Commuting Elements in q-Deformed Heisenberg Algebras

Commuting Elements in *q*-Deformed Heisenberg Algebras

Lars Hellström
Umeå University, Sweden

Sergei D. Silvestrov
Lund Institute of Technology, Sweden

World Scientific
Singapore • New Jersey • London • Hong Kong

Published by

World Scientific Publishing Co. Pte. Ltd.

P O Box 128, Farrer Road, Singapore 912805

USA office: Suite 1B, 1060 Main Street, River Edge, NJ 07661

UK office: 57 Shelton Street, Covent Garden, London WC2H 9HE

British Library Cataloguing-in-Publication Data
A catalogue record for this book is available from the British Library.

COMMUTING ELEMENTS IN q-DEFORMED HEISENBERG ALGEBRAS

ISBN 981-02-4403-7

Printed in Singapore by World Scientific Printers

Commuting Elements in q-Deformed Heisenberg Algebras

Preface

The main objects studied in this book are q-deformed Heisenberg algebras. More specifically the monograph is about commuting elements in q-deformed Heisenberg algebras. These algebras have rich properties as mathematical objects and have many important applications in physics and beyond.

The structure of commuting elements in an algebra is of fundamental importance for its structure and representation theory as well as for its applications. In this book the structure of commuting elements in q-deformed Heisenberg algebras is studied in a systematic way. Many new results are presented with complete proofs. One of the major achievements in this monograph is the result saying that commuting elements in a q-deformed Heisenberg algebra must be algebraically dependent for deformation parameters q of free type. We believe that this result will have a fundamental role for developing algebraic-geometrical methods in investigations of linear and non-linear q-difference and q-integral equations and in spectral theory of q-difference and q-integral operators.

In addition to new results the book contains some basic definitions and facts on q-deformed Heisenberg algebras. Several appendices also include some general theory used in other parts of the book. In the first appendix the Diamond Lemma and some related definitions and results are presented. In the second appendix we develop a theory of degree functions in arbitrary associative algebras. Finally, the third appendix contains a definition and several properties of some basic q-combinatorial functions over an arbitrary field. Some notations appearing in the book are discussed in the fourth appendix.

The bibliography contains in addition to references on q-deformed Heisenberg algebras some selected references on related subjects and on existing and potential applications. Subjects considered in each reference are indicated.

The book is self contained both as far as proofs and the background material is concerned. In addition to research and reference purposes, it can also be used in a special course or a series of lectures on the subject or as a complementary literature to a general course on algebra. A standard linear algebra course and some basic course in algebra should be enough to be able to read this book. We thus hope that the book will be useful for specialists as well as doctoral and advanced undergraduate students.

Acknowledgments

Several people and institutions have supported our work. Firstly and most of all we are grateful to the Department of Mathematics at Umeå University for constant support from the very beginning of the project. We are specially grateful to Professor Hans Wallin and Professor Roland Häggkvist for their support and encouragement.

This book has been written while Lars Hellström has been a doctoral student at the Department of Mathematics at Umeå University, and while Sergei Silvestrov has been working at the Department of Mathematics at Umeå University and at the Department of Mathematics at the Royal Institute of Technology in Stockholm, and also during his research visit to the Department of Mathematics and the Obermann Center for Advanced Studies at the University of Iowa. The support of those institutions is gratefully acknowledged.

Sergei Silvestrov would like to extend his special thanks to Professor Dan Laksov, Professor Håkan Eliasson, Professor Michael Benedicks, Professor Ari Laptev and Professor Thomas Höglund for their support and encouragement of his research and pedagogical activities at the Department of Mathematics at Royal Institute of Technology in Stockholm. Sergei Silvestrov is grateful also to Professor Palle Jørgensen for helping to make the visit to the University of Iowa comfortable and interesting as far as both research and every day live is concerned, to Professor Florin Radulescu, Professor Paul Muhly, Professor Raul Curto, Professor Tuong Ton-That, Professor Fred Goodman, Professor Philip Kutzko, and other participants of the seminars on Operator Theory, Mathematical Physics and Representation Theory at the University of Iowa for creating an inspiring research environment, and to Jay Semel, Lorna Olson and Karla Tonella from the Obermann Center for Advanced Studies for opportunity to use centers facilities and for their kind help. Sergei Silvestrov is also very grateful to STINT foundation for supporting financially his visit to the University of Iowa.

We wish to thank Nicke Sjödin for supplying us with an English translation of one of the texts we quoted.

Lars Hellström is grateful to his parents Bo Hellström och Birgitta Hellström, and Sergei Silvestrov is grateful to his parents Evelina Silvestrova and Dmitrii Silvestrov and to his wife Zhiyi Silvestrov Liang for their encouragement and support.

Contents

Preface vii

Chapter 1 Introduction **1**

1.1 q-Deformed Heisenberg algebras 1
1.2 Some references and motivation 5
1.3 Contents by chapters 14
1.4 Conventions and notations 15

Chapter 2 Immediate consequences of the commutation relations **19**

2.1 The values of $\mathbf{q}$ 19
2.2 Reordering formulae 20
2.3 Simplifying commutation relations 30

Chapter 3 Bases and normal form in $\mathcal{H}(q)$ and $\mathcal{H}(\mathbf{q}, J)$ **35**

3.1 The definition of $\mathcal{H}(\mathbf{q}, J)$ 36
3.2 Three bases for $\mathcal{H}(\mathbf{q}, J)$ 38
3.3 Comparison of the three bases 48
3.4 Computational aspects 49

Chapter 4 Degree in and gradation of $\mathcal{H}(\mathbf{q}, J)$ **53**

4.1 Degree in $\mathcal{H}(\mathbf{q}, J)$ 54
4.2 Grading $\mathcal{H}(\mathbf{q}, J)$ 62
4.3 Some useful properties 68

Chapter 5 Centralisers of elements in $\mathcal{H}(\mathbf{q}, J)$ **73**

5.1 General definitions and theorems 74
5.2 Classification of $\mathcal{H}(\mathbf{q}, J)$ 77
5.3 The case q_j of torsion type for some $j \in J$ 81
5.4 The centre of $\mathcal{H}(\mathbf{q}, J)$ when $\mathbf{q}$ is of strictly direct type on J . . 84
5.5 $\mathcal{H}(\mathbf{q}, J)$ for $\mathbf{q} \in \mathcal{Q}(J, \mathcal{K})$ 86

Chapter 6 Centralisers of elements in $\mathcal{H}(q)$ **95**

6.1 Classification of $\mathcal{H}(1)$ 95
6.2 $\mathcal{H}(q)$ when q is of free type 99
6.3 $\mathcal{H}(q)$ when q is of torsion type 104

Chapter 7 Algebraic dependence of commuting elements in $\mathcal{H}(q)$ and $\mathcal{H}(\mathbf{q}, n)$ **117**

Chapter 8 Representations of $\mathcal{H}(\mathbf{q}, J)$ by q-difference operators **129**

Appendix A The Diamond Lemma **141**

A.1 Definitions and proofs 142
A.2 A short key to the notations 149
A.3 A few extra results 151

Appendix B Degree functions and gradations **155**

B.1 General theory of degree functions 156
B.1.1 A generalisation of the Bernstein filtration 156
B.1.2 The basic properties 160
B.2 Degree and free algebras 162
B.2.1 Additivity of degree in free algebras 163
B.2.2 Some technical lemmas 168
B.2.3 Degree in a free algebra versus degree in a quotient . . . 175
B.3 Gradations 179

Appendix C q-special combinatorics **183**

C.1 Definitions and existence 184
C.2 Other properties 186
C.3 q-Stirling numbers 196
C.4 Extending the q-combinatorial functions 215

Appendix D Notes on notations **221**

Bibliography 223

Index 255

Chapter 1

Introduction

"What are we going to do today?"
"My boy," said the adult, seeing a universe of possibilities, "there is nothing that we are *not* going to do today."
— from *Q-squared* by P. DAVID

This book is about commuting elements in q-deformed Heisenberg algebras. These algebras are known to have rich properties as mathematical objects as well as to be important in a variety of physical models.

1.1 q-Deformed Heisenberg algebras

Linear operators A and B satisfying the Heisenberg canonical commutation relation

$$AB - BA = I, \tag{1.1}$$

where I denotes the identity operator, occupy a key place in the operator formulation of Quantum Mechanics. Relation (1.1) is satisfied for example by the operators of creation and annihilation for systems with one degree of freedom subject to Bose statistics. For systems with more then one degree of freedom one considers instead finite or infinite families of pairs of linear operators $\{A_j, B_j\}_{j\in J}$ satisfying the Heisenberg canonical commutation relations

$$A_jB_j - B_jA_j = I, \tag{1.2a}$$
$$A_iB_j - B_jA_i = 0 \quad \text{for } i \neq j, \tag{1.2b}$$
$$A_iA_j - A_jA_i = 0, \quad B_iB_j - B_jB_i = 0 \tag{1.2c}$$

for all $i, j \in J$. For systems subject to Fermi statistics a special role is played, in the case of a single degree of freedom, by pairs of linear operators A and B satisfying the canonical anticommutation relation

$$AB + BA = I, \tag{1.3}$$

and in the case of more then one degree of freedom, by finite or infinite families of pairs of linear operators $\{A_j, B_j\}_{j\in J}$ satisfying

$$A_jB_j + B_jA_j = I, \tag{1.4a}$$
$$A_iB_j - B_jA_i = 0 \quad \text{for } i \neq j, \tag{1.4b}$$
$$A_iA_j - A_jA_i = 0, \quad B_iB_j - B_jB_i = 0. \tag{1.4c}$$

The relations (1.2) and (1.4) belong to a family of commutation relations

$$A_jB_j - q_jB_jA_j = I, \tag{1.5a}$$
$$A_iB_j - B_jA_i = 0 \quad \text{for } i \neq j, \tag{1.5b}$$
$$A_iA_j - A_jA_i = 0, \quad B_iB_j - B_jB_i = 0 \tag{1.5c}$$

parameterised by a vector $\mathbf{q} = (q_j)_{j\in J}$ of parameters from the field $\mathcal{K}$ over which the linear operators are defined. In the case of a single pair of operators A, B this family consists of a single relation

$$AB - qBA = I \tag{1.6}$$

depending on one parameter q from $\mathcal{K}$. This relation reduces to (1.1) when $q = 1$ and to (1.3) when $q = -1$. More generally, the q-deformed commutation relations (1.5) become the canonical commutation relations (1.2) if $q_j = 1$ for all $j \in J$, and the canonical anticommutation relations if $q_j = -1$ for all $j \in J$. With this in mind we refer to both (1.6) and (1.5) as q-deformed Heisenberg canonical commutation relations.

Any set of linear operators satisfying the above commutation relations is called a representation of these commutation relations. There are many kinds of representations with different applications and significance in physics and with different mathematical properties. At the same time there are many properties which are the same for all representations. Often such properties have algebraic nature and can be studied via passing to another level of abstraction by considering A's and B's not as linear operators but as formal generators of the algebra consisting of non-commutative polynomial expressions, that is finite linear combinations of finite products of A's and B's. Two non-commutative polynomials in this algebra are said to be equal if one of them can be obtained from the other by a finite number of operations of addition, multiplication and substitution using the commutation relations satisfied by A's and B's. More systematic definition of this algebra can be given as a quotient of the free algebra by the ideal defined by the commutation relations. We will refer to these algebras as q-deformed Heisenberg algebras. There are many linear space bases in q-deformed Heisenberg algebras each having its advantages and disadvantages. Important bases consisting of products of generators taken in a certain specific order can be constructed using the Diamond Lemma for rings and algebras. These bases prove to be very useful in understanding the structure and representation theory of the q-deformed Heisenberg algebras as well as in their numerous applications.

If $A = D = \frac{d}{dx} : f(x) \mapsto (Df)(x) = \frac{df}{dx} = f'(x)$ is the operator of differentiation and $B = M_x : f(x) \mapsto xf(x)$ is the operator of multiplication by the indeterminate x both acting for example on the linear space of infinitely differentiable functions, or on the linear spaces of formal power series or polynomials in x, then they satisfy the Heisenberg canonical commutation relation (1.1) with $I : f(x) \mapsto f(x)$ denoting the identity operator. In other words the pair $(A, B) = (D, M)$ is a representation of the commutation relation (1.1). Similarly, the relations (1.2) are satisfied if, for all $j \in J = \{1, \ldots, n\}$,

$$A_j = \partial_j = \partial_{x_j} : f(x_1, \ldots x_n) \mapsto (\partial_{x_j} f)(x_1, \ldots x_n)$$

is the operator of partial differentiation with respect to the indeterminate x_j, and

$$B_j = M_{x_j} : f(x_1, \ldots x_n) \mapsto x_j f(x_1, \ldots x_n)$$

is the operator of multiplication by x_j acting for example on the linear space of infinitely differentiable functions, or on the linear spaces of formal power series or polynomials in $x_1, \ldots, x_n$. These at first sight innocent observations make the Heisenberg canonical commutation relations fundamentally important for differentiation and integration theory, and thus for physics and many other subjects where integration and differentiation are involved.

The situation is very similar for the q-deformed Heisenberg commutation relations. If A is the operator of q-differentiation defined for nonzero $q \neq 1$ and $x \neq 0$ as

$$A = D_{x,q} = D_q : f(x) \mapsto (D_q f)(x) = \frac{f(x) - f(qx)}{(1-q)x},$$

and if

$$B = M_x : f(x) \mapsto x f(x)$$

is the operator of multiplication by x both acting for example on the linear space of infinitely differentiable functions, or on the linear space of formal power series or polynomials in nonzero x, then they satisfy the q-deformed Heisenberg canonical commutation relation (1.6). Similarly, the relations (1.5) are satisfied by the operators of partial q-differentiation defined for all $j \in \{1, \ldots, n\}$, nonzero $q_j \neq 1$ and $x_j \neq 0$ as

$$A_j = \partial_{j,q_j} = \partial_{x_j,q_j} : f(x_1, \ldots x_n) \mapsto \frac{f(x_1, \ldots, x_n) - f(x_1, \ldots, q_j x_j, \ldots, x_n)}{(1-q_j)x_j},$$

and the operators of multiplication defined as

$$B_j = M_{x_j} : f(x_1, \ldots x_n) \mapsto x_j f(x_1, \ldots x_n),$$

all acting for example on the linear space of infinitely differentiable functions, or on the linear spaces of formal power series or polynomials in nonzero $x_1, \ldots, x_n$. Moreover, by definition of derivative, $D_q(f)$ tends to $D(f)$ and $\partial_{x_j,q_j}(f)$ tends to $\partial_{x_j}(f)$ when q tends to 1. A consequence of all these observations is that the whole calculus based on the operators D_q and ∂_{x_j,q_j} can be considered in a natural way both as a deformation and a prelimit discretisation of the usual differential and integral calculus. The q-deformed Heisenberg commutation relations play the same fundamental role in this q-deformed differential and integral calculus as the Heisenberg canonical commutation relations do in the usual undeformed one.

1.2 Some references and motivation

Leaving more detailed definitions and discussions of relevant notions to later chapters we would like to mention now some works concerned specifically or in part with q-deformed Heisenberg algebras, also called sometimes, for various reasons, q-deformed Heisenberg–Weyl algebras or q-deformed Weyl algebras.

Various physically important representations and diverse applications of q-deformed Heisenberg algebras have been considered in the works of M. Arik and D. D. Coon [22], M. Arik [20; 21], V. V. Kuryshkin [249], O. W. Greenberg [164; 165], H. Morikawa [289], A. Kempf [225; 226], J. Hruby [188], S. Chaturvedi and V. Srinivasan [68], V. P. Spiridonov [365], S. Skorik and V. P. Spiridonov [359], K. N. Ilinski, G. V. Kalinin and A. S. Stepanenko [194], M. Chaichian, H. Grosse and P. Prešnajder [60], M. Chaichian, F. R. Gonzalez and P. Prešnajder [60], M. Chaichian, M. N. Mnatsakanova and Yu. S. Vernov [63], J. Wess [391; 392], J. Schwenk [346; 347], J. Schwenk and J. Wess [349], A. Hebecker, S. Schreckenberg, J. Schwenk, W. Weich and J. Wess [179], A. Hebecker and W. Weich [178], B. L. Cerchiai, R. Hinterding, J. Madore and J. Wess [56], A. S. Zhedanov [408], A. I. Solomon [362], R. J. McDermott and A. I. Solomon [260; 261; 262; 263], A. S. Zhedanov [408; 409], F. H. L. Essler and V. Rittenberg [108], C. Quesne and N. Vansteenkiste [320], C. Delbecq and C. Quesne [90; 91], W. Pusz and S. L. Woronowicz [318], W. Pusz [317], R. Speicher [363; 364], M. Bożejko and R. Speicher [43; 44; 45], D. Zagier [401], A. N. Kochubei [233], D. Bonatsos [39], D. Bonatsos, C. Daskaloyannis and A. Faessler [40], S. V. Shabanov [352; 353], S. S Avancini and G Krein [25], Z. Chang, H.-Y. Guo and H. Yan [66], H.-Y. Fan and S.-C. Jing [112; 113; 206], M. Rausch de Traubenberg [322; 323], L. Ma, Z. Tang and Y.-D. Zhang [258; 259], O. R. Jensen [204], G. Fiore [129], D. I. Fivel [130], and also M. Aspenberg and S. D. Silvestrov [24]. Useful reordering formulas for elements in q-deformed Heisenberg algebras and in their generalisations and extentions have been obtained in the articles by J. Cigler [73], Ph. Feinsilver [114; 115; 116], T. H. Koornwinder [237], A. Turbiner [373], A. Turbiner and G. Post [374], N. Fleury and A. Turbiner [132], G. Post [312], R. Berger [31], W. A. Al-Salam and E. H. Ismail [11], J. S. Møller [295], and L. Hellström and S. D. Silvestrov [182; 183; 184; 356].

The q-deformed Heisenberg relations alone or accompanied by some additional commutation relations play a key role in the definitions, theory and applications of such physically important objects as quantum groups, quantum spaces, deformed harmonic oscillator algebras, q-analogues of Virasoro algebra and of other important algebras, q-analogues of various objects from homological algebra, and in investigations on braided geometry, noncommutative differential and integral calculus, deformation quantization, q-special functions, q-orthogonal polynomials, q-Fourier analysis and umbral and q-umbral calculus as described for instance by L. C. Biedenharn [35], L. C. Biedenharn and M. A. Lohe [36], M. Chaichian and P. P. Kulish [61], M. Chaichian, A. P. Demichev, P. P. Kulish [58], M. Chaichian and A. P Demichev [57], M. Chaichian, P. P. Kulish and J. Lukierski [62], E. V. Damaskinsky and P. P. Kulish [86; 87; 88; 89], P. P. Kulish [248], A. J. Macfarlane [265], W.-S. Chung [71], W.-S. Chung and A. U. Klimyk [72], T. L. Curtright [84], I. M. Gelfand and D. B. Fairlie [156], D. B. Fairlie [110], D. B. Fairlie and C. K. Zachos [111], C. Zachos [400], A. Jannussis [200], R. J. Finkelstein [117; 118; 119; 120; 121; 122; 123; 124; 125; 126], R. J. Finkelstein and E. Marcus [127], A. C. Cadavid and R. J. Finkelstein [49; 50], F. L. Chan and R. J. Finkelstein [64], F. L. Chan, R. J. Finkelstein and V. Oganesyan [65], R. Floreanini, V. P. Spiridonov and L. Vinet [134; 135], R. Floreanini, J. LeTourneux and L. Vinet [136; 137], R. Floreanini and L. Vinet [138; 139; 140; 141; 145; 146], P. Furlan, L. K. Hadjiivanov and I. T. Todorov [150], T. Hayashi [177], S. Jing and J. J. Xu [207], S. Rodriguez - Romo and D. W. Ebner [324], G. Fiore [128], J. Schwenk [346; 347; 348], M. R. Ubriaco [377], C.-Z. Zha and W.-Z. Zhao [406], J. - H. Dai, H. -Y. Guo and H. Yan [85], G. Kaniadakis, A. Lavagno and P. Quarati [220], S. Chaturvedi, R. Jagannathan, R. Sridhar and V. Srinivasan [67], U. Carow-Watamura, M. Schlieker and S. Watamura [54], U. Carow-Watamura and S. Watamura [55], J. Cigler [73; 74; 75; 78; 77; 76; 79; 80; 81], T. H. Koornwinder [236; 237], A. Dimakis, F. Müller-Hoissen and T. Striker [94], S. Roman [327; 328; 329; 330; 331], S. Roman and G.-C. Rota [332], G.-C. Rota, D. Kahaner and A. Odlyzko [335], C. Kassel [223; 224], E. C. Ihrig and M. E. H. Ismail [189], R. S. Dunne, A. J. Macfarlane, J. A. de Azcárraga and J. C. Pérez Bueno [104; 105], J. A. de Azcárraga and A. J. Macfarlane [26], M. Dubois-Violette and R. Kerner [101], S. Durand [106], B. Y. Hou, B. Y. Hou and Z. Q. Ma [187], K. Aomoto and Y. Kato [19], K. Aomoto [18], F. Bonechi, N. Ciccoli, R. Giachetti, E. Sorace and M. Tarlini [41], J. Wess and B. Zumino [393; 394], and B. Zumino [410], J. Bertrand, M. Irac-Astaud [34], S. Majid [268; 269; 270] Yu. I. Manin [272], and S. L. Woronowicz [398]. The articles of E. V. Damaskinsky and P. P. Kulish [86; 87; 88; 89; 248] contain, together with important results, also a review of many publications related to q-deformed Heisenberg algebras and their representations and applications. The q-deformed Heisenberg algebras arise also as an important example in Santilli's theory of Lie-admissible algebras described for example by R. M. Santilli [338; 339], and R. M. Santilli and H. C. Myung [294], and in non-canonical mechanics as discussed in the articles by A. Jannussis, L. Papaloucas and P. Siafarikas [203], A. Jannussis, G. Brodimas, D. Sourlas, A. Streclas, P. Siafarikas, L. Papaloucas, and N. Tsangas [201], A. Jannussis, G. Brodimas, D. Sourlas, K. Vlachos, P. Siafarikas and L. Papaloucas [202], and A. Jannussis [200]. The C^*-algebras and $*$-algebras associated with q-deformed Heisenberg relations as well as their representations by bounded and unbounded operators in a Hilbert space have been considered by K. Dykema and A. Nica [107], A. L. Rosenberg [333; 334], P. E. T. Jørgensen [212], P. E. T. Jørgensen and R. F. Werner [219], P. E. T. Jørgensen, L. M. Schmitt and R. F. Werner [217; 218], D. P. Proskurin [316], P. E. T. Jørgensen, D. P. Proskurin and Yu. S. Samoĭlenko [216], V. Mazorchuk and L. B. Turowska [277], E. C. Lance [251], V. L. Ostrovskyĭ and Yu. S. Samoĭlenko [302; 303; 304], and K. Schmüdgen [343; 344]. The q-deformed Heisenberg commutation relations has been mentioned also in the monographs by A. U. Klimyk and K. Schmüdgen, [231], M. Chaichian and A. P. Demichev [57], A. Joseph [208] and S. Majid [269] in relation to quantum groups, braided Lie algebras, braided geometry and noncommutative differential calculus, in the monographs by H. Exton [109], and N. J. Vilenkin and A. U. Klimyk [386] in connection to q-analysis, q-difference equations, and q-special functions, and in the monographs by V. P. Maslov [275], M. V. Karasev, V. P. Maslov [221], and V. E. Nazaikinskii, V. E. Shatalov and B. Yu. Sternin [296] as an important example in Maslov's noncommutative operational calculus.

The q-difference operator $D_{x,q}$ forming together with multiplication by x a representation of the q-deformed Heisenberg relation, or its slight modification, the operator $xD_{x,q}$, has appeared actually much earlier then the q-deformed Heisenberg relations themselves in the works by E. Heine, J. Thomae, L. J. Rogers and F. H. Jackson. This operators are the basic operators for the theory of q-difference equations, q-analysis and their applications, playing the same role as the ordinary diferential operator D_x or the operator xD_x in the differentiation and integration theory, analysis and their applications. The q-difference operators appear in important ways in many of the cited in the book works on quanum groups, algebras and spaces, deformed oscillator algebras, noncommutative geometry and noncommutative differential calculus. The theory of linear q-difference equations, applications of q-difference equations in analysis, theory of q-special functions and combinatorics, q-difference analogues of variouse physically important linear and non-linear differential equations, and numerouse applications of q-difference equations and q-difference operators in variouse parts of mathematics and physics has been also considered by E. Heine [180; 181], J. Thomae [368; 369; 370], L. J. Rogers [325], F. H. Jackson [195; 196; 197; 198; 199], G. D. Birkhoff [37], R. D. Carmichael [53], C. R. Adams [3; 4; 5; 6], T. E. Mason [276], W. J. Trjitzinsky [372], W. Hahn [166; 167; 168; 169; 170; 171; 172; 173; 174], N. E. Nörlund [300], F. Ryde [336], J. Le Caine [250], G. W. Starcher [366], W. H. Abdi [1; 2], G. E. Andrews [15; 16], K.-W. Yang [399], P. A. Hendriks [185], S.-C. Jing and H.-Y. Fan [205], S. C. Milne [280], V. K. Dobrev [96; 97], V. K. Dobrev, H. D. Doebner and R. Twarock [98], V. K. Dobrev and B. S. Kostadinov [99], E. Papp [306; 307; 308], R. Twarock [375; 376], R. Floreanini and L. Vinet [142; 143; 144], M. Pillin [311], P. Lesky [255], R. F. Swarttouw and H. G. Meijer [367], F. Marotte and C. Zhang [274], M. Klimek [229; 230], A. Schirrmacher [341], U. Meyer [278], I. B. Frenkel and N. Yu. Reshetikhin [148], K. Aomoto [17], I. M. Gelfand, M. I. Graev and V. S. Retakh [157], K. Mimachi [281; 282; 283; 284; 285], K. Mimachi and M. Noumi [286], M. Nishizawa [298], M. Nishizawa and K. Ueno [299], R. K. Saxena and R. Kumar [340] I. Mukhopadhaya and A. R. Chowdhury [291], J. M. Thuswaldner [371], C. Zhang [407], W. Miller [279], A. K. Agarwal, E. G. Kalnins and W. Miller [9], E. Horikawa [186], R. Wallisser [390], R. P. Agarwal [8], W. A. Al-Salam [10], W. A. Al-Salam and A. Verma [12; 13], M. Upadhyay [378; 379; 380], R. Askey and J. Wilson [23], H. Exton [109], and G. Gasper and M. Rahman [153].

We have mentioned in the beginning of the introduction, and it follows from the title, that the main subject of the study in the book is the structure of the commuting elements in the q-deformed Heisenberg algebras. The commuting elements in algebras, rings and groups occupy a very special and important place in the theory and applications.

In representation theory, operators of representations are often conveniently expressed in the bases consisting of joint eigenvectors or generalised eigenvectors for operators representing the elements in some sets of commuting elements. Such descriptions of representations often have explicit physical interpretation and are useful from computational as well as theoretical points of view. As a consequence the problem of description of commuting elements in algebras, rings and groups becomes important not only for representation theory itself, but also for such mathematical subjects as operator algebras and operator theory, non-commutative geometry, theory of special functions, probability theory, dynamical systems as well as for many applications within classical and quantum physics, chemistry and other subjects employing representation theory as a tool or using it as their axiomatic base. The literature where commuting operators appear in connection to representations of groups, rings and algebras is huge, constantly growing and far beyond the scope of this introduction. Relevant material and references can be found for example in the articles by L. Gårding and A. Wightman [151; 152], V. Ya. Golodets [160], G. W. Mackey [267], A. M. Vershik, I. M. Gelfand and M. I. Graev [384], Yu. M. Berezanskij, V. L. Ostrovskyĭ and Yu. S. Samoĭlenko [30], V. I. Gorbachuk, Yu. S. Samoĭlenko and G. F. Us [161], E. Ye. Vaysleb and Yu. S. Samoĭlenko [382] E. Ye. Vaysleb [381], M. V. Karasev and E. Novikova [222], B. Fuglede [149], P. E. T. Jørgensen [209; 210; 211], S. Pedersen [310], P. E. T. Jørgensen and S. Pedersen, [215], J. F. Van Diejen [93], K. Mimachi and M. Noumi [286], L. Lapointe and L. Vinet [252], in the books by I. M. Gelfand and N. Ya. Vilenkin [158], Yu. M. Berezanskij [28], Yu. M. Berezanskij and Yu. G. Kondrat'ev [29], Yu. S. Samoĭlenko [337], V. L. Ostrovskyĭ and Yu. S. Samoĭlenko [304], P. E. T. Jørgensen [213], and P. E. T. Jørgensen and R. T. More [214], K. Schmüdgen [342], G. K. Pedersen [309], C. R. Putnam [319], I. G. Macdonald [264], and A. Connes [83], and also in the articles by S. D. Silvestrov and H. Wallin [358], S. D. Silvestrov and V. L. Ostrovskyĭ [305], S. D. Silvestrov and L. B. Turowska [357], and S. D. Silvestrov [354; 355].

Another main concrete motivation for this book came from extensive applications of commuting differential, difference, integral and other operators to solution of Korteweg–de Vries (KdV) equation, Kadomtsev–Petviashvili (KP) equation, Gardner equation, Nizhnik–Veselov–Novikov equation, Yang–Mills equations, Landau–Lifshitz equation, Einstein equation, Novikov and Dubrovin type equations, Zakharov–Shabat type equations, Lax type equations, and of many other nonlinear and linear equations arising in physics, as has been demonstrated in fundamental pioneering works of V. E. Zakharov, A. B. Shabat [404; 405], B. A. Dubrovin, V. B. Matveev, and S. P. Novikov [103], B. A. Dubrovin [102], S. P. Novikov [301], I. M. Krichever and S. P. Novikov [244; 245; 246; 247], I. M. Krichever

[239; 240; 241; 242; 243], V. E. Zakharov and L. D. Faddeev [402], L. Dickey [92], I. M. Gelfand and L. Dickey [154; 155], G. Wilson [395; 396; 397], G. B. Segal and G. Wilson [351], P. D. Lax [253; 254], D. Mumford [293], D. Mumford and P. van Moerbeke [287], H. P. McKean [266], Yu. I. Manin [271], J.-L. Verdier [383], V. G. Drinfeld [100], V. V. Sokolov [360; 361], A. P. Veselov [385], O. I. Mokhov [288], A. R. Chowdhury and N. D. Gupta [70], and in the books by V. E. Zakharov, S. V. Manakov, S. P. Novikov and L. P. Pitayevsky [403], by B. G. Konopelchenko [235], by J. Moser [290], A. Pressley and G. Segal [313], where many further references, results and applications can be found. The applications of commuting operators considered in these publications as well as other relevant results have been also discussed in the articles by J. Harnad [176] and E. Previato [314; 315]. Many relevant results and extensive bibliography can also be found in the semi-review article by F. Gesztesy and R. Weikard [159].

A distinctive common feature of all these works is that commuting operators, and as a consequence the corresponding solutions of equations, are described in terms of algebraic curves, surfaces and functions on them using methods of algebraic geometry. A fundamental role in these developments is played by the results of J. L. Burchnall and T. W. Chaundy [46; 47; 48; 69] connecting algebraic geometry to commuting differential operators. Closely related to investigations of nonlinear differential equations and to the results of J. L. Burchnall and T. W. Chaundy are important applications of commuting operators, and hence of the methods of algebraic geometry and the theory of Riemann surfaces, in the theory of colligations, and in investigations on non-selfadjoint operators and representations of semigroups as developed by M. S. Livšic, N. Kravitsky, A. S. Markus, V. Vinnikov and L. L. Waksman [256; 257]. We have been motivated also by the works of G. Floquet [133], I. Schur [345], G. Wallenberg [389], and H. Flanders [131], S. A. Amitsur [14] and J. Dixmier [95], R. C. Carlson and K. R. Goodearl [52], where commuting differential operators and commuting elements in Heisenberg (Weyl) algebras and their applications have been considered.

The direct relevance of commuting differential operators to solution of non-linear differential equations is probably seen best in the case of the celebrated Korteweg–de Vries (KdV) equation. This equation has been shown by D. Korteweg and G. de Vries in [238] to be the equation describing propagation of solitary water waves in a channel, the phenomenon qualitatively described in a published form for example by J. Scott-Russell [350]. This phenomenon has been also studied by a number of people before appearance of the work of Korteweg and de Vries [238]. In particular, the Korteweg–de Vries equation has been actually already discovered much earlier by M. J. Boussinesq [42] in his study of solitary waves. Since the discovery of the Korteweg–de Vries equation in connection to propagation of solitary waves, almost magically it has been shown also to describe many other phenomena investigated in different parts of physics. The KdV equation is the partial differential equation for the function of two real variables $u = u(x,t)$, and is often presented in the following form:

$$\partial_t u - 6u\partial_x u + \partial_x^3 u = 0. \tag{1.7}$$

The key observation is that this equation can be reformulated as the condition of commutativity of differential operators

$$\begin{aligned} Z &= \partial_t u - P = \partial_t u + 4\partial_x^3 - 6u\partial_x - 3\partial_x u, \\ L &= -\partial_x^2 + u. \end{aligned}$$

In other words $[L, \partial_t - P]$ if and only if $u(x,t)$ is a solution of the KdV equation (1.7). Thus a description of commuting differential operators P and L gives also a description of solutions of the KdV equation (1.7).

At the start of the project leading to this book we anticipated that q-difference equations analogous to nonlinear differential equations can be constructed, and that many methods of solution of the corresponding nonlinear differential equations can be extended to their q-anologues. Indeed, while the book has been written, in the articles by E. Frenkel [147], P. Iliev [190; 191; 192; 193], L. Haine and P. Iliev [175], B. Khesin, V. Lyubashenko and C. Roger [228], M. Adler, E. Horozov and P. van Moerbeke [7] several q-analogs of KdV and KP equations have been introduced and some of their properties and solutions have been studied. The commuting q-difference operators can effectively be used in investigation of these and many other linear and nonlinear q-difference equations in a similar way as commuting differential operators are employed in the study of differential equations. The description of q-analogues of the KdV equations requires more preliminaries, than would be reasonable to have in this introduction. An easy example, nevertheless showing equally well the connection between commuting q-difference operators and q-difference equations, is obtained for instance by making the following observation. If q-difference operators are acting on a space of functions where the equality of the linear q-difference operators is equivalent to the equality of their coefficients, then the operators $u\partial_{x,q} = u(x,t)\partial_{x,q}$ and $v\partial_{x,q} = v(x,t)\partial_{x,q}$ commute, that is satisfy

$$[u\partial_{x,q}, v\partial_{x,q}] = (u\partial_{x,q})(v\partial_{x,q}) - (v\partial_{x,q})(u\partial_{x,q}) = 0,$$

if and only if the functions $u = u(x,y)$ and $v = v(x,y)$ solve the q-difference equation

$$u(\partial_{x,q}v) - v(\partial_{x,q}u) = 0.$$

For instance, if $v = \partial_{x,q}u$ then this equation becomes the nonlinear q-difference equation

$$u(\partial_{x,q}^2 u) - (\partial_{x,q}u)^2 = 0.$$

Of course, in this simple case, solutions of these equations are easily described by noting that, in order to be solutions, the functions v and u must satisfy $v(x,t) = k(x,t)u(x,t)$ at all points (x,t) where $u(x,t) \neq 0$ and $u(qx,t) \neq 0$ for some function $k(x,t)$ such that $k(qx,t) = k(x,t)$. However, solving q-difference equations, obtained from commutativity of the q-difference operators of higher order then one, is a significantly more complicated task. Description of commuting q-difference operators, or more generally of commuting elements in q-deformed Heisenberg algebras is a way to approach this problem.

We feel that the methods of algebraic geometry so successfully applied to nonlinear differential equations can also be very effective in the study of the q-analogues of these equations. A key fact connecting algebraic geometry to differential equations is the result of J. L. Burchnall and T. W. Chaundy [46; 47; 48; 69] saying that commuting differential operators α and β are algebraically dependent, that is satisfy an algebraic equation

$$P(\alpha,\beta) = \sum_{\substack{0\leq j\leq n\\ 0\leq k\leq m}} p_{jk}\alpha^j\beta^k = 0,$$

where $P(x,y) = \sum_{\substack{0\leq j\leq n\\ 0\leq k\leq m}} p_{jk}x^jy^k$ is a polynomial in two variables. This result and its multi-dimensional extensions allow one to study spectral properties and solutions to eigenvalue problems for commuting differential operators, and hence the solutions to nonlinear differential equations, in terms of the corresponding algebraic curve $P(x,y) = 0$, or in case of several operators or of higher dimensions in terms of the corresponding algebraic surfaces. An important observation is that Burchnall's and Chaundy's result on algebraic dependence of commuting differential operators is actually not specific for differential operators. It holds in general for commuting elements in the Heisenberg (Weyl) algebra defined as an algebra with abstract generators $\{A_j, B_j\}_{j\in J}$ and defining commutation relations (1.2), or even if necessary in a proper extention of the Heisenberg algebra allowing some power series in generators to be involved, as for example needed for studying differential operators with coefficients expressible as power series. A consequence of this observation is that much of the theory and applications based on this result remain true also for other representations of Heisenberg algebras and not just for the standard representation generated by differentiation operators and operators of multiplication by independent variables.

Similarly to the fundamental role played in study of solutions of nonlinear differential operators by Burchnall's and Chaundy's result on algebraic dependence of commuting differential operators, the analogous result for q-difference operators ought to be as important for study of the solutions of the corresponding nonlinear q-difference equations. In Chapter 7 we show that commuting elements in the q-deformed Heisenberg algebra with two generators must be algebraically dependent if the deformation parameter q is of free type, which in the field of characteristic zero means that either $q = 1$ or q is not a root of 1. When q is a root of 1, we show that commuting elements are algebraically dependent over the center of the algebra. Generalisations of this result to general q-deformed Heisenberg algebras are also considered. In Chapter 8 we consider the representation of q-deformed Heisenberg algebras generated by q-difference and multiplication operators, lying in the foundation of the theory of q-differentiation, q-integration, q-difference operators and q-difference equations. This representation is a natural q-deformation of the standard representation of the Heisenberg algebra by differentiation and multiplication operators. The algebra of q-difference operators is generated by the operators of this representation. Algebraic dependence of commuting elements in a q-deformed Heisenberg algebra thus imply that commuting q-differential operators are algebraically dependent provided q is of free type. This corollary is presented at the end of Chapter 8.

1.3 Contents by chapters

In order to make the book self-contained and useful not only for specialists but also for graduate and advanced undergraduate students, some relevant necessary basic definitions and background material on q-deformed Heisenberg algebras have been included in Chapter 2 and Chapter 3. In Chapter 2 the q-deformed Heisenberg algebras are defined and several formulas for reordering of elements are proved. In Chapter 3 several bases in these algebras are described using the Diamond Lemma. The last four chapters—Appendices A, B, and C—contain some general material which has been used throughout the book but which is not specifically concerned with the q-deformed Heisenberg algebras, and also a discussion of some notations appearing in other chapters. Appendix A is devoted to the Diamond Lemma and some related notions and results. In Appendix B we develop a general theory of degree functions that can be used in any associative algebra. In Appendix C we define basic q-combinatorial functions for q from any field and describe some of their properties used in the book. Finally, in Appendix D we discuss some notations appearing in the book.

In Chapter 4 we describe properties of the degree functions and introduce and study certain gradations. Chapter 5 is about centralisers of elements in general q-deformed Heisenberg algebras. In Chapter 6 we study the centralisers of elements in the q-deformed Heisenberg algebras with two generators. Chapter 7 is devoted to the question of algebraic dependence of commuting elements in the q-deformed Heisenberg algebras. In Chapter 8 we consider the representations of the q-deformed Heisenberg algebras by q-difference operators.

The material in the main chapters of the book, primary concerned with the q-deformed Heisenberg algebras, can also serve as an illustration and application of the general notions and concepts described in the appendices. The main chapters of the book can however be read successfully independently of the appendices. We thus feel that a good way to read this book from research as well as pedagogical and pleasure points of view would be to proceed directly with the main chapters and to turn to the appendices only when such a necessity arises.

Finally we wish to mention that the bibliography at the end of the book, comprising the references mentioned in this introduction and other parts, is most probably far from exhausting the constantly growing enormous literature mentioning or specifically devoted to q-deformed Heisenberg algebras and their applications and generalisations. We hope however that these references would provide a rather good overview of the subject and will encourage the reader to pursue further study of q-deformed Heisenberg algebras discovering new references, applications and interconnections with other parts of mathematics and physics.

We wish to mention for the record that the preprint [184] contains a preliminary version of the book. Selected results from the book have been also announced in [183].

1.4 Conventions and notations

We frequently use notation with parentheses, braces, and brackets with index, e.g.

$$(a_i)_{i=1}^{n} \qquad \{a_i\}_{i=1}^{n} \qquad \left[a_{ij}\right]_{i=1;j=1}^{m;n}$$

The first of these denotes the vector (ordered n-tuple) $(a_1, a_2, \ldots, a_n)$, the second denotes the set $\{a_1, a_2, \ldots, a_n\}$, and the third denotes the $m \times n$ matrix

$$\begin{pmatrix} a_{11} & a_{12} & \cdots & a_{1n} \\ a_{21} & a_{22} & \cdots & a_{2n} \\ \vdots & \vdots & \ddots & \vdots \\ a_{m1} & a_{m2} & \cdots & a_{mn} \end{pmatrix}$$

If there is only one index on a pair of brackets then these denote an $m \times 1$ matrix.

Our main reason for using this notation is that we often have a need for sets, vectors, and matrices where the indices are not simply numbers—indeed, the set of indices may well be infinite. This makes the simple '...' notation for leaving out elements a bit too far-fetched to be practical in many cases.

Throughout the book, a boldfaced letter such as $\mathbf{s}$ or $\mathbf{k}$ will denote a vector $(s_1, s_2, \ldots)$ (or $(k_1, k_2, \ldots)$ respectively) where the names of the components are formed by attaching an index from some index set to the corresponding non-boldfaced letter. Thus if we say $\mathbf{s} \in \mathbb{N}^n$, we mean $(s_1, \ldots, s_n) \in \mathbb{N}^n$. If we then say $c_{\mathbf{s}}$, we mean $c_{s_1,\ldots,s_n}$.

The vectors need not have finitely (or even denumerably) many components. If a vector $\mathbf{q}$ can be infinite then we will write something like $\mathbf{q} \in \mathcal{K}^J$ to specify that J is the set used to index the components of the vector and these components will be members of the set $\mathcal{K}$. If no index set has been explicitly specified however (as when we write $\mathbf{s} \in \mathbb{N}^n$), the index set will be $\{1, 2, \dots, n\}$ for some positive integer n.

A generalisation of this notation that will come in handy is that it is often convenient to consider vectors which has components for every element in a set J, but which may also contain arbitrarily many other components. In such a case we will write the set* of such vectors as $\mathcal{K}^{\square \supseteq J}$, where the square represents the unnamed and unspecified superset of J which will act as index set, rather than $\mathcal{K}^J$. A concrete example of this is that $\mathbf{q} \in \mathcal{K}^{\square \supseteq \{1\}}$ allows the case $\mathbf{q} = (q_1, q_2) \in \mathcal{K}^{\{1,2\}}$ since $\{1, 2\} \supseteq \{1\}$.

If the components of a vector $\mathbf{s}$ belong to some set S for which the concept of absolute value is meaningful (for example, S might be the integers), we use $|\mathbf{s}|$ to denote the l^1 norm of $\mathbf{s}$, i.e. if $\mathbf{s} \in S^J$ then

$$|\mathbf{s}| = \sum_{j \in J} |s_j|.$$

This practice is common for multiindices, and our vectors $\mathbf{s}$ will often serve as such.

In some places, there are $\sum$-sums of no terms or $\prod$-products of no factors. The convention in such cases is that

$$\sum_{i=n}^{n-1} a_i = \sum_{i \in \emptyset} a_i = 0, \qquad \prod_{i=n}^{n-1} a_i = \prod_{i \in \emptyset} a_i = 1.$$

In some cases we use $\prod$ on factors which do not commute. In these cases we mean

$$\prod_{i=1}^{n} \alpha_i = \alpha_1 \alpha_2 \cdots \alpha_n.$$

*At least in some axiomatizations of set-theory, this will actually *not* be a set, but this is nothing that affects our proofs.

The symbol δ_{ij} (a delta with two indices) will be used to denote the Kronecker delta, i.e.,

$$\delta_{ij} = \begin{cases} 1 & \text{if } i = j, \\ 0 & \text{if } i \neq j. \end{cases}$$

The indices may come from any set, all that matters is whether they are equal or not. The 1 and 0 are the unit and zero respectively of the field under consideration.

Chapter 2

Immediate consequences of the commutation relations

immediate (i mē´di *a*t) *adj.* [...] **2** Close, with nothing intervening.

WEBSTER'S NEW ILLUSTRATED DICTIONARY

In this chapter, we define generalised q-deformed Heisenberg algebras and prove some reordering formulae, which are used throughout the book.

2.1 The values of q

Theorem 2.1 *Let $\mathcal{A}$ be an associative algebra with unit over a field $\mathcal{K}$ and denote the unit of $\mathcal{A}$ by I. Let J be a nonempty set and let $q_{ij} \in \mathcal{K}$ be arbitrary for all $i, j \in J$.*

If there exist some elements $\{A_j, B_j\}_{j\in J} \subseteq \mathcal{A}$ that satisfy

$$[A_i, A_j] = 0, \qquad [B_i, B_j] = 0,$$
$$A_i B_j - q_{ij} B_j A_i = \delta_{ij} I$$

for all $i, j \in J$, then $q_{ij} = 1$ whenever $i \neq j$.

Proof. Let $i, j \in J$ satisfy $i \neq j$. Then

$$\begin{aligned} A_i &= q_{jj} A_i B_j A_j + A_i - q_{jj} A_i B_j A_j = A_i A_j B_j - q_{jj} q_{ij} B_j A_i A_j = \\ &= A_j A_i B_j - q_{ij} q_{jj} B_j A_j A_i = q_{ij} A_j B_j A_i - q_{ij} q_{jj} B_j A_j A_i = \\ &= q_{ij} (A_j B_j - q_{jj} B_j A_j) A_i = q_{ij} A_i, \end{aligned}$$

which is equivalent to $(1 - q_{ij}) A_i = 0$. Since $A_i B_i - q_{ii} B_i A_i = I$ implies that A_i is nonzero, $q_{ij} = 1$. □

It can be shown however, that as soon as $q_{ij} = 1$ for all $i \neq j$, then there does exist an algebra $\mathcal{A}$ that satisfies the conditions in the theorem. This motivates the following definition.

Definition 2.1 Let J be a nonempty set and $\mathcal{K}$ a field. An associative algebra over $\mathcal{K}$ with unit I that is generated by the elements $\{A_j, B_j\}_{j \in J}$ is called a **generalised q-deformed Heisenberg algebra** by J and over $\mathcal{K}$, or alternatively a **J-generalised q-deformed Heisenberg algebra** over $\mathcal{K}$, if

- $[A_i, A_j] = [B_i, B_j] = 0$ for all $i, j \in J$,
- $[A_i, B_j] = 0$ for all $i, j \in J$ such that $i \neq j$, and
- $A_j B_j - q_j B_j A_j = I$ for all $j \in J$ and some fixed $\{q_j\}_{j \in J} \subseteq \mathcal{K}$.

We will give a general construction of such algebras in Chapter 3. Though most of that chapter is about finding bases for the algebras, it is that work which will show that the construction really gives the generalised q-deformed Heisenberg algebras we claim it gives.

2.2 Reordering formulae

The reordering formulae stated in this section are valid in all generalised q-deformed Heisenberg algebras by J and over $\mathcal{K}$. As generators with different indices always commute in these algebras, it is a trivial problem to work out a formula for reordering A_i and B_j when $i \neq j$, namely

$$A_i^k B_j^l = B_j^l A_i^k$$

for all $k, l \in \mathbb{N}$. Hence the formulae we will present are for two elements A and B which satisfy the relation

$$AB = qBA + I, \tag{2.1}$$

where $q \in \mathcal{K}$ is arbitrary. This corresponds to the situation with generators with the same indices in a generalised q-deformed Heisenberg algebra.

Some later considerations will also show that it is interesting to know some reordering formulae involving the monomial BA. It is in fact of such interest that we often give this monomial the special name C. Thus, $C = BA$ in this section (and quite often in the rest of the book too).

To state these formulae in a reasonably short way, we have used five functions which are not standard, namely

$$\{n\}_q, \qquad \{n\}_q!, \qquad \binom{n}{k}_q, \qquad \begin{bmatrix} n \\ k \end{bmatrix}_q, \quad \text{and} \quad \left\{ \begin{matrix} n \\ k \end{matrix} \right\}_q .$$

The definitions of these can be found in Appendix C; note in particular that it is not $\left[\begin{smallmatrix} n \\ k \end{smallmatrix}\right]_q$ which denotes the q-binomial coefficients. To make this text self-contained, that appendix also contains proofs of all properties of these functions that we use in the book.

Theorem 2.2 *If $n \in \mathbb{Z}^+$ then*

$$AB^n = q^n B^n A + \{n\}_q B^{n-1} \tag{2.2}$$
$$A^n B = q^n BA^n + \{n\}_q A^{n-1} \tag{2.3}$$

Proof. The proof is by induction on n. If $n = 1$ then both equations are exactly the commutation relation (2.1). In the case $n > 1$, assume that (2.2) and (2.3) hold for $n - 1$. Then it follows that

$$\begin{aligned} AB^n &= qBAB^{n-1} + B^{n-1} = \\ &= qB\left(q^{n-1}B^{n-1}A + \{n-1\}_q B^{n-2}\right) + B^{n-1} = \\ &= q^n B^n A + \{n\}_q B^{n-1} \\ A^n B &= qA^{n-1}BA + A^{n-1} = \\ &= q\left(q^{n-1}BA^{n-1} + \{n-1\}_q A^{n-2}\right)A + A^{n-1} = \\ &= q^n BA^n + \{n\}_q A^{n-1} \end{aligned}$$

which completes the proof. □

Theorem 2.3 *Let $i, j \in \mathbb{N}$, then*

$$A^i B^j = \sum_{k=0}^{\min(i,j)} q^{(i-k)(j-k)} \{k\}_q! \binom{i}{k}_q \binom{j}{k}_q B^{j-k} A^{i-k} \tag{2.4}$$

Proof. The method of the proof is induction on $\min(i,j)$. The case $\min(i,j)=0$ is obvious since k then must be 0.

Let $\min(i,j)=n>0$ and assume (2.4) holds for $\min(i,j)<n$. There are two cases to consider:

Case 1: $i=\min(i,j)$. Here it is known that i can be substituted for $\min(i,j)$, thus (2.4) for this case can be shown through the calculations in (2.5) below. When, in some step, a numbered equation from elsewhere in this book is applied, the number of that equation is placed above the corresponding equality sign. Similarly, the equality sign at which the induction hypothesis is applied is marked with $(*)$.

$$
\begin{aligned}
&A^iB^j \overset{(2.2)}{=} A^{i-1}(q^jB^jA+\{j\}_qB^{j-1}) = \\
&= q^jA^{i-1}B^jA+\{j\}_qA^{i-1}B^{j-1} \overset{(*)}{=} \\
&\overset{(*)}{=} q^j\sum_{k=0}^{i-1} q^{(i-1-k)(j-k)}\{k\}_q!\binom{i-1}{k}_q\binom{j}{k}_q B^{j-k}A^{i-k}+ \\
&\quad+\{j\}_q\sum_{k=0}^{i-1} q^{(i-1-k)(j-1-k)}\{k\}_q!\binom{i-1}{k}_q\binom{j-1}{k}_q B^{j-1-k}A^{i-1-k} = \\
&= q^j\left(\sum_{k=1}^{i-1} q^{(i-1-k)(j-k)}\{k\}_q!\binom{i-1}{k}_q\binom{j}{k}_q B^{j-k}A^{i-k}+\right. \\
&\quad\left.+q^{(i-1)j}B^jA^i\right)+\{j\}_q\left(\{i-1\}_q!\binom{i-1}{i-1}_q\binom{j-1}{i-1}_q B^{j-i}+\right. \\
&\quad\left.+\sum_{k=1}^{i-1} q^{(i-k)(j-k)}\{k-1\}_q!\binom{i-1}{k-1}_q\binom{j-1}{k-1}_q B^{j-k}A^{i-k}\right) = \\
&= \sum_{k=1}^{i-1}\left(q^{(i-k)(j-k)}q^k\{k\}_q!\binom{i-1}{k}_q\binom{j}{k}_q+\right. \\
&\quad\left.+q^{(i-k)(j-k)}\{j\}_q\{k-1\}_q!\binom{i-1}{k-1}_q\binom{j-1}{k-1}_q\right)B^{j-k}A^{i-k}+ \\
&\quad+q^{ij}B^jA^i+\{i\}_q!\binom{i}{i}_q\binom{j}{i}_q B^{j-i} \overset{(C.21)}{=}
\end{aligned}
$$

$$\overset{(C.21)}{=} \sum_{k=1}^{i-1} q^{(i-k)(j-k)} \{k\}_q! \binom{i}{k}_q \binom{j}{k}_q B^{j-k} A^{i-k} +$$

$$+ q^{ij} B^j A^i + \{i\}_q! \binom{i}{i}_q \binom{j}{i}_q B^{j-i} =$$

$$= \sum_{k=0}^{i} q^{(i-k)(j-k)} \{k\}_q! \binom{i}{k}_q \binom{j}{k}_q B^{j-k} A^{i-k} \tag{2.5}$$

Case 2: $j = \min(i, j)$. In analogy with the first case, j can be substituted for $\min(i, j)$, thus (2.4) for this case can be shown through the calculations in (2.6) below. The reference system is the same as in the first case.

$$A^i B^j \overset{(2.3)}{=} (q^i B A^i + \{i\}_q A^{i-1}) B^{j-1} =$$

$$= q^i B A^i B^{j-1} + \{i\}_q A^{i-1} B^{j-1} \overset{(*)}{=}$$

$$\overset{(*)}{=} q^i \sum_{k=0}^{j-1} q^{(i-k)(j-1-k)} \{k\}_q! \binom{i}{k}_q \binom{j-1}{k}_q B^{j-k} A^{i-k} +$$

$$+ \{i\}_q \sum_{k=0}^{j-1} q^{(i-1-k)(j-1-k)} \{k\}_q! \binom{i-1}{k}_q \binom{j-1}{k}_q B^{j-1-k} A^{i-1-k} =$$

$$= q^i \left(\sum_{k=1}^{j-1} q^{(i-k)(j-1-k)} \{k\}_q! \binom{i}{k}_q \binom{j-1}{k}_q B^{j-k} A^{i-k} + \right.$$

$$\left. + q^{i(j-1)} B^j A^i \right) + \{i\}_q \left(\{j-1\}_q! \binom{i-1}{j-1}_q \binom{j-1}{j-1}_q A^{i-j} + \right.$$

$$\left. + \sum_{k=1}^{j-1} q^{(i-k)(j-k)} \{k-1\}_q! \binom{i-1}{k-1}_q \binom{j-1}{k-1}_q B^{j-k} A^{i-k} \right) =$$

$$= \sum_{k=1}^{j-1} \left(q^{(i-k)(j-k)} q^k \{k\}_q! \binom{i}{k}_q \binom{j-1}{k}_q + \right.$$

$$\left. + q^{(i-k)(j-k)} \{i\}_q \{k-1\}_q! \binom{i-1}{k-1}_q \binom{j-1}{k-1}_q \right) B^{j-k} A^{i-k} +$$

$$+ q^{ij} B^j A^i + \{j\}_q! \binom{i}{j}_q \binom{j}{j}_q A^{i-j} \overset{(C.21)}{=}$$

$$\overset{(C.21)}{=} \sum_{k=1}^{j-1} q^{(i-k)(j-k)} \{k\}_q! \binom{i}{k}_q \binom{j}{k}_q B^{j-k} A^{i-k} +$$
$$+ q^{ij} B^j A^i + \{j\}_q! \binom{i}{j}_q \binom{j}{j}_q A^{i-j} =$$
$$= \sum_{k=0}^{j} q^{(i-k)(j-k)} \{k\}_q! \binom{i}{k}_q \binom{j}{k}_q B^{j-k} A^{i-k} \tag{2.6}$$

Together the two cases form the inductive step and thus, by the induction principle, the theorem follows. □

This book only treats nonnegative powers of A and B, but we do sometimes comment on how results generalise to such cases, or more precisely, what generalisations there are if we additionally assume that there exists inverses of A and B. In the case of Theorem 2.3, that generalisation is straightforward. Provided (i) that the definition of the q-binomial coefficients $\binom{n}{k}_q$ is extended to negative n and non-negative k, as described in Section C.4, and (ii) that the sum in (2.4) is taken over all $k \in \mathbb{N}$, that reordering formula actually holds for all $i, j \in \mathbb{Z}$. If only one of i and j is negative then only finitely many terms in the sum are nonzero and one can even use the above proof to prove that the formula holds—only the base for the induction needs to be replaced! When both i and j are negative things get trickier, as in general the sum will become a series with infinitely many nonzero terms, but it can still be shown using elementary methods that the formula holds as an equality of formal power series.

If one applies the special properties of $\{n\}_1$ described in Section C.2, the formula (2.4) can be slightly rewritten when $q = 1$.

Corollary 2.4 *If $q = 1$ then*

$$A^i B^j = \sum_{k=0}^{\min(i,j)} \left\{ k! \binom{i}{k} \binom{j}{k} \right\}_1 B^{j-k} A^{i-k} \tag{2.7}$$

for all $i, j \in \mathbb{N}$.

It is possible to prove a formula for rewriting $B^i A^j$ as a sum of multiples of $A^{j-k} B^{i-k}$ in the same way as (2.4) was proved, but that would be pretty long. As it turns out, there is a simpler way.

Theorem 2.5 *Let $i, j \in \mathbb{N}$. If $q \neq 0$ then*

$$B^i A^j = \sum_{k=0}^{\min(i,j)} (-1)^k q^{-(i-k)(j-k)-k} \{k\}_{q^{-1}}! \binom{i}{k}_{q^{-1}} \binom{j}{k}_{q^{-1}} A^{j-k} B^{i-k} = \tag{2.8}$$

$$= \sum_{k=0}^{\min(i,j)} (-1)^k q^{\binom{k}{2}-ij} \{k\}_q! \binom{i}{k}_q \binom{j}{k}_q A^{j-k} B^{i-k} \tag{2.9}$$

Proof. Let $A' = -qB$, $B' = A$, and $q' = q^{-1}$. If (A', B', q') are substituted for (A, B, q) in (2.1) then that equation still holds. Hence by Theorem 2.3,

$$(-q)^i B^i A^j = (-qB)^i A^j = A'^i B'^j =$$
$$= \sum_{k=0}^{\min(i,j)} q'^{(i-k)(j-k)} \{k\}_{q'}! \binom{i}{k}_{q'} \binom{j}{k}_{q'} B'^{j-k} A'^{i-k} =$$
$$= \sum_{k=0}^{\min(i,j)} q^{-(i-k)(j-k)} \{k\}_{q^{-1}}! \binom{i}{k}_{q^{-1}} \binom{j}{k}_{q^{-1}} A^{j-k} (-qB)^{i-k} =$$
$$= (-q)^i \sum_{k=0}^{\min(i,j)} (-q)^{-k} q^{-(i-k)(j-k)} \{k\}_{q^{-1}}! \binom{i}{k}_{q^{-1}} \binom{j}{k}_{q^{-1}} A^{j-k} B^{i-k},$$

which proves (2.8).

Showing (2.9) is then just a matter of rewriting the coefficients. Note that by Theorem C.10,

$$(-1)^k q^{-(i-k)(j-k)-k} \{k\}_{q^{-1}}! \binom{i}{k}_{q^{-1}} \binom{j}{k}_{q^{-1}} =$$
$$= (-1)^k q^{-(i-k)(j-k)-k} \cdot q^{-\binom{k}{2}} \{k\}_q! \cdot q^{-k(i-k)} \binom{i}{k}_q \cdot q^{-k(j-k)} \binom{j}{k}_q =$$
$$= (-1)^k q^{-i(j-k)+k(j-k)-k-\frac{1}{2}(k^2-k)-k(i-k)-k(j-k)} \{k\}_q! \binom{i}{k}_q \binom{j}{k}_q =$$
$$= (-1)^k q^{-ij+ik-\frac{1}{2}k+\frac{1}{2}k^2-ik} \{k\}_q! \binom{i}{k}_q \binom{j}{k}_q =$$
$$= (-1)^k q^{\binom{k}{2}-ij} \{k\}_q! \binom{i}{k}_q \binom{j}{k}_q .$$

Thus (2.9) holds as well. □

The formulae in the following corollary are the important special cases $i = 1$ and $j = 1$ respectively of (2.8) and (2.9), obtained by using (C.10). They can alternatively be deduced from (2.2) and (2.3) with the method used in the proof of Theorem 2.5.

Corollary 2.6 *If $i, j \in \mathbb{Z}^+$ and $q \neq 0$ then*

$$BA^j = q^{-j}A^jB - q^{-1}\{j\}_{q^{-1}}A^{j-1} = \tag{2.10}$$

$$= q^{-j}A^jB - q^{-j}\{j\}_qA^{j-1}, \tag{2.11}$$

$$B^iA = q^{-i}AB^i - q^{-1}\{i\}_{q^{-1}}B^{i-1} = \tag{2.12}$$

$$= q^{-i}AB^i - q^{-i}\{i\}_qB^{i-1}. \tag{2.13}$$

Theorem 2.7 *Recall that $C = BA$. Let $i, j \in \mathbb{N}$. Then*

$$A^iC^j = \left(q^iC + \{i\}_qI\right)^jA^i, \tag{2.14}$$

$$C^iB^j = B^j\left(q^jC + \{j\}_qI\right)^i, \tag{2.15}$$

and if $q \neq 0$ then

$$C^iA^j = q^{-ij}A^j\left(C - \{j\}_qI\right)^i, \tag{2.16}$$

$$B^iC^j = q^{-ij}\left(C - \{i\}_qI\right)^jB^i. \tag{2.17}$$

Proof. By Theorem 2.2,

$$A^iC = A^iBA = q^iBA^{i+1} + \{i\}_qA^i = (q^iC + \{i\}_qI)A^i. \tag{2.18}$$

It is obvious that (2.14) holds for $j = 0$. When $n > 0$ and assuming that (2.14) holds for $j = n - 1$, it follows from

$$A^iC^j = (q^iC + \{i\}_qI)A^iC^{j-1} = (q^iC + \{i\}_qI)(q^iC + \{i\}_qI)^{j-1}A^i = \\ = (q^iC + \{i\}_qI)^jA^i$$

that (2.14) holds for $j = n$. Thus by induction it holds for all $n \in \mathbb{N}$.

Theorem 2.2 also implies that

$$CB^j = BAB^j = q^jB^{j+1}A + \{j\}_qB^j = q^jB^jC + \{j\}_qB^j, \tag{2.19}$$

from which follows that $CB^j = B^j\left(q^jC + \{j\}_qI\right)$. Then (2.15) follows by induction on i, just as did (2.14).

Assume $q \neq 0$. Then (2.18) and (2.19) also imply that

$$CA^j = q^{-j} A^j \big(C - \{j\}_q I\big),$$
$$B^i C = q^{-i} \big(C - \{i\}_q I\big) B^i$$

respectively. It is obvious that (2.16) and (2.17) hold for $i = 0$ and $j = 0$ respectively. When $n > 0$ and assuming that these two equations hold for $i = n-1$ and $j = n-1$ respectively, it follows from

$$C^i A^j = q^{-j} C^{i-1} A^j \big(C - \{j\}_q I\big) = q^{-ij} A^j \big(C - \{j\}_q I\big)^j,$$
$$B^i C^j = q^{-i} (C - \{i\}_q I) B^i C^{j-1} = q^{-ij} (C - \{i\}_q I)^j B^i$$

that they hold for $i = n$ and $j = n$ respectively as well. Thus by induction they hold for all $i, j \in \mathbb{N}$. □

Theorem 2.8 *Note that $C = BA$. If $n \in \mathbb{N}$ then*

$$A^{n+1} B^{n+1} = A^n B^n \big(q^{n+1} C + \{n+1\}_q I\big), \tag{2.20}$$

hence

$$A^n B^n = \prod_{k=1}^{n} \big(q^k C + \{k\}_q I\big) \tag{2.21}$$

for all $n \in \mathbb{N}$. Furthermore, if $q \neq 0$ and $n \in \mathbb{N}$ then

$$B^{n+1} A^{n+1} = q^{-n} B^n A^n \big(C - \{n\}_q I\big), \tag{2.22}$$

hence

$$B^n A^n = \prod_{k=0}^{n-1} q^{-k} \big(C - \{k\}_q I\big) \tag{2.23}$$

for all $n \in \mathbb{N}$.

Proof. Let $n \in \mathbb{N}$ be arbitrary. Then (2.2) implies that

$$\begin{aligned} A^{n+1} B^{n+1} &= q^n A^n B^n AB + \{n\}_q A^n B^n = A^n B^n (q^n AB + \{n\}_q I) = \\ &= A^n B^n (q^{n+1} BA + q^n I + \{n\}_q I) = \\ &= A^n B^n (q^{n+1} C + \{n+1\}_q I), \end{aligned}$$

which is exactly (2.20). Equation (2.21) follows from (2.20) by induction.

It can be deduced from (2.3) that

$$B^nA^nBA = q^nB^{n+1}A^{n+1} + \{n\}_qB^nA^n.$$

Hence $q^nB^{n+1}A^{n+1} = B^nA^n\big(BA - \{n\}_qI\big)$, which implies (2.22). Then a simple induction gives (2.23). □

Theorem 2.9 *For all $n \in \mathbb{N}$ and $q \in \mathcal{K}$,*

$$C^n = \sum_{k=0}^{n} \left\{ {n \atop k} \right\}_q q^{\binom{k}{2}} B^kA^k, \tag{2.24}$$

and if $q \neq 0$ then

$$B^nA^n = q^{-\binom{n}{2}} \sum_{k=0}^{n} \begin{bmatrix} n \\ k \end{bmatrix}_q (-1)^{n-k}C^k, \tag{2.25}$$

$$A^nB^n = q^{\binom{n}{2}} \sum_{k=0}^{n} \begin{bmatrix} n+1 \\ k+1 \end{bmatrix}_{q^{-1}} q^kC^k, \tag{2.26}$$

$$C^n = q^{-n} \sum_{k=0}^{n} \left\{ {n+1 \atop k+1} \right\}_{q^{-1}} (-1)^{n-k}q^{-\binom{k}{2}}A^kB^k. \tag{2.27}$$

Proof. The formula (2.25) is a direct corollary to (2.23), since by that equation and Definition C.2,

$$B^nA^n = \prod_{k=0}^{n-1} q^{-k}\big(C - \{k\}_qI\big) = q^{-\binom{n}{2}}F_n(C;q) = q^{-\binom{n}{2}} \sum_{k=0}^{n} \begin{bmatrix} n \\ k \end{bmatrix}_q (-1)^{n-k}C^k.$$

The formula (2.24) could be inferred from this and Corollary C.19, but it is more instructive to do it directly. By Lemma 2.2 and Theorem C.13, for all $n \in \mathbb{N}$,

$$\begin{aligned} C\sum_{k=0}^{n} \left\{ {n \atop k} \right\}_q q^{\binom{k}{2}} B^kA^k &= \sum_{k=0}^{n} \left\{ {n \atop k} \right\}_q q^{\binom{k}{2}} BAB^kA^k = \\ &= \sum_{k=0}^{n} \left\{ {n \atop k} \right\}_q q^{\binom{k}{2}} B\big(q^kB^kA + \{k\}_qB^{k-1}\big)A^k = \\ &= \sum_{i=1}^{n+1} \left\{ {n \atop i-1} \right\}_q q^{\binom{i}{2}} B^iA^i + \sum_{k=0}^{n} \{k\}_q \left\{ {n \atop k} \right\}_q q^{\binom{k}{2}} B^kA^k = \end{aligned}$$

$$= \sum_{k=1}^{n+1} \begin{Bmatrix} n \\ k-1 \end{Bmatrix}_q q^{\binom{k}{2}} B^k A^k + \sum_{k=1}^{n+1} \{k\}_q \begin{Bmatrix} n \\ k \end{Bmatrix}_q q^{\binom{k}{2}} B^k A^k =$$

$$= \sum_{k=1}^{n+1} \begin{Bmatrix} n+1 \\ k \end{Bmatrix}_q q^{\binom{k}{2}} B^k A^k.$$

Hence by induction on n,

$$C^{n+1} = C \sum_{k=0}^{n} \begin{Bmatrix} n \\ k \end{Bmatrix}_q q^{\binom{k}{2}} B^k A^k =$$

$$= \sum_{k=1}^{n+1} \begin{Bmatrix} n+1 \\ k \end{Bmatrix}_q q^{\binom{k}{2}} B^k A^k = \sum_{k=0}^{n+1} \begin{Bmatrix} n+1 \\ k \end{Bmatrix}_q q^{\binom{k}{2}} B^k A^k.$$

The formula (2.26) is a corollary to (2.21), but here some trickery is needed. By that formula, Theorem C.10, and Definition C.2,

$$qCA^nB^n = qC \prod_{k=1}^{n} \left(q^k C + \{k\}_q I\right) = qC \prod_{k=1}^{n} q^{k-1} \left(qC + \{k\}_{q^{-1}} I\right) =$$

$$= q^{\binom{n}{2}} \prod_{k=0}^{n} -\left(-qC - \{k\}_{q^{-1}} I\right) =$$

$$= q^{\binom{n}{2}} (-1)^{n+1} F_{n+1}(-qC; q^{-1}) =$$

$$= q^{\binom{n}{2}} (-1)^{n+1} \sum_{k=0}^{n+1} \begin{bmatrix} n+1 \\ k \end{bmatrix}_{q^{-1}} (-1)^{n+1-k} (-qC)^k =$$

$$= q^{\binom{n}{2}} \sum_{k=1}^{n+1} \begin{bmatrix} n+1 \\ k \end{bmatrix}_{q^{-1}} q^k C^k$$

for all $n \in \mathbb{N}$. Dividing by qC here would give the wanted result, but that is not necessarily a valid step. One can however combine the result with an induction to get the same effect. Clearly (2.26) holds for $n = 0$. Let $m > 0$ and assume that it holds for $n = m - 1$. Then

$$q^{-\binom{n+1}{2}} A^{n+1} B^{n+1} = q^{-\binom{n}{2}} \left(qBA^{n+1}B^n + q^{-n}\{n+1\}_q A^n B^n\right) =$$

$$= q^{-\binom{n}{2}} \left(qCA^nB^n + \{n+1\}_{q^{-1}} A^n B^n\right) =$$

$$= \sum_{k=1}^{n+1} \begin{bmatrix} n+1 \\ k \end{bmatrix}_{\frac{1}{q}} q^k C^k + \{n+1\}_{q^{-1}} \sum_{k=0}^{n} \begin{bmatrix} n+1 \\ k+1 \end{bmatrix}_{\frac{1}{q}} q^k C^k =$$

$$= \{n+1\}_{q^{-1}} \begin{bmatrix} n+1 \\ 1 \end{bmatrix}_{\frac{1}{q}} I + \begin{bmatrix} n+1 \\ n+1 \end{bmatrix}_{\frac{1}{q}} q^{n+1}C^{n+1} +$$

$$+ \sum_{k=1}^{n} \left(\begin{bmatrix} n+1 \\ k \end{bmatrix}_{\frac{1}{q}} + \{n+1\}_{q^{-1}} \begin{bmatrix} n+1 \\ k+1 \end{bmatrix}_{\frac{1}{q}} \right) q^k C^k =$$

$$= \sum_{k=0}^{n+1} \begin{bmatrix} n+2 \\ k+1 \end{bmatrix}_{\frac{1}{q}} q^k C^k,$$

and thus the validity of (2.26) for all $n \in \mathbb{N}$ follows by induction.

The formula (2.27), finally, is shown the same way as (2.24) was. We leave the proof to the reader. □

Some of these formulae are already known in the literature. The formula (2.4) in the case $j \leq i$, and the formulae (2.23), (2.24), and (2.25) can be found in [73], where they are deduced for an operator algebra which satisfies (2.1). The formula (2.7) is a special case of a formula in [387], and Theorem 2.2 appears in [373].

2.3 Simplifying commutation relations

It is not unusual that one encounters various generalised forms of the basic q-deformed Heisenberg commutation relation $AB = qBA + I$. One of the simplest of these are

$$AB = qBA + pI, \tag{2.28}$$

where $p \in \mathcal{K} \setminus \{0\}$. This can easily be rewritten in the form (2.1) through the substitution $B' = p^{-1}B$, since

$$AB' = p^{-1}AB = p^{-1}(qBA + pI) = q(p^{-1}B)A + I = qB'A + I.$$

Conversely this implies that the reordering formulae deduced in Section 2.2 have easily deduced counterparts for the commutation relation (2.28). Indeed, assuming that A, B, q, and p satisfy (2.28), and that $B = pB'$, one quickly sees that

$$\begin{aligned} A^iB^j &= p^j A^i (B')^j = \\ &= p^j \sum_{k=0}^{\min(i,j)} q^{(i-k)(j-k)} \{k\}_q! \binom{i}{k}_q \binom{j}{k}_q (B')^{j-k} A^{i-k} = \\ &= \sum_{k=0}^{\min(i,j)} q^{(i-k)(j-k)} \{k\}_q! \binom{i}{k}_q \binom{j}{k}_q p^j p^{-j+k} B^{j-k} A^{i-k} = \\ &= \sum_{k=0}^{\min(i,j)} q^{(i-k)(j-k)} \{k\}_q! \binom{i}{k}_q \binom{j}{k}_q p^k B^{j-k} A^{i-k} \end{aligned} \tag{2.29}$$

by Theorem 2.3. Similarly by Theorem 2.5,

$$\begin{aligned} B^iA^j &= p^i (B')^i A^j = \\ &= p^i \sum_{k=0}^{\min(i,j)} (-1)^k q^{-(i-k)(j-k)-k} \{k\}_{q^{-1}}! \binom{i}{k}_{q^{-1}} \binom{j}{k}_{q^{-1}} A^{j-k} (B')^{i-k} = \\ &= \sum_{k=0}^{\min(i,j)} (-1)^k q^{-(i-k)(j-k)-k} \{k\}_{q^{-1}}! \binom{i}{k}_{q^{-1}} \binom{j}{k}_{q^{-1}} p^k A^{j-k} B^{i-k} = \\ &= \sum_{k=0}^{\min(i,j)} (-1)^k q^{\binom{k}{2}-ij} \{k\}_q! \binom{i}{k}_q \binom{j}{k}_q p^k A^{j-k} B^{i-k}. \end{aligned} \tag{2.30}$$

The same idea can be applied to the results in Theorems 2.7–2.9. One only has to remember that since $C = BA = pB'A$, it is suitable to introduce a $C' = B'A$ for which the formulae in those theorems apply. Thus

$$A^iC^j = p^j A^i (C')^j = p^j \left(q^i C' + \{i\}_q I\right)^j A^i = \left(q^i C + \{i\}_q pI\right)^j A^i, \tag{2.31}$$

and

$$\begin{aligned} C^iB^j = p^{i+j}(C')^i(B')^j &= p^{i+j}(B')^j \left(q^j C' + \{j\}_q I\right)^i = \\ &= B^j \left(q^j C + \{j\}_q pI\right)^i . \end{aligned} \tag{2.32}$$

The remaining formulae in these theorems become

$$C^iA^j = q^{-ij} A^j \left(C - \{j\}_q pI\right)^i \tag{2.33}$$

$$B^iC^j = q^{-ij} \left(C - \{i\}_q pI\right)^j B^i \tag{2.34}$$

$$A^nB^n = \prod_{k=1}^{n} \left(q^k C + p\{k\}_q I\right) = q^{\binom{n}{2}} \sum_{k=0}^{n} \begin{bmatrix} n+1 \\ k+1 \end{bmatrix}_{\frac{1}{q}} p^{n-k} q^k C^k \tag{2.35}$$

$$B^n A^n = q^{-\binom{n}{2}} \prod_{k=0}^{n-1} (C - p\{k\}_q I) = q^{-\binom{n}{2}} \sum_{k=0}^{n} \begin{bmatrix} n \\ k \end{bmatrix}_q (-p)^{n-k} C^k \tag{2.36}$$

$$C^n = \sum_{k=0}^{n} \begin{Bmatrix} n \\ k \end{Bmatrix}_q p^{n-k} q^{\binom{k}{2}} B^k A^k = \tag{2.37}$$

$$= q^{-n} \sum_{k=0}^{n} \begin{Bmatrix} n+1 \\ k+1 \end{Bmatrix}_{\frac{1}{q}} (-p)^{n-k} q^{\binom{k}{2}} A^k B^k \tag{2.38}$$

Another family of commutation relations which generalise (2.1) even further concerns an algebra with three elements A, B, and D, which satisfy

$$AB = qBA + D \qquad \text{and} \qquad [A, D] = 0 = [B, D] \tag{2.39}$$

for some $q \in \mathcal{K}$. Clearly, (2.28) is a special case of (2.39), but just as results for algebras satisfying (2.1) could be lifted to results for algebras satisfying (2.28), results for algebras satisfying (2.28) can be lifted to results for algebras satisfying (2.39). In this case though, the connection is not as immediate as before, and it therefore seems best to start with a fake proof, since the technicalities would otherwise obscure the idea.

Fake proof: Start by letting $\mathcal{K}' = \mathcal{K}(D)$—the extension of the field $\mathcal{K}$ by D, which may be algebraic or transcendental over $\mathcal{K}$ (it works either way). Since $\mathcal{K}'$ is a field itself, there is no problem with picking the scalars q and p for use in (2.28) from $\mathcal{K}'$. Furthermore, if $p = D$ then (2.39) for an algebra over $\mathcal{K}$ is equivalent to (2.28) for an algebra over $\mathcal{K}'$. This in turn implies that generalisations of (2.29)–(2.38) that hold in an algebra over $\mathcal{K}$ which satisfies (2.39) can be deduced from exactly those equations! The generalisations in question are

$$A^i B^j = \sum_{k=0}^{\min(i,j)} q^{(i-k)(j-k)} \{k\}_q! \binom{i}{k}_q \binom{j}{k}_q B^{j-k} A^{i-k} D^k, \tag{2.40}$$

$$A^i C^j = \left(q^i C + \{i\}_q D\right)^j A^i, \tag{2.41}$$

$$C^i B^j = B^j \left(q^j C + \{j\}_q D\right)^i, \tag{2.42}$$

$$A^n B^n = \prod_{k=1}^{n} \left(q^k C + \{k\}_q D\right), \tag{2.43}$$

$$C^n = \sum_{k=0}^{n} \left\{ {n \atop k} \right\}_q q^{\binom{k}{2}} B^k A^k D^{n-k} \tag{2.44}$$

and for the case $q \neq 0$,

$$B^i A^j = \sum_{k=0}^{\min(i,j)} (-1)^k q^{\binom{k}{2}-ij} \{k\}_q! \binom{i}{k}_q \binom{j}{k}_q A^{j-k} B^{i-k} D^k, \tag{2.45}$$

$$C^i A^j = q^{-ij} A^j \left(C - \{j\}_q D\right)^i, \tag{2.46}$$

$$B^i C^j = q^{-ij} \left(C - \{i\}_q D\right)^j B^i, \tag{2.47}$$

$$A^n B^n = \prod_{k=1}^{n} \left(q^k C + \{k\}_q D\right) = q^{\binom{n}{2}} \sum_{k=0}^{n} \begin{bmatrix} n+1 \\ k+1 \end{bmatrix}_{\frac{1}{q}} q^k C^k D^{n-k}, \tag{2.48}$$

$$B^n A^n = \prod_{k=0}^{n-1} q^{-k} \left(C - \{k\}_q D\right) = q^{-\binom{n}{2}} \sum_{k=0}^{n} \begin{bmatrix} n \\ k \end{bmatrix}_q (-1)^{n-k} C^k D^{n-k}, \tag{2.49}$$

$$C^n = q^{-n} \sum_{k=0}^{n} \left\{ {n+1 \atop k+1} \right\}_{\frac{1}{q}} (-1)^{n-k} q^{\binom{k}{2}} A^k B^k D^{n-k}. \tag{2.50}$$

Now what is wrong with this proof? Only that D might satisfy some equation which contradicts that D is invertible—it might for example be the case that $D^n = 0$ for some positive integer n. If that is the case then it isn't possible to construct $\mathcal{K}(D)$, and so the entire argument fails. There is however a way around this.

Let $\mathcal{A}$ be the $\mathcal{K}$-algebra in which (2.40)–(2.50) are to be proved. Let $\mathcal{K}' = \mathcal{K}(x)$, for some x which really is transcendental over $\mathcal{K}$. Construct a $\mathcal{K}'$-algebra $\mathcal{B}$ in which there are elements A' and B' which satisfy $A'B' = qB'A' + xI$, but in which no other defining relations hold. This is a non-trivial step, but one may choose $\mathcal{B} = \mathcal{H}(q, \mathcal{K}')$ from Chapter 3, $A' = A$ (the A of $\mathcal{H}(q, \mathcal{K}')$, not the A in $\mathcal{A}$), and $B' = xB$ (similarly). By (2.29), this implies that

$$(A')^i (B')^j = \sum_{k=0}^{\min(i,j)} q^{(i-k)(j-k)} \{k\}_q! \binom{i}{k}_q \binom{j}{k}_q x^k (B')^{j-k} (A')^{i-k}.$$

Now let $\mathcal{C}$ be the $\mathcal{K}$-subalgebra of $\mathcal{B}$ that is generated by A', B', and xI. Clearly

$$(A')^i(B')^j = \sum_{k=0}^{\min(i,j)} q^{(i-k)(j-k)}\{k\}_q! \binom{i}{k}_q \binom{j}{k}_q (xI)^k(B')^{j-k}(A')^{i-k} \tag{2.51}$$

must hold in $\mathcal{C}$.

The trick is that there exists a $\mathcal{K}$-algebra homomorphism $\phi\colon \mathcal{C} \longrightarrow \mathcal{A}$ such that $A' \mapsto A$, $B' \mapsto B$, and $xI \mapsto D$. This is also non-trivial, but it is true because there are no defining relations in $\mathcal{C}$ which does not hold in $\mathcal{A}$ as well. Applying ϕ to both sides of (2.51) yields

$$A^iB^j = \sum_{k=0}^{\min(i,j)} q^{(i-k)(j-k)}\{k\}_q! \binom{i}{k}_q \binom{j}{k}_q D^kB^{j-k}A^{i-k},$$

which is the wanted equality. The same procedure can be repeated for the other formulae to produce (2.41)–(2.50).

Alternatively, one may prove (2.40)–(2.50) by induction. The reader who prefers that method will find it a trivial, even if rather boring, task to make proofs of (2.40)–(2.50) out of the proofs of Theorems 2.2–2.9.

Chapter 3

Bases and normal form in $\mathcal{H}(q)$ and $\mathcal{H}(\mathrm{q}, J)$

'Twas brillig, and the slithy toves
Did gyre and gimble in the wabe:
All mimsy were the borogoves
And the mome raths outgrabe.
— from *Jabberwocky* by L. CARROLL

For most of us, the above piece of poetry would not make any sense. The reason for this is that many words in it lack meaning—or at least they do at the moment of the first reading. There are two obvious ways to resolve this problem. One is to skip this poem altogether. Another way is to try to somehow find, or to assign, some meaning to those words that are unclear, so that the poem would make sense. A good thing is that both approaches are acceptable and it is a matter of personal taste which of them to take.

Before reading this chapter, the reader should be aware that it employs a rather comprehensive machinery to get the results proved. This machinery is connected to a theorem known as the Diamond Lemma for Ring Theory, a theorem which is well known in the literature. The reader will find the complete definitions of all concepts we use that are connected to the Diamond Lemma, together with a proof of the Lemma itself, in Appendix A.

It is perfectly possible to understand most of this book, in particular the new results, without learning the Diamond Lemma formalism. Hence we have seen it best to provide readers who would prefer not to make any deeper studies in the Diamond Lemma formalism with some kind of a guide on what to read and what to skip. Such readers should however be aware that this requires that most of the basic theorems in this book must be accepted on a "without proof" basis, as the proofs we give often rely heavily on the Diamond Lemma and the formalism connected to it. Luckily, these basic theorems are often almost intuitive; what we need the Diamond Lemma for is to prove in general things which seem almost obviously true for any given example.

What then, should one read to be able to continue with the rest of this book without first having to get into the Diamond Lemma formalism? Start by reading the comments to Definitions 3.2–3.4, these should explain enough about the algebras $\mathcal{H}(\mathbf{q}, J, \mathcal{K})$ to form a working metaphor for them. Then read the Note to Theorem 3.1, this should put the working metaphor on more solid ground. Finally read everything after the Note except the proofs.

3.1 The definition of $\mathcal{H}(\mathbf{q}, J)$

Leaving suggestions for a first reading through aside, we now return to the strict order of mathematical reasoning and continue with some important, however technical, definitions.

Definition 3.1 Let $\mathcal{R}$ be an associative and commutative ring with unit. Let $\mathcal{A}$ be an $\mathcal{R}$-algebra with unit and let $X \subseteq \mathcal{A}$ be a nonempty set. We then define

$$\mathrm{Alg}_{\mathcal{R}}\langle X\rangle = \bigcap_{\substack{\mathcal{B}\supseteq X \\ \mathcal{B}\text{ is a subalgebra of }\mathcal{A}\text{ with unit}}} \mathcal{B},$$

and call $\mathrm{Alg}_{\mathcal{R}}\langle X\rangle$ the $\mathcal{R}$-subalgebra of $\mathcal{A}$ with unit **generated** by X.

Definition 3.2 Let J be a nonempty set and $\mathcal{K}$ be a field. Then $\mathcal{F}(J, \mathcal{K})$ denotes $\mathcal{K}\big\langle\{\mathsf{a}_j, \mathsf{b}_j\}_{j\in J}\big\rangle$, the free associative $\mathcal{K}$-algebra with unit 1 whose set of generators is $\{\mathsf{a}_j, \mathsf{b}_j\}_{j\in J}$. If there is no doubt or irrelevant which the field $\mathcal{K}$ is then it might be dropped from the notation, leaving $\mathcal{F}(J)$.

Note: Given a nonempty set J, a field $\mathcal{K}$, and a any nonempty subset $J' \subseteq J$, we will identify the free algebra $\mathcal{F}(J', \mathcal{K})$ with the algebra

$$\mathrm{Alg}_{\mathcal{K}}\langle \{\mathsf{a}_j, \mathsf{b}_j\}_{j\in J'} \rangle \subseteq \mathcal{F}(J, \mathcal{K}).$$

Definition 3.3 Let J be a nonempty set and $\prec$ be some fixed total ordering of J. Let $\mathcal{K}$ be a field and let $\mathbf{q} \in \mathcal{K}^J$, or more generally let $\mathbf{q} \in \mathcal{K}^{\square \supseteq J}$, be arbitrary. Consider the algebra $\mathcal{F}(J, \mathcal{K})$.

Now define the reduction system $S(\mathbf{q}, J)$ to be the reduction system that consists exactly of those rules (μ_s, α_s) that fit some of the patterns

$$(\mathsf{a}_i\mathsf{a}_j, \mathsf{a}_j\mathsf{a}_i),\ (\mathsf{b}_i\mathsf{a}_j, \mathsf{a}_j\mathsf{b}_i),\ (\mathsf{a}_i\mathsf{b}_j, \mathsf{b}_j\mathsf{a}_i),\ (\mathsf{b}_i\mathsf{b}_j, \mathsf{b}_j\mathsf{b}_i) \quad \text{where } i \succ j, \tag{3.1a}$$

$$(\mathsf{a}_j\mathsf{b}_j, q_j\mathsf{b}_j\mathsf{a}_j + 1) \tag{3.1b}$$

for $i, j \in J$.

Note: It may seem peculiar to take the deformation vector $\mathbf{q}$ from $\mathcal{K}^{\square \supseteq J}$ rather than from $\mathcal{K}^J$, since none of the additional elements are ever used. The reason we do this is that we now and then consider not only the main reduction system $S(\mathbf{q}, J)$ but also subsystems of the form $S(\mathbf{q}, J')$, where J' is some subset of J. If we do this while requiring that $\mathbf{q} \in \mathcal{K}^J$ for $\mathbf{q}$ used in $S(\mathbf{q}, J)$, then we would also have to introduce an abbreviated deformation vector $\mathbf{q}' = (q_j)_{j\in J'}$ for use in $S(\mathbf{q}', J')$. We have decided against that approach since it would only conceal the actual structure.

The above definitions are mainly technical (but handy, once the Diamond Lemma formalism is comprehended). The actual purpose of (3.1) is to specify which equalities, in addition to the ones that hold in every associative algebra, that should hold in $\mathcal{H}(\mathbf{q}, J, \mathcal{K})$.

Definition 3.4 Let J be a nonempty set and $\mathcal{K}$ be a field. Let $\mathbf{q} \in \mathcal{K}^{\square \supseteq J}$ be arbitrary. Define

$$\mathcal{H}(\mathbf{q}, J, \mathcal{K}) = \mathcal{F}(J, \mathcal{K}) / \mathcal{I}\big(S(\mathbf{q}, J)\big) \tag{3.2}$$

and

$$A_j = \mathsf{a}_j + \mathcal{I}\big(S(\mathbf{q}, J)\big), \qquad B_j = \mathsf{b}_j + \mathcal{I}\big(S(\mathbf{q}, J)\big),$$
$$I = 1 + \mathcal{I}\big(S(\mathbf{q}, J)\big)$$

for all $j \in J$. This clearly makes $\mathcal{H}(\mathbf{q}, J, \mathcal{K})$ a generalised q-deformed Heisenberg algebra by J.

As with $\mathcal{F}(J, \mathcal{K})$, the $\mathcal{K}$ will often be left out from $\mathcal{H}(\mathbf{q}, J, \mathcal{K})$, leaving only $\mathcal{H}(\mathbf{q}, J)$. Furthermore if $J = \{1, \ldots, n\}$ then one may write $\mathcal{H}(\mathbf{q}, n)$ for $\mathcal{H}(\mathbf{q}, J)$. Finally, if $|J| = 1$ then one can drop the J too from $\mathcal{H}(\mathbf{q}, J)$ and identify $\mathbf{q}$ with its only component, leaving only $\mathcal{H}(q)$ to denote the algebra. As this calls for a corresponding simplification in the notation for the generators of the algebra, the indices are dropped from A_j and B_j, leaving only A and B.

Note: Given a nonempty set J, a field $\mathcal{K}$, a deformation vector $\mathbf{q} \in \mathcal{K}^{\Box \supseteq J}$, and a nonempty subset $J' \subseteq J$, we will identify the generalised q-deformed Heisenberg algebra $\mathcal{H}(\mathbf{q}, J', \mathcal{K})$ with the algebra

$$\mathrm{Alg}_{\mathcal{K}}\langle \{A_j, B_j\}_{j \in J'} \rangle \subseteq \mathcal{H}(\mathbf{q}, J, \mathcal{K}).$$

The $\mathcal{H}(\mathbf{q}, J)$ algebras are defined constructively, with a construction that happens to simplify some of our proofs, and not axiomatically or through some other more descriptive method. We will therefore give such a description. Let J, $\mathcal{K}$, and $\mathbf{q}$ be given. The algebra $\mathcal{H}(\mathbf{q}, J, \mathcal{K})$ will then be the freest generalised q-deformed Heisenberg algebra by J and over $\mathcal{K}$, for that particular choice of $\mathbf{q} = (q_j)_{j \in J}$. Every generalised q-deformed Heisenberg algebra $\mathcal{H}$ that is by J, over $\mathcal{K}$, and has the same choice of $\mathbf{q} = (q_j)_{j \in J}$ will either be isomorphic to $\mathcal{H}(\mathbf{q}, J, \mathcal{K})$ itself, or else isomorphic to a quotient algebra of $\mathcal{H}(\mathbf{q}, J, \mathcal{K})$.

3.2 Three bases for $\mathcal{H}(\mathbf{q}, J)$

In this section we will consider the question of finding a basis for $\mathcal{H}(\mathbf{q}, J)$. By using the Diamond Lemma we are able to derive no less than three different bases.

Theorem 3.1 *Let J be a nonempty set and $\prec$ be a total ordering of J. Let $\mathbf{q} \in \mathcal{K}^{\Box \supseteq J}$ be arbitrary. Let $\phi\colon \mathcal{F}(J) \longrightarrow \mathcal{F}(J)/\mathcal{I}\big(S(\mathbf{q}, J)\big)$ be the natural algebra homomorphism. Then the following hold:*

(i) $\mathrm{Red}\big(S(\mathbf{q}, J)\big) = \mathcal{F}(J)$.

(ii) *The restriction of ϕ to $\mathrm{Irr}\big(S(\mathbf{q}, J)\big)$ is a vector space isomorphism.*

(iii) *The set Y, which is defined by*

$$Y = \bigcup_{n=1}^{\infty} \left\{ \prod_{i=1}^{n} \mathsf{b}_{j_i}^{k_i} \mathsf{a}_{j_i}^{l_i} \,\middle|\, \begin{array}{l} \{j_i\}_{i=1}^{n} \subseteq J, \\ j_1 \prec j_2 \prec \cdots \prec j_n, \\ \{k_i\}_{i=1}^{n} \subset \mathbb{N}, \text{ and } \{l_i\}_{i=1}^{n} \subset \mathbb{N} \end{array} \right\}, \tag{3.3}$$

is a basis for $\operatorname{Irr}\bigl(S(\mathbf{q}, J)\bigr)$.

(iv) $\mathcal{H}(\mathbf{q}, J)$ *is a nontrivial generalised q-deformed Heisenberg algebra by J.*

(v) *The set*

$$\phi(Y) = \bigcup_{n=1}^{\infty} \left\{ \prod_{i=1}^{n} B_{j_i}^{k_i} A_{j_i}^{l_i} \,\middle|\, \begin{array}{l} \{j_i\}_{i=1}^{n} \subseteq J, \\ j_1 \prec j_2 \prec \cdots \prec j_n, \\ \{k_i\}_{i=1}^{n} \subset \mathbb{N}, \text{ and } \{l_i\}_{i=1}^{n} \subset \mathbb{N} \end{array} \right\} \tag{3.4}$$

is a basis for $\mathcal{H}(\mathbf{q}, J)$.

Note: The basis (3.4) looks worse than it actually is. It is a set because that is what a basis is, by definition. The union is there because if J is infinite then there is no upper bound on how many generators might be needed to express a monomial.

In the very simple case of $\mathcal{H}(q)$, the above expression just becomes

$$\{\, B^k A^l \mid k, l \in \mathbb{N} \,\}. \tag{3.5}$$

For $\mathcal{H}(\mathbf{q}, n)$ it becomes

$$\left\{ B_1^{k_1} A_1^{l_1} B_2^{k_2} A_2^{l_2} \cdots B_n^{k_n} A_n^{l_n} \,\middle|\, \{k_i\}_{i=1}^{n}, \{l_i\}_{i=1}^{n} \subset \mathbb{N} \right\}, \tag{3.6}$$

if the total ordering used is $<$. The total ordering is, by the way, completely arbitrary, and just has to exist for the proof to go through. It is actually possible to prove that every total ordering of J will yield the same basis (3.4), but since $S(\mathbf{q}, J)$, and hence $\operatorname{Irr}\bigl(S(\mathbf{q}, J)\bigr)$, depends on the ordering, it still makes sense to explicitly name the ordering.

There is also an alternative way to write the basis (3.6) which might be of interest. It is a trivial matter to see that this is the same thing as

$$\left\{ B_1^{k_1} \cdots B_n^{k_n} A_1^{l_1} \cdots A_n^{l_n} \,\middle|\, \{k_i\}_{i=1}^{n}, \{l_i\}_{i=1}^{n} \subset \mathbb{N} \right\}, \tag{3.7}$$

an expression which is perhaps more useful in some applications. For the purposes of this book however, the first form is the most useful.

Proof. The main task is to prove (i). Then all the other claims will follow rather easily.

Begin with the case that J is finite. Let $X = \{\mathsf{a}_j, \mathsf{b}_j\}_{j \in J'}$ be the set of generators of $\mathcal{F}(J)$, let $X_j = \{\mathsf{a}_j, \mathsf{b}_j\}$ for all $j \in J$, and let $\mathcal{X} = \{X_j\}_{j \in J}$ be the corresponding partition of X. The total ordering $\prec$ of J naturally defines a total ordering $\prec$ of $\mathcal{X}$ by letting $X_i \prec X_j$ if and only if $i \prec j$.

The reduction system $S(\mathbf{q}, J)$ can be partitioned as

$$S(\mathbf{q}, J) = C \cup \left(\bigcup_{j \in J} S_j \right),$$

where C contains all the pairs (μ_s, α_s) that fit pattern (3.1a), and

$$S_j = \{(\mathsf{a}_j \mathsf{b}_j, q_j \mathsf{b}_j \mathsf{a}_j + 1)\}$$

for all $j \in J$. Thus, by Lemma A.4, all ambiguities of the reduction system $S(\mathbf{q}, J)$ are resolvable if and only if all ambiguities of all the reduction systems S_j are.

None of these reduction systems contain any ambiguities, however, because there is only one element in each reduction system S_j, and the left hand side of that element, $\mathsf{a}_j \mathsf{b}_j$, can not form an ambiguity with itself. Hence all ambiguities of $S(\mathbf{q}, J)$ are resolvable. It is furthermore easy to see that $S(\mathbf{q}, J)$ is compatible with the degree lexicographic ordering $<$ that is built on the following ordering of the generators in X:

- $\mathsf{a}_i < \mathsf{a}_j$, $\mathsf{a}_i < \mathsf{b}_j$, $\mathsf{b}_i < \mathsf{a}_j$, and $\mathsf{b}_i < \mathsf{b}_j$ for $i, j \in J$ such that $i \prec j$;
- $\mathsf{b}_j < \mathsf{a}_j$ for all $j \in J$.

The degree lexicographic ordering satisfies the descending chain condition and thus the conditions in the Diamond Lemma (Theorem A.3) are fulfilled. It follows that

$$\operatorname{Red}\big(S(\mathbf{q}, J)\big) = \mathcal{F}(J),$$

and so (i) has been shown for finite J.

Now consider the case that J is infinite. Let $\alpha \in \mathcal{F}(J)$ be arbitrary, then there exists a finite subset J' of J such that $\alpha \in \mathcal{F}(J')$. Hence, by the above, $\alpha \in \operatorname{Red}\big(S(\mathbf{q}, J')\big)$.

Note that $S(\mathbf{q}, J')$ is a subset of $S(\mathbf{q}, J)$ with the property that an $s \in S(\mathbf{q}, J)$ belongs to $S(\mathbf{q}, J')$ if and only if $\mu_s \in \mathcal{F}(J')$. Thus the reduction systems $S(\mathbf{q}, J')$ and $S(\mathbf{q}, J)$ fulfill the conditions of Lemma A.5, and therefore

$$\operatorname{Red}\big(S(\mathbf{q}, J')\big) \subseteq \operatorname{Red}\big(S(\mathbf{q}, J)\big).$$

This implies that $\alpha \in \operatorname{Red}\big(S(\mathbf{q}, J)\big)$, and since α was arbitrary, it follows that $\mathcal{F}(J) \subseteq \operatorname{Red}\big(S(\mathbf{q}, J)\big)$. The reverse inclusion is true by definition, and hence $\operatorname{Red}\big(S(\mathbf{q}, J)\big) = \mathcal{F}(J)$, i.e., (i) is proved for infinite J too.

By the Note to the Diamond Lemma, it follows from (i) that

$$\operatorname{Irr}\big(S(\mathbf{q}, J)\big) \oplus \mathcal{I}\big(S(\mathbf{q}, J)\big) = \mathcal{F}(J).$$

Then (ii) is an easy consequence of that $\mathcal{I}\big(S(\mathbf{q}, J)\big)$ is the kernel of ϕ.

As for (iii), it is easy to see that Y is the set of all monomials in $\operatorname{Irr}\big(S(\mathbf{q}, J)\big)$. The set of all monomials in $\mathcal{F}(J)$ is a basis for $\mathcal{F}(J)$, hence Y is linearly independent. Finally every element of $\operatorname{Irr}\big(S(\mathbf{q}, J)\big)$ must be possible to write as a linear combination of monomials in $\operatorname{Irr}\big(S(\mathbf{q}, J)\big)$, since otherwise it would not be irreducible. Thus (iii) is a basis for $\operatorname{Irr}\big(S(\mathbf{q}, J)\big)$.

The algebra $\mathcal{H}(\mathbf{q}, J) = \mathcal{F}(J)/\mathcal{I}\big(S(\mathbf{q}, J)\big)$ is by definition a generalised q-deformed Heisenberg algebra by J. It is nontrivial because it is isomorphic, as a vector space, to $\operatorname{Irr}\big(S(\mathbf{q}, J)\big)$, which has dimension > 0. Hence (iv) is proved. Claim (v) follows from (ii) and (iii). □

Definition 3.5 If an element in $\mathcal{H}(\mathbf{q}, J)$ is expressed as a linear combination of elements in the basis (3.4), then that element is said to be expressed in **normal form**.

Note: It does not matter much which total ordering of J is used, since it is a trivial task to rewrite a monomial from one ordering to another, and hence little attention will be paid to what ordering is used in the rest of the book. The important thing is that for all $j \in J$, there is no B_j after an A_j.

Theorem 3.2 *Let J be a nonempty set and $\prec$ be a total ordering of J. Let $D \subseteq J$ be arbitrary and let $\mathbf{q} \in \mathcal{K}^{\square \supseteq J}$ satisfy $q_j \neq 0$ for all $j \in D$.*

Define $P_D(j, k, l)$ to be

$$P_D(j,k,l) = \begin{cases} A_j^l B_j^k & \text{if } j \in D, \\ B_j^k A_j^l & \text{if } j \notin D. \end{cases}$$

Then the set

$$\bigcup_{n=1}^{\infty}\left\{\prod_{i=1}^{n} P_D(j_i, k_i, l_i) \,\middle|\, \begin{array}{l}\{j_i\}_{i=1}^{n} \subseteq J,\\ j_1 \prec j_2 \prec \cdots \prec j_n,\\ \{k_i\}_{i=1}^{n} \subset \mathbb{N},\ \textit{and}\ \{l_i\}_{i=1}^{n} \subset \mathbb{N}\end{array}\right\} \tag{3.8}$$

is a basis for $\mathcal{H}(\mathbf{q}, J)$.

Proof. This is proved using the same technique as in the proof of Theorem 3.1. The only difference is that another reduction system is used. This reduction system will be called S_D.

The system S_D is similar to $S(\mathbf{q}, J)$ in that it can be partitioned in one part C containing only rules saying that two generators commute and a family $\{S_j\}_{j\in J}$ of parts, each of which is a reduction system on $\mathcal{H}(q_j)$. The C parts of S_D and $S(\mathbf{q}, J)$ respectively are identical, as are the S_j parts for $j \notin D$. The difference lies in the remaining S_j. These are

$$S_j = \left\{\left(\mathsf{b}_j\mathsf{a}_j, \tfrac{1}{q_j}\mathsf{a}_j\mathsf{b}_j - \tfrac{1}{q_j}1\right)\right\} \tag{3.9}$$

for all $j \in D$.

There is also some difference in the ordering of the generators on which the degree lexicographic ordering is built. As before, $i \prec j$ will imply $\mathsf{a}_i < \mathsf{a}_j$, $\mathsf{a}_i < \mathsf{b}_j$, $\mathsf{b}_i < \mathsf{a}_j$, and $\mathsf{b}_i < \mathsf{b}_j$, but then there are differences. If $j \in D$ then $\mathsf{a}_j < \mathsf{b}_j$ and if $j \notin D$ then $\mathsf{b}_j < \mathsf{a}_j$. These changes are necessary to make the ordering compatible with the reduction system.

This will of course lead to some slight differences in the intermediate results. The basis for the set of irreducible elements in $\mathcal{F}(J)$ will for example be

$$\bigcup_{n=1}^{\infty}\left\{\prod_{i=1}^{n} Q_D(j_i, k_i, l_i) \,\middle|\, \begin{array}{l}\{j_i\}_{i=1}^{n} \subseteq J,\\ j_1 \prec j_2 \prec \cdots \prec j_n,\\ \{k_i\}_{i=1}^{n} \subset \mathbb{N},\ \text{and}\ \{l_i\}_{i=1}^{n} \subset \mathbb{N}\end{array}\right\},$$

where

$$Q_D(j, k, l) = \begin{cases} \mathsf{a}_j^l\mathsf{b}_j^k & \text{if } j \in D \\ \mathsf{b}_j^k\mathsf{a}_j^l & \text{if } j \notin D. \end{cases}$$

These differences are however small. A formal proof is simply a repetition of the technicalities in the proof of Theorem 3.1. □

Theorem 3.3 *Let J be a nonempty set and $\prec$ be a total ordering of J. Let $\mathbf{q} \in \mathcal{K}^{\Box \supseteq J}$ be such that $q_j \neq 0$ for all $j \in J$. Let $C_j = B_jA_j$ for all*

$j \in J$ and let

$$P(j,k,l) = \begin{cases} C_j^k B_j^l & \text{if } l > 0 \\ C_j^k & \text{if } l = 0 \\ C_j^k A_j^{-l} & \text{if } l < 0 \end{cases}.$$

Then the set

$$\bigcup_{n=1}^{\infty} \left\{ \prod_{i=1}^{n} P(j_i, k_i, l_i) \,\middle|\, \begin{array}{l} \{j_i\}_{i=1}^n \subseteq J, \\ j_1 \prec j_2 \prec \cdots \prec j_n, \\ \{k_i\}_{i=1}^n \subset \mathbb{N}, \text{ and } \{l_i\}_{i=1}^n \subset \mathbb{Z} \end{array} \right\} \tag{3.10}$$

is a basis for $\mathcal{H}(\mathbf{q}, J)$.

Proof. This proof is also similar to the proof of Theorem 3.1, but there is a lot more to it than that. The main idea is to construct an algebra, that is isomorphic to $\mathcal{H}(\mathbf{q}, J)$, in such a way that the basis (3.10) is what one gets from doing the same thing as in Theorem 3.1. This also defines a natural partition of the proof into two steps—the first step establishes an isomorphism which is used in the second step to interpret the result of an application of the Diamond Lemma.

Step 1: For brevity, put

$$\mathcal{E}(J) = \mathcal{K}\langle \{\mathsf{a}_j, \mathsf{b}_j, \mathsf{c}_j\}_{j \in J} \rangle.$$

The algebra $\mathcal{F}(J) = \mathcal{K}\langle \{\mathsf{a}_j, \mathsf{b}_j\}_{j \in J} \rangle$ should be considered a subalgebra of $\mathcal{E}(J)$. Let

$$\begin{aligned} P_{ij} &= \{\, (\lambda_2\lambda_1, \lambda_1\lambda_2) \mid \lambda_1 \in \{\mathsf{a}_i, \mathsf{b}_i, \mathsf{c}_i\} \text{ and } \lambda_2 \in \{\mathsf{a}_j, \mathsf{b}_j, \mathsf{c}_j\} \,\}, \\ S_j &= \{ (\mathsf{b}_j\mathsf{a}_j, \mathsf{c}_j), (\mathsf{a}_j\mathsf{b}_j, q_j\mathsf{c}_j + 1), (\mathsf{a}_j\mathsf{c}_j, q_j\mathsf{c}_j\mathsf{a}_j + \mathsf{a}_j), \\ &\qquad (\mathsf{b}_j\mathsf{c}_j, \tfrac{1}{q_j}\mathsf{c}_j\mathsf{b}_j - \tfrac{1}{q_j}\mathsf{b}_j) \} \end{aligned}$$

for all $i, j \in J$. Then

$$S = \Bigl(\bigcup_{\substack{i,j \in J \\ i \prec j}} P_{ij} \Bigr) \cup \Bigl(\bigcup_{j \in J} S_j \Bigr),$$

$$S_0 = \{ (\mathsf{c}_j, \mathsf{b}_j\mathsf{a}_j) \}_{j \in J}$$

will be reduction systems for $\mathcal{E}(J)$.

The reduction systems S and S_0 define two two-sided ideals: $\mathcal{I}(S)$ and $\mathcal{I}(S_0)$. It is easy to see that $\mathcal{I}(S_0) \subset \mathcal{I}(S)$, and hence there is also a third ideal $\mathcal{I}_0 = \mathcal{I}(S)/\mathcal{I}(S_0) \subseteq \mathcal{E}(J)/\mathcal{I}(S_0)$. These ideals make it possible to set up the following diagram

$$\begin{array}{ccccc}
\mathcal{E}(J) & \xrightarrow{\phi_1} & \mathcal{E}(J)/\mathcal{I}(S_0) & \xrightarrow{\nu_2} & \mathcal{F}(J) \\
{\scriptstyle\phi_2}\downarrow & & {\scriptstyle\phi_3}\downarrow & & {\scriptstyle\phi_4}\downarrow \\
\mathcal{E}(J)/\mathcal{I}(S) & \xrightarrow{\nu_1} & (\mathcal{E}(J)/\mathcal{I}(S_0))/\mathcal{I}_0 & \xrightarrow{\nu_3} & \mathcal{H}(q)
\end{array}$$

where ϕ_1, ϕ_2, ϕ_3, and ϕ_4 are the natural homomorphisms from an algebra to a quotient. The other homomorphisms are

$$\nu_1 : \left\{ \begin{array}{rcl}
\mathcal{E}(J)/\mathcal{I}(S) & \longrightarrow & (\mathcal{E}(J)/\mathcal{I}(S_0))/\mathcal{I}_0 \\
\mathsf{a}_j + \mathcal{I}(S) & \longmapsto & (\mathsf{a}_j + \mathcal{I}(S_0)) + \mathcal{I}_0 \\
\mathsf{b}_j + \mathcal{I}(S) & \longmapsto & (\mathsf{b}_j + \mathcal{I}(S_0)) + \mathcal{I}_0 \\
\mathsf{c}_j + \mathcal{I}(S) & \longmapsto & (\mathsf{c}_j + \mathcal{I}(S_0)) + \mathcal{I}_0
\end{array} \right.$$

$$\nu_2 : \left\{ \begin{array}{rclcrcl}
\mathcal{E}(J)/\mathcal{I}(S_0) & \longrightarrow & \mathcal{F}(J) & & \mathsf{a}_j + \mathcal{I}(S_0) & \longmapsto & \mathsf{a}_j \\
\mathsf{b}_j + \mathcal{I}(S_0) & \longmapsto & \mathsf{b}_j & & \mathsf{c}_j + \mathcal{I}(S_0) & \longmapsto & \mathsf{b}_j\mathsf{a}_j
\end{array} \right.$$

$$\nu_3 : \left\{ \begin{array}{rclcl}
(\mathcal{E}(J)/\mathcal{I}(S_0))/\mathcal{I}_0 & \longrightarrow & \mathcal{F}(J)/\mathcal{I}(S(\mathbf{q}, J)) & = & \mathcal{H}(\mathbf{q}, J) \\
(\mathsf{a}_j + \mathcal{I}(S_0)) + \mathcal{I}_0 & \longmapsto & \mathsf{a}_j + \mathcal{I}(S(\mathbf{q}, J)) & = & A_j \\
(\mathsf{b}_j + \mathcal{I}(S_0)) + \mathcal{I}_0 & \longmapsto & \mathsf{b}_j + \mathcal{I}(S(\mathbf{q}, J)) & = & B_j \\
(\mathsf{c}_j + \mathcal{I}(S_0)) + \mathcal{I}_0 & \longmapsto & \mathsf{b}_j\mathsf{a}_j + \mathcal{I}(S(\mathbf{q}, J)) & = & C_j
\end{array} \right.$$

The diagram is in fact commutative and ν_1, ν_2, and ν_3 are isomorphisms. All that is left in the first step of this proof is to prove this. Commutativity of the left square and bijectivity of ν_1 is easy, as this is precisely the claim of one of the Isomorphism Theorems for algebras.

The system S_0 is clearly a reduction system without ambiguities. Using the degree functions defined in Section B.1 of Appendix B, a semigroup partial ordering that is compatible with S_0 can be defined as follows: Let $V = \{\mathsf{c}_j\}_{j \in J}$ and define $\mu_1 < \mu_2$ if and only if $d_V(\mu_1) < d_V(\mu_2)$. What it does is simply that it counts the number of c's in the monomials (the indices are ignored) and compares these numbers—the fewer the c's, the smaller the monomial.

This means that the Diamond Lemma applies and it follows that $\mathcal{E}(J) = \mathcal{I}(S_0) \oplus \mathrm{Irr}(S_0)$. Note that $\mathrm{Irr}(S_0) = \mathcal{F}(J)$, hence in particular $\phi_1\big|_{\mathcal{F}(J)}$ is a vector space isomorphism. It is in fact an isomorphism of algebras since $\mathcal{F}(J)$ is an algebra. Finally, ν_2 happens to be the inverse of $\phi_1\big|_{\mathcal{F}(J)}$, so it is an isomorphism as well.

Commutativity of the right square and bijectivity of ν_3 will follow immediately once it has been shown that $\nu_2(\mathcal{I}_0) = \mathcal{I}\bigl(S(\mathbf{q}, J)\bigr)$. This is equivalent to showing that $(\nu_2 \circ \phi_1)\bigl(\mathcal{I}(S)\bigr) = \mathcal{I}\bigl(S(\mathbf{q}, J)\bigr)$, which will now be done. To begin with, $(\nu_2 \circ \phi_1)(\mu_s - \alpha_s) \in \mathcal{I}\bigl(S(\mathbf{q}, J)\bigr)$ for all $s \in S$, because

$$\begin{aligned}
\mu_s - \alpha_s &\mapsto (\nu_2 \circ \phi_1)(\mu_s - \alpha_s) \\
\mathsf{a}_j\mathsf{b}_i - \mathsf{b}_i\mathsf{a}_j &\mapsto \mathsf{a}_j\mathsf{b}_i - \mathsf{b}_i\mathsf{a}_j \\
\mathsf{a}_j\mathsf{c}_i - \mathsf{c}_i\mathsf{a}_j &\mapsto \mathsf{a}_j\mathsf{b}_i\mathsf{a}_i - \mathsf{b}_i\mathsf{a}_i\mathsf{a}_j = \\
&= (\mathsf{a}_j\mathsf{b}_i - \mathsf{b}_i\mathsf{a}_j)\mathsf{a}_i + \mathsf{b}_i(\mathsf{a}_j\mathsf{a}_i - \mathsf{a}_i\mathsf{a}_j) \\
\mathsf{c}_j\mathsf{c}_i - \mathsf{c}_i\mathsf{c}_j &\mapsto \mathsf{b}_j\mathsf{a}_j\mathsf{b}_i\mathsf{a}_i - \mathsf{b}_i\mathsf{a}_i\mathsf{b}_j\mathsf{a}_j = \\
&= \mathsf{b}_j(\mathsf{a}_j\mathsf{b}_i - \mathsf{b}_i\mathsf{a}_j)\mathsf{a}_i + (\mathsf{b}_j\mathsf{b}_i - \mathsf{b}_i\mathsf{b}_j)\mathsf{a}_j\mathsf{a}_i + \\
&\qquad + \mathsf{b}_i\mathsf{b}_j(\mathsf{a}_j\mathsf{a}_i - \mathsf{a}_i\mathsf{a}_j) + \mathsf{b}_i(\mathsf{b}_j\mathsf{a}_i - \mathsf{a}_i\mathsf{b}_j)\mathsf{a}_j \\
&\text{(the other reductions in } P_{ij} \text{ are handled similarly)} \\
\mathsf{b}_j\mathsf{a}_j - \mathsf{c}_j &\mapsto \mathsf{b}_j\mathsf{a}_j - \mathsf{b}_j\mathsf{a}_j = 0 \\
\mathsf{a}_j\mathsf{b}_j - q_j\mathsf{c}_j - 1 &\mapsto \mathsf{a}_j\mathsf{b}_j - q_j\mathsf{b}_j\mathsf{a}_j - 1 \\
\mathsf{a}_j\mathsf{c}_j - q_j\mathsf{c}_j\mathsf{a}_j - \mathsf{a}_j &\mapsto \mathsf{a}_j\mathsf{b}_j\mathsf{a}_j - q_j\mathsf{b}_j\mathsf{a}_j\mathsf{a}_j - \mathsf{a}_j = \\
&= (\mathsf{a}_j\mathsf{b}_j - q_j\mathsf{b}_j\mathsf{a}_j - 1)\mathsf{a}_j \\
\mathsf{b}_j\mathsf{c}_j - \tfrac{1}{q_j}\mathsf{c}_j\mathsf{b}_j + \tfrac{1}{q_j}\mathsf{b}_j &\mapsto \mathsf{b}_j\mathsf{b}_j\mathsf{a}_j - \tfrac{1}{q_j}\mathsf{b}_j\mathsf{a}_j\mathsf{b}_j + \tfrac{1}{q_j}\mathsf{b}_j = \\
&= -\tfrac{1}{q_j}\mathsf{b}_j(\mathsf{a}_j\mathsf{b}_j - q_j\mathsf{b}_j\mathsf{a}_j - 1)
\end{aligned}$$

for arbitrary $j \in J$ and $i \prec j$. The rightmost parts obviously belong to $\mathcal{I}\bigl(S(\mathbf{q}, J)\bigr)$ and hence $(\nu_2 \circ \phi_1)\bigl(\mathcal{I}(S)\bigr) \subseteq \mathcal{I}\bigl(S(\mathbf{q}, J)\bigr)$. The other way is even simpler, since the only rules in $S(\mathbf{q}, J)$ that do not appear in S are the ones of the form $(\mathsf{a}_j\mathsf{b}_j, q_j\mathsf{b}_j\mathsf{a}_j + 1)$, and

$$\mathsf{a}_j\mathsf{b}_j - q_j\mathsf{c}_j - 1 \mapsto \mathsf{a}_j\mathsf{b}_j - q_j\mathsf{b}_j\mathsf{a}_j - 1.$$

Hence $\nu_2(\mathcal{I}_0) = \mathcal{I}\bigl(S(\mathbf{q}, J)\bigr)$, and ν_3 is an isomorphism.

The interesting fact is that $\mathcal{E}(J)/\mathcal{I}(S)$ is isomorphic to $\mathcal{H}(\mathbf{q}, J)$, because this makes it possible to get a basis for $\mathcal{H}(\mathbf{q}, J)$ from each basis of $\mathcal{E}(J)/\mathcal{I}(S)$.

Step 2: First consider the case when J is finite. Let $X = \{\mathsf{a}_j, \mathsf{b}_j, \mathsf{c}_j\}_{j \in J}$ be the set of generators of $\mathcal{E}(J)$, let $X_j = \{\mathsf{a}_j, \mathsf{b}_j, \mathsf{c}_j\}$ for all $j \in J$, and let $\mathcal{X} = \{X_j\}_{j \in J}$ be the corresponding partition of X. The total ordering $\prec$ of J naturally defines a total ordering $\prec$ of $\mathcal{X}$ by letting $X_i \prec X_j$ if and only if $i \prec j$.

Thus the conditions in Lemma A.4 are fulfilled, and hence all ambiguities of the reduction system S are resolvable if and only if all ambiguities of all the reductions systems S_j are resolvable. This is easy to verify, since every S_j contains exactly four rules and since all the S_j are, in a sense, identical.

It is easy to see that there are no inclusion ambiguities in S_j, but there are some overlap ambiguities. To specify them, let $s_1 = (\mathsf{b}_j\mathsf{a}_j, \mathsf{c}_j)$, $s_2 = (\mathsf{a}_j\mathsf{b}_j, q_j\mathsf{c}_j + 1)$, $s_3 = (\mathsf{a}_j\mathsf{c}_j, q_j\mathsf{c}_j\mathsf{a}_j + \mathsf{a}_j)$, and $s_4 = (\mathsf{b}_j\mathsf{c}_j, \frac{1}{q_j}\mathsf{c}_j\mathsf{b}_j - \frac{1}{q_j}\mathsf{b}_j)$. Then the overlap ambiguities are

$$(s_2, s_1, \mathsf{a}_j, \mathsf{b}_j, \mathsf{a}_j), \qquad (s_2, s_4, \mathsf{a}_j, \mathsf{b}_j, \mathsf{c}_j),$$
$$(s_1, s_2, \mathsf{b}_j, \mathsf{a}_j, \mathsf{b}_j), \text{ and } \qquad (s_1, s_3, \mathsf{b}_j, \mathsf{a}_j, \mathsf{c}_j).$$

These ambiguities can be resolved as follows:

$$\mathsf{a}_j\mathsf{b}_j\mathsf{a}_j \xmapsto{\quad s_2 \quad} q_j\mathsf{c}_j\mathsf{a}_j + \mathsf{a}_j$$
$$\mathsf{a}_j\mathsf{b}_j\mathsf{a}_j \xmapsto{s_1} \mathsf{a}_j\mathsf{c}_j \xmapsto{s_3} q_j\mathsf{c}_j\mathsf{a}_j + \mathsf{a}_j$$

$$\mathsf{a}_j\mathsf{b}_j\mathsf{c}_j \xmapsto{\quad s_2 \quad} q_j\mathsf{c}_j^2 + \mathsf{c}_j$$
$$\mathsf{a}_j\mathsf{b}_j\mathsf{c}_j \xmapsto{s_4} \frac{1}{q_j}\mathsf{a}_j\mathsf{c}_j\mathsf{b}_j - \frac{1}{q_j}\mathsf{a}_j\mathsf{b}_j \xmapsto{s_3} \mathsf{c}_j\mathsf{a}_j\mathsf{b}_j \xmapsto{s_2} q_j\mathsf{c}_j^2 + \mathsf{c}_j$$

$$\mathsf{b}_j\mathsf{a}_j\mathsf{b}_j \xmapsto{\quad s_1 \quad} \mathsf{c}_j\mathsf{b}_j$$
$$\mathsf{b}_j\mathsf{a}_j\mathsf{b}_j \xmapsto{s_2} q_j\mathsf{b}_j\mathsf{c}_j + \mathsf{b}_j \xmapsto{s_4} \mathsf{c}_j\mathsf{b}_j$$

$$\mathsf{b}_j\mathsf{a}_j\mathsf{c}_j \xmapsto{\quad s_1 \quad} \mathsf{c}_j^2$$
$$\mathsf{b}_j\mathsf{a}_j\mathsf{c}_j \xmapsto{s_3} q_j\mathsf{b}_j\mathsf{c}_j\mathsf{a}_j + \mathsf{b}_j\mathsf{a}_j \xmapsto{s_4} \mathsf{c}_j\mathsf{b}_j\mathsf{a}_j \xmapsto{s_1} \mathsf{c}_j^2$$

The reduction system S is compatible with the degree lexicographic ordering $<$ that is built on the following ordering of the generators in X:

- $x_1 < x_2$ for all $x_1 \in X_i$ and $x_2 \in X_j$ such that $i \prec j$.
- $\mathsf{c}_j < \mathsf{b}_j < \mathsf{a}_j$ for all $j \in J$.

This ordering satisfies the descending chain condition and thus the conditions in the Diamond Lemma (Theorem A.3) are fulfilled. It follows that

$$\mathrm{Red}(S) = \mathcal{E}(J)$$

for finite J.

Now let $\alpha \in \mathcal{E}(J)$ be arbitrary, then there exists a finite subset J' of J such that $\alpha \in \mathcal{E}(J')$. Let

$$S' = \Biggl(\bigcup_{\substack{i,j \in J' \\ i \prec j}} P_{ij} \Biggr) \cup \Biggl(\bigcup_{j \in J'} S_j \Biggr).$$

By the above, $\operatorname{Red}\bigl(S', \mathcal{E}(J')\bigr) = \mathcal{E}(J')$ and hence $\alpha \in \operatorname{Red}\bigl(S', \mathcal{E}(J')\bigr)$.

It is then easy to see that S' is a subset of S with the property that an $s \in S$ belongs to S' if and only if $\mu_s \in \mathcal{E}(J')$. Therefore the reduction systems S' and S fulfill the conditions of Lemma A.5, and hence $\operatorname{Red}\bigl(S', \mathcal{E}(J')\bigr) \subseteq \operatorname{Red}\bigl(S, \mathcal{E}(J)\bigr)$. Thus $\alpha \in \operatorname{Red}(S)$, and since α was arbitrary, it follows that $\mathcal{E}(J) \subseteq \operatorname{Red}(S)$. The reverse inclusion is true by definition, and hence $\operatorname{Red}(S) = \mathcal{E}(J)$. In particular, $\phi_2\big|_{\operatorname{Irr}(S)}$ is a vector space isomorphism.

It is easily seen that S can reduce a monomial if and only if it meets any of the following criteria: (i) there is an a_j to the left of a b_j; (ii) there is an a_j to the left of a c_j; (iii) there is a b_j to the left of a c_j; (iv) there is a b_j to the left of an a_j. Thus $\operatorname{Irr}(S)$ must be spanned by the monomials which do not meet any of criteria (i)–(iv), and it follows that

$$\bigcup_{n=1}^{\infty} \left\{ \prod_{i=1}^{n} Q(j_i, k_i, l_i) \;\middle|\; \begin{array}{l} \{j_i\}_{i=1}^{n} \subseteq J, \\ j_1 \prec j_2 \prec \cdots \prec j_n, \\ \{k_i\}_{i=1}^{n} \subset \mathbb{N}, \text{ and } \{l_i\}_{i=1}^{n} \subset \mathbb{Z} \end{array} \right\}$$

is a basis for $\operatorname{Irr}(S)$ with

$$Q(j,k,l) = \begin{cases} \mathsf{c}_j^k \mathsf{b}_j^l & \text{if } l > 0 \\ \mathsf{c}_j^k & \text{if } l = 0 \\ \mathsf{c}_j^k \mathsf{a}_j^{-l} & \text{if } l < 0 \end{cases}.$$

Our conclusion of these steps is that the mapping $\nu_3 \circ \nu_1 \circ \bigl(\phi_2\big|_{\operatorname{Irr}(S)}\bigr)$ from $\operatorname{Irr}(S)$ to $\mathcal{H}(\mathbf{q}, J)$ is a vector space isomorphism. Since $\nu_3 \circ \nu_1 \circ \phi_2 \circ Q$ is P (defined in the statement of the theorem), it follows that (3.10) is indeed a basis for $\mathcal{H}(\mathbf{q}, J)$. $\square$

3.3 Comparison of the three bases

All three bases (3.4), (3.8), and (3.10) are sets of the form

$$\bigcup_{n=1}^{\infty} \left\{ \prod_{i=1}^{n} P_t(j_i, k_i, l_i) \,\middle|\, \begin{array}{l} \{j_i\}_{i=1}^{n} \subseteq J, \\ j_1 \prec j_2 \prec \cdots \prec j_n, \\ \{k_i\}_{i=1}^{n} \subset \mathbb{N}, \text{ and } \{l_i\}_{i=1}^{n} \subset \mathbb{N} \end{array} \right\}. \tag{3.11}$$

The difference lies in the monomials $P_t(j,k,l)$, which are respectively

$$P_1(j,k,l) = B_j^k A_j^l \qquad \text{(Theorem 3.1)}$$

$$P_2(j,k,l) = \begin{cases} A_j^l B_j^k & \text{if } j \in D \\ B_j^k A_j^l & \text{if } j \notin D \end{cases} \qquad \text{(Theorem 3.2)}$$

$$P_3(j,k,l) = \begin{cases} C_j^l B_j^{k-l} & \text{if } k > l \\ C_j^k & \text{if } k = l \\ C_j^k A_j^{l-k} & \text{if } k < l \end{cases} \qquad \text{(Theorem 3.3)}$$

Note that Theorem 3.3 uses another indexation of the monomials—the l there equals $k-l$ here and the k there equals $\min(k,l)$ here. Also, recall that $C_j = B_j A_j$, just as in that theorem.

Note that only the BA set (the set based on P_1) is a basis for all $\mathcal{H}(\mathbf{q}, J)$ algebras. For the other two sets to become bases in $\mathcal{H}(\mathbf{q}, J)$, it is necessary that $q_j \neq 0$ for some or all $j \in J$. If these conditions are not fulfilled then neither of the two latter sets are linearly independent. On the other hand, the $\mathcal{H}(\mathbf{q}, J)$ algebras where some $q_j = 0$ are best treated as a special case—this book treats them only when it does not complicate the proofs—so this disadvantage of those two sets is not as large as it might seem.

Apart from that, there is little theoretical difference between the BA basis and the AB basis (the basis based on P_2 with $D = J$). We have chosen to use the BA basis as normal form, but almost all proofs would (after minor adjustments) have worked just as fine with the AB basis instead. A counterexample to this rule of thumb is Theorem 4.11, where the proof needs to mix BA and AB ordering of the generators.

The practical viewpoint however, is likely to favour the BA basis before the AB basis. The reason for this is (see Chapter 8) that many applications interpret B as a 'multiplication by variable' operator and A as a differentiation or difference type operator. Since additional factors tend to make differentiation messier with more terms it is often preferable to make differentiations first.

The remaining basis is the one that is based on P_3—we will call it the *CBA* basis after the ordering of generators that was used to deduce it. This basis portends Chapter 4 in the sense that it has a much closer relation to the very important concept of homogeneity that is introduced there than the other bases do. As a result of this, the *CBA* basis is sometimes much simpler to use, in spite of its complex form, than the other two. A good example of this is the proof of Lemma 6.19.

3.4 Computational aspects

If any one of these bases is used systematically, computations in $\mathcal{H}(\mathbf{q}, n)$ will be much simpler to implement. The reason for this is that an element in one of these bases can be uniquely specified with $2n$ integers, in the case of the *BA* and *AB* bases these integers suitably being the exponents for B_1, A_1, ..., B_n, and A_n. This makes the basic task of *storing* elements in the algebra dramatically simpler than if arbitrary monomials had to be kept track of (even to uniquely describe a general monomial is a nightmare in most computer languages).

Storage using a basis for the algebra is completely indifferent to addition and multiplication by a scalar—both work monomial-wise regardless of any linear independence—and it solves the problem of testing for equality, which is a most laborious task if general monomials can appear. Still, there is a price to pay, namely that a multiplication will require more effort than it would if the products were allowed to be general monomials. On the other hand, this additional effort can be made relatively small.

If the *BA* basis is used then the canonical problem is to rewrite the product

$$B_1^{r_1} A_1^{s_1} \cdots B_n^{r_n} A_n^{s_n} \cdot B_1^{k_1} A_1^{l_1} \cdots B_n^{k_n} A_n^{l_n} \tag{3.12}$$

into a linear combination of elements in the basis. Since generators with different indices commute, this can be rewritten as

$$\left(B_1^{r_1} A_1^{s_1} \cdot B_1^{k_1} A_1^{l_1}\right) \cdots \left(B_n^{r_n} A_n^{s_n} \cdot B_n^{k_n} A_n^{l_n}\right).$$

Now (2.4) can be applied in each parenthesis, resulting in something that looks like

$$\Big(\sum_i a_{1i} B_1^{r_1+k_1-i} A_1^{s_1+l_1-i}\Big) \cdots \Big(\sum_i a_{ni} B_n^{r_n+k_n-i} A_n^{s_n+l_n-i}\Big).$$

Expanding this will produce the required linear combination of elements in the basis.

The AB basis can be handled in the same way, but one would use (2.9) to rewrite the products rather than (2.4). Using mixed BA and AB ordering is probably a bad idea since it requires implementing both the BA reordering and the AB reordering.

The CBA basis is, hardly surprising, more troublesome in most cases, but the general outline is the same. First collect C_j, B_j, and A_j according to index so that the product is put into the form

$$\big(P_3(1, r_1, s_1) \cdot P_3(1, k_1, l_1)\big) \cdots \big(P_3(n, r_n, s_n) \cdot P_3(n, k_n, l_n)\big).$$

Then rewrite each product $P_3(j, r_j, s_j) \cdot P_3(j, k_j, l_j)$ as a linear combination of $P_3(j, u_{ji}, v_{ji})$ factors using the reordering formulae (3.13). Finally expand the whole product. The phase that is different from the BA and AB cases is the rewriting phase.

It is not only that (2.4) cannot be used that is a novelty. Because the proof of Theorem 3.3 lists four (slightly overlapping) situations in which the reduction system can change a monomial, there ought to be four, not one, reordering formulae to resolve the respective situations. These formulae were derived in Section 2.2, but it might be best to repeat them here. The formulae in question are

$$A_j^n C_j^m = \big(q_j^n C_j + \{n\}_{q_j} I\big)^m A_j^n \tag{3.13a}$$

$$B_j^n C_j^m = q_j^{-nm} \big(C_j - \{n\}_{q_j} I\big)^m B_j^n \tag{3.13b}$$

$$A_j^n B_j^n = \prod_{k=1}^{n} \big(q_j^k C_j + \{k\}_{q_j} I\big) \tag{3.13c}$$

$$B_j^n A_j^n = \prod_{k=0}^{n-1} q_j^{-k} \big(C_j - \{k\}_{q_j} I\big). \tag{3.13d}$$

With these, all noncommutative calculations can be handled in a few steps and no more than four integers are needed to uniquely describe every monomial that appears.

Admittedly, these formulae are not as straight on as (2.4), but there is one advantage—none of them requires that a q-factorial or a q-binomial coefficient is computed. Whether or not this is enough to compensate for the increased complexity in the algorithm must be decided from case to case.

To some readers, the whole idea of computations with elements in $\mathcal{H}(\mathbf{q}, J)$ might seem a little esoteric; we all know that it is possible, but might find it hard to imagine a context where any larger amount of computations needs to be performed. The comments to Theorem 7.3 do however give an example of such a context.

Chapter 4

Degree in and gradation of $\mathcal{H}(\mathrm{q}, J)$

> I have had my results for a long time, but I do not yet know how to arrive at them.
>
> K. F. GAUSS

We hope the reader will quickly find that the formulations of most results in this chapter are easily anticipated. As is annoyingly often the case with such easily formulated results however, proving them turns out to be a completely different matter indeed. It was not until we had done some rather extensive formalisation of the basic concepts involved that we could go on to construct the rigorous proofs on which we base most of our work in subsequent chapters.

Most of that formalisation has been put in Appendix B, which relates to the present chapter very much like Appendix A to Chapter 3. Logically this entire chapter is based on the material in Appendix B, but one can develop an understanding of the concepts defined without first understanding the formal constructions behind.

4.1 Degree in $\mathcal{H}(\mathbf{q}, J)$

Definition 4.1 The following short names for degree functions in $\mathcal{H}(\mathbf{q}, J)$ are used:

$$\begin{array}{lll} \deg_j \alpha & \text{means} & d_{\{A_j,B_j\}}(\alpha) \\ \deg_{\mathrm{A}_j} \alpha & \text{means} & d_{\{A_j\}}(\alpha) \\ \deg_{\mathrm{B}_j} \alpha & \text{means} & d_{\{B_j\}}(\alpha) \end{array}$$

for $j \in J$. Furthermore deg with no index will just as usual denote the total degree, i.e.,

$$\deg \alpha = d_X(\alpha),$$

where $X = \{A_j, B_j\}_{j \in J}$.

Considering the vast amount of material that is at work behind this definition, the resulting concept is surprisingly intuitive, as the next theorem will show.

Theorem 4.1 *Let $\{j_i\}_{i=1}^{n} \subseteq J$ be n distinct elements. Let $\{k_{ir}\}_{i=1;r=1}^{n;m} \subset \mathbb{N}$ and $\{l_{ir}\}_{i=1;r=1}^{n;m} \subset \mathbb{N}$ be such that if $k_{ir} = k_{is}$ and $l_{ir} = l_{is}$ for all i then $r = s$. Let $\{a_r\}_{r=1}^{m} \subseteq \mathcal{K} \setminus \{0\}$ and let $V \subseteq \{A_j, B_j\}_{j \in J}$. Then*

$$d_V\left(\sum_{r=1}^{m} a_r B_{j_1}^{k_{1r}} A_{j_1}^{l_{1r}} \cdots B_{j_n}^{k_{nr}} A_{j_n}^{l_{nr}}\right) = \max_{1 \leqslant r \leqslant m}\left(\sum_{\substack{1 \leqslant i \leqslant n \\ A_{j_i} \in V}} l_{ir} + \sum_{\substack{1 \leqslant i \leqslant n \\ B_{j_i} \in V}} k_{ir}\right). \quad (4.1)$$

Note: What this theorem says is that if an element of $\mathcal{H}(\mathbf{q}, J)$ is expressed in normal form, i.e, if it is written as a linear combination of elements in the basis (3.4), then the degree of that element can be determined by inspection. All one needs to do is to sum the exponents of the relevant generators in every term separately and then find the largest such sum.

Proof. Let $\prec$ be a total ordering of J such that $j_1 \prec j_2 \prec \cdots \prec j_n$. Let $W = \{\mathsf{a}_j \mid A_j \in V\} \cup \{\mathsf{b}_j \mid B_j \in V\}$. Let

$$\alpha = \sum_{r=1}^{m} a_r B_{j_1}^{k_{1r}} A_{j_1}^{l_{1r}} \cdots B_{j_n}^{k_{nr}} A_{j_n}^{l_{nr}},$$

$$\hat{\alpha} = \sum_{r=1}^{m} a_r \mathsf{b}_{j_1}^{k_{1r}} \mathsf{a}_{j_1}^{l_{1r}} \cdots \mathsf{b}_{j_n}^{k_{nr}} \mathsf{a}_{j_n}^{l_{nr}}.$$

Then it is easy to see that by Theorem B.18, $d_V(\alpha) = d_W(\hat{\alpha})$, since the defining expression for α is in normal form and $\hat{\alpha}$ is the corresponding element in $\operatorname{Irr}(S(\mathbf{q}, J))$.

By Lemma B.8 and Corollary B.11,

$$d_W(\hat{\alpha}) = \max_{1 \leqslant r \leqslant m} d_W(\mathsf{b}_{j_1}^{k_{1r}} \mathsf{a}_{j_1}^{l_{1r}} \cdots \mathsf{b}_{j_n}^{k_{nr}} \mathsf{a}_{j_n}^{l_{nr}}) =$$

$$= \max_{1 \leqslant r \leqslant m} \Bigg(\sum_{\substack{1 \leqslant i \leqslant n \\ \mathsf{a}_{j_i} \in W}} l_{ir} + \sum_{\substack{1 \leqslant i \leqslant n \\ \mathsf{b}_{j_i} \in W}} k_{ir} \Bigg) = \max_{1 \leqslant r \leqslant m} \Bigg(\sum_{\substack{1 \leqslant i \leqslant n \\ A_{j_i} \in V}} l_{ir} + \sum_{\substack{1 \leqslant i \leqslant n \\ B_{j_i} \in V}} k_{ir} \Bigg).$$

Thus (4.1) is proved. □

The formula (4.1) can be used as an alternative definition of degree in $\mathcal{H}(\mathbf{q}, J)$. The reason for not doing this here is that it would make some important properties of the degree functions more cumbersome to prove. A context where the necessary properties can be presented without proof, and where the inner works of the degree concept are of no interest, could well benefit from using Theorem 4.1 as a definition of degree instead of Definition 4.1.

The next theorem is very technical in nature, but it answers the important question "*When multiplying two polynomials α and β, will I be able to predict what the most significant term in the product will be simply by looking at the most significant terms of α and β?*". The answer is "*Yes, you will.*" if all $q_j \neq 0$ and "*No, you will not.*" otherwise.

The first point in this that needs to be explained is what the most significant term of a polynomial is. In a polynomial algebra such as $\mathcal{K}[x]$, the most significant term is simply the term with highest degree, i.e., the term in which the formal variable x is raised to the highest power. This carries over to other commutative polynomial algebras, but there is a complication. For example $x + y$ is a perfectly good polynomial in $\mathcal{K}[x, y]$, but which is the most significant term? Our method of answering this question is to fix a sequence of degree functions, and for each function in the sequence to keep the terms that obtain a maximum for the function and throw away the others, a process that is formalised in Definitions B.4 and B.5. If the functions are chosen correctly, only one term will remain when this procedure has been applied with all the functions in the sequence, and that term will be the most significant term with respect to that particular sequence of degree functions.

The simplest example of a case where the above question is answered negatively occurs in $\mathcal{H}(0)$. Take $\alpha = A + I$ and $\beta = B$. It will be enough in this case to use the total degree to determine the most significant term. Note that $\alpha\beta = (A+I)B = AB + B = I + B$. The most significant terms of α and β respectively are A and B, but their contribution to $\alpha\beta$ is not the most significant term of the product.

The positive answer in the first case is not possible to verify by examples, but it should be observed that it is what is to be expected in a commutative polynomial algebra. Thus the following theorem, whose main claim is (4.2), indicates that most of the q-deformed Heisenberg algebras $\mathcal{H}(\mathbf{q}, J)$ behave like commutative polynomial algebras in this matter.

Theorem 4.2 *Let J be a nonempty set and $\prec$ be a total ordering of J. Let $\mathbf{q} \in \mathcal{K}^{\Box \supseteq J}$ be arbitrary. Let $\alpha, \beta \in \operatorname{Irr}(S(\mathbf{q}, J)) \setminus \{0\}$. Then there exists a finite $J' \subseteq J$ such that*

$$d_{\{\mathsf{b}_j\}}(\alpha) = d_{\{\mathsf{a}_j\}}(\alpha) = d_{\{\mathsf{b}_j\}}(\beta) = d_{\{\mathsf{a}_j\}}(\beta) = 0$$

for all $j \in J \setminus J'$. Choose such a J'. Let $s = |J'|$, and let $\{j_1, \dots, j_s\} = J'$ be such that $j_1 \prec j_2 \prec \cdots \prec j_s$.

Let $\{W_i\}_{i=1}^r$ be a family of subsets of $\{\mathsf{a}_j, \mathsf{b}_j\}_{j \in J'}$ such that

$$\{W_i\}_{i=1}^r \supseteq \bigl\{\{\mathsf{a}_{j_1}\}, \{\mathsf{b}_{j_1}\}, \dots, \{\mathsf{a}_{j_s}\}, \{\mathsf{b}_{j_s}\}\bigr\}.$$

Let $\{f_i\}_{i=1}^r$ be defined by

$$f_i = d_{W_i} \circ \mathrm{pr}_2$$

for $i = 1, \dots, r$. Let $\mathbf{f} = (f_1, \dots, f_r)$.

Then there exists pairs (a, μ) and (b, ν) such that

$$\bigl\{(a,\mu)\bigr\} = \operatorname{Maj}\bigl(\operatorname{Mon}(\alpha), \mathbf{f}\bigr)$$
$$\bigl\{(b,\nu)\bigr\} = \operatorname{Maj}\bigl(\operatorname{Mon}(\beta), \mathbf{f}\bigr).$$

Furthermore, if $q_j \neq 0$ for all $j \in J'$ then

$$\operatorname{Maj}\Bigl(\operatorname{Mon}\bigl(t^S(\alpha\beta)\bigr), \mathbf{f}\Bigr) = \bigl\{(c, \lambda)\bigr\}, \tag{4.2}$$

where

$$\lambda = \mathsf{b}_{j_1}^{k_1}\mathsf{a}_{j_1}^{l_1}\cdots\mathsf{b}_{j_s}^{k_s}\mathsf{a}_{j_s}^{l_s},$$
$$c = ab\prod_{i=1}^{s} q_{j_i}^{m_i},$$
$$k_i = d_{\{\mathsf{b}_{j_i}\}}(\mu) + d_{\{\mathsf{b}_{j_i}\}}(\nu),$$
$$l_i = d_{\{\mathsf{a}_{j_i}\}}(\mu) + d_{\{\mathsf{a}_{j_i}\}}(\nu),$$
$$m_i = d_{\{\mathsf{a}_{j_i}\}}(\mu)\, d_{\{\mathsf{b}_{j_i}\}}(\nu).$$

Proof. To keep it short, denote a_{j_i} by a_i, b_{j_i} by b_i, A_{j_i} by A_i, B_{j_i} by B_i, and q_{j_i} by q_i for $i = 1,\ldots,s$. Also let $\phi\colon \mathcal{F}(J) \longrightarrow \mathcal{H}(\mathbf{q}, J)$ be the natural algebra homomorphism and let $S = S(\mathbf{q}, J)$.

The first claim of the theorem—that J' exists—is trivial. There is a finite number (two, to be exact) of nonzero polynomials involved. Each of these polynomials can be written as a linear combination of a finite number of monomials, and each monomial is the product of a finite number of a_j's and b_j's. Hence the degree functions can be nonzero only for a finite number of j's.

The second claim of the theorem—that

$$U = \operatorname{Maj}\big(\operatorname{Mon}(\alpha), \mathbf{f}\big) \qquad \text{and} \qquad V = \operatorname{Maj}\big(\operatorname{Mon}(\beta), \mathbf{f}\big)$$

contain exactly one element each—is also easy. They cannot be empty, since $\operatorname{Mon}(\alpha)$ and $\operatorname{Mon}(\beta)$ are both nonempty. It is furthermore the case that $f_i(t_1) = f_i(t_2)$ for all $i = 1,\ldots,r$ and $t_1, t_2 \in U$. This means in particular that $d_{\{\mathsf{a}_j\}} \circ \operatorname{pr}_2$ and $d_{\{\mathsf{b}_j\}} \circ \operatorname{pr}_2$ are known and equal for every μ such that $(a, \mu) \in U$. Now fix a μ such that $(a, \mu) \in U$. Since the monomial $\mu \in \operatorname{Irr}(S)$, and μ thus must be an element in the basis (3.3), it follows that the degrees of μ in the generators uniquely determine μ. Hence there is exactly one element in U. Also, the argument remains true if U is replaced by V.

Now for the final claim, equation (4.2). Let

$$u_i = d_{\{\mathsf{b}_i\}}(\mu), \qquad v_i = d_{\{\mathsf{a}_i\}}(\mu),$$
$$w_i = d_{\{\mathsf{b}_i\}}(\nu), \text{ and} \qquad x_i = d_{\{\mathsf{a}_i\}}(\nu)$$

for $i = 1,\ldots,s$. Then $\mu = \mathsf{b}_1^{u_1}\mathsf{a}_1^{v_1}\cdots\mathsf{b}_s^{u_s}\mathsf{a}_s^{v_s}$ and $\nu = \mathsf{b}_1^{w_1}\mathsf{a}_1^{x_1}\cdots\mathsf{b}_s^{w_s}\mathsf{a}_s^{x_s}$.

The method of the proof will be induction on s. If $s = 1$ then by Theorem 2.3,

$$\phi(\mu\nu) = B_1^{u_1} A_1^{v_1} B_1^{w_1} A_1^{x_1} = \\ = \sum_{i=0}^{\min(v_1,w_1)} q_1^{(v_1-i)(w_1-i)} \{i\}_{q_1}! \binom{v_1}{i}_{q_1} \binom{w_1}{i}_{q_1} B_1^{k_1-i} A_1^{l_1-i}.$$

Hence

$$t^S(\mu\nu) = \sum_{i=0}^{\min(v_1,w_1)} q_1^{(v_1-i)(w_1-i)} \{i\}_{q_1}! \binom{v_1}{i}_{q_1} \binom{w_1}{i}_{q_1} \mathsf{b}_1^{k_1-i} \mathsf{a}_1^{l_1-i}.$$

Corollary B.11 implies that if a single term in a polynomial simultaneously maximises all $d_{\{\mathsf{a}_j\}}$ and all $d_{\{\mathsf{b}_j\}}$ for $j \in J$, then that term will also maximise d_{W_i} for all $i = 1, \ldots, r$. Thus

$$\mathrm{Maj}\Big(\mathrm{Mon}\big(t^S(ab\mu\nu)\big), \mathbf{f}\Big) = \big\{(abq_1^{v_1w_1}, \mathsf{b}_1^{k_1}\mathsf{a}_1^{l_1})\big\} = \big\{(c,\lambda)\big\}.$$

The conditions in Lemma B.15 are now fulfilled and therefore

$$\mathrm{Maj}\Big(\mathrm{Mon}\big(t^S(\alpha\beta)\big), \mathbf{f}\Big) = \mathrm{Maj}\Big(\mathrm{Mon}\big(t^S(ab\mu\nu)\big), \mathbf{f}\Big) = \big\{(c,\lambda)\big\}.$$

as required. Thus the base for the induction has been laid.

Now let $s > 1$ and assume that (4.2) holds for all $\alpha, \beta \in \mathrm{Irr}(S)$ for which a set J' (with properties as above) can be found that satisfies $|J'| = s - 1$. It will be shown that the α and β from the statement of the theorem also satisfy (4.2).

One has that

$$\phi(\mu\nu) = B_1^{u_1} A_1^{v_1} \cdots B_s^{u_s} A_s^{v_s} B_1^{w_1} A_1^{x_1} \cdots B_s^{w_s} A_s^{x_s} = \\ = B_1^{u_1} A_1^{v_1} \cdots B_{s-1}^{u_{s-1}} A_{s-1}^{v_{s-1}} B_1^{w_1} A_1^{x_1} \cdots B_{s-1}^{w_{s-1}} A_{s-1}^{x_{s-1}} B_s^{u_s} A_s^{v_s} B_s^{w_s} A_s^{x_s} = \\ = \phi(\gamma\delta\epsilon),$$

where

$$\gamma = \mathsf{b}_1^{u_1}\mathsf{a}_1^{v_1} \cdots \mathsf{b}_{s-1}^{u_{s-1}}\mathsf{a}_{s-1}^{v_{s-1}}, \\ \delta = \mathsf{b}_1^{w_1}\mathsf{a}_1^{x_1} \cdots \mathsf{b}_{s-1}^{w_{s-1}}\mathsf{a}_{s-1}^{x_{s-1}}, \\ \epsilon = \mathsf{b}_s^{u_s}\mathsf{a}_s^{v_s}\mathsf{b}_s^{w_s}\mathsf{a}_s^{x_s}.$$

Hence $t^S(\mu\nu) = t^S(\gamma\delta\epsilon) = t^S\big(t^S(\gamma\delta)\, t^S(\epsilon)\big)$.

The inductive assumption can be applied to $\gamma\delta$. $\mathrm{Maj}\big(\mathrm{Mon}(\gamma), \mathbf{f}\big) = \big\{(1,\gamma)\big\}$ and $\mathrm{Maj}\big(\mathrm{Mon}(\delta), \mathbf{f}\big) = \big\{(1,\delta)\big\}$. Hence one can set $d = \prod_{i=1}^{s-1} q_i^{m_i}$ and $\kappa = \mathsf{b}_1^{k_1}\mathsf{a}_1^{l_1}\cdots\mathsf{b}_{s-1}^{k_{s-1}}\mathsf{a}_{s-1}^{l_{s-1}}$ and get

$$\mathrm{Maj}\Big(\mathrm{Mon}\big(t^S(\gamma\delta)\big), \mathbf{f}\Big) = \big\{(d,\kappa)\big\},$$

because $d_{\{\mathsf{b}_i\}}(\gamma) + d_{\{\mathsf{b}_i\}}(\delta) = u_i + w_i = k_i$, $d_{\{\mathsf{a}_i\}}(\gamma) + d_{\{\mathsf{a}_i\}}(\delta) = v_i + x_i = l_i$, and $d_{\{\mathsf{a}_i\}}(\gamma)\, d_{\{\mathsf{b}_i\}}(\delta) = v_i w_i = m_i$.

The factor $t^S(\epsilon)$ can be computed explicitly—just as was done above for $t^S(\mu\nu)$ in the case $s = 1$—and this yields

$$t^S(\epsilon) = \sum_{i=0}^{\min(v_s,w_s)} q_s^{(v_s-i)(w_s-i)}\{i\}_{q_s}! \binom{v_s}{i}_{q_s} \binom{w_s}{i}_{q_s} \mathsf{b}_s^{k_s-i}\mathsf{a}_s^{l_s-i}.$$

Hence

$$\mathrm{Maj}\Big(\mathrm{Mon}\big(t^S(\epsilon)\big), \mathbf{f}\Big) = \big\{(q_s^{m_s}, \mathsf{b}_s^{k_s}\mathsf{a}_s^{l_s})\big\}.$$

Let $e = q_s^{m_s}$ and $\iota = \mathsf{b}_s^{k_s}\mathsf{a}_s^{l_s}$.

Now

$$\phi(\kappa\iota) = B_1^{k_1}A_1^{l_1}\cdots B_{s-1}^{k_{s-1}}A_{s-1}^{l_{s-1}}B_s^{k_s}A_s^{l_s} = B_1^{k_1}A_1^{l_1}\cdots B_s^{k_s}A_s^{l_s} = \phi(\lambda),$$

and hence $t^S(\kappa\iota) = \lambda$. These clearly fulfill the conditions in Lemma B.15. Therefore

$$\begin{aligned}\mathrm{Maj}\left(\mathrm{Mon}\Big(t^S\big(t^S(\gamma\delta)\,t^S(\epsilon)\big)\Big), \mathbf{f}\right) &= \mathrm{Maj}\Big(\mathrm{Mon}\big(t^S(de\kappa\iota)\big), \mathbf{f}\Big) = \\ &= \mathrm{Maj}\big(\mathrm{Mon}(de\lambda), \mathbf{f}\big) = \big\{(de,\lambda)\big\}.\end{aligned}$$

Now the conditions are fulfilled for another application of Lemma B.15 giving

$$\begin{aligned}&\mathrm{Maj}\Big(\mathrm{Mon}\big(t^S(\alpha\beta)\big), \mathbf{f}\Big) = \mathrm{Maj}\Big(\mathrm{Mon}\big(t^S(ab\mu\nu)\big), \mathbf{f}\Big) = \\ &\quad = \mathrm{Maj}\left(\mathrm{Mon}\Big(t^S\big(ab\,t^S(\gamma\delta)\,t^S(\epsilon)\big)\Big), \mathbf{f}\right) = \big\{(abde,\lambda)\big\} = \big\{(c,\lambda)\big\}.\end{aligned}$$

The inductive step is now completed, and the last claim of the theorem follows from an application of the principle of induction. □

The following theorem is more or less the final step in proving that degree in $\mathcal{H}(\mathbf{q}, J)$ behaves as it does in the ordinary commutative polynomial algebras: when two elements are multiplied, the degree of the product will be the sum of the degrees of the two factors.

Theorem 4.3 *Let $V \subseteq \{A_j, B_j\}_{j\in J}$. Let $\alpha, \beta \in \mathcal{H}(\mathbf{q}, J)$. If $q_j \neq 0$ for all $j \in J$ then*

$$d_V(\alpha\beta) = d_V(\alpha) + d_V(\beta). \tag{4.3}$$

Proof. If $\alpha = 0$ or $\beta = 0$ then the theorem is trivial. Therefore assume that $\alpha, \beta \neq 0$.

Let $W_1 = \{\, \mathsf{a}_j \mid A_j \in V\,\} \cup \{\, \mathsf{b}_j \mid B_j \in V\,\}$. Let $S = S(\mathbf{q}, J)$ and let ϕ be the natural algebra homomorphism $\mathcal{F}(J) \longrightarrow \mathcal{H}(\mathbf{q}, J)$. Let $\hat{\alpha}, \hat{\beta} \in \mathrm{Irr}(S)$ be the unique elements that satisfy $\alpha = \phi(\hat{\alpha})$ and $\beta = \phi(\hat{\beta})$ respectively. It then follows from Theorem B.18 that

$$\begin{aligned} d_V(\alpha) &= d_{W_1}(\hat{\alpha}), \\ d_V(\beta) &= d_{W_1}(\hat{\beta}), \\ d_V(\alpha\beta) &= d_{W_1}\big(t^S(\hat{\alpha}\hat{\beta})\big). \end{aligned}$$

It will now be convenient to adopt much of the notation from Theorem 4.2. Let J' be a finite subset of J such that

$$d_{\{\mathsf{b}_j\}}(\hat{\alpha}) = d_{\{\mathsf{a}_j\}}(\hat{\alpha}) = d_{\{\mathsf{b}_j\}}(\hat{\beta}) = d_{\{\mathsf{a}_j\}}(\hat{\beta}) = 0$$

for all $j \in J \setminus J'$. Let $s = |J'|$, and let $\{j_1, \ldots, j_s\} = J'$ be such that $j_1 \prec j_2 \prec \cdots \prec j_s$. Denote a_{j_i} by a_i and b_{j_i} by b_i for $i = 1, \ldots, s$.

Let $W_{i+1} = \{\mathsf{a}_i\}$ and $W_{i+s+1} = \{\mathsf{b}_i\}$ for $i = 1, \ldots, s$. Define $\{f_i\}_{i=1}^{2s+1}$ by

$$f_i = d_{W_i} \circ \mathrm{pr}_2$$

for $i = 1, \ldots, 2s+1$. Let $\mathbf{f} = (f_1, \ldots, f_{2s+1})$. Let (a, μ), (b, ν), and (c, λ) be the pairs that satisfy

$$\begin{aligned} \{(a, \mu)\} &= \mathrm{Maj}\big(\mathrm{Mon}(\alpha), \mathbf{f}\big), \\ \{(b, \nu)\} &= \mathrm{Maj}\big(\mathrm{Mon}(\beta), \mathbf{f}\big), \\ \{(c, \lambda)\} &= \mathrm{Maj}\Big(\mathrm{Mon}\big(t^S(\hat{\alpha}\hat{\beta})\big), \mathbf{f}\Big). \end{aligned}$$

Note that $d_{W_1}(\hat{\alpha}) = d_{W_1}(\mu)$, $d_{W_1}(\hat{\beta}) = d_{W_1}(\nu)$, and $d_{W_1}\big(t^S(\hat{\alpha}\hat{\beta})\big) = d_{W_1}(\lambda)$.

Moreover let

$$u_i = d_{\{\mathsf{b}_i\}}(\mu), \qquad v_i = d_{\{\mathsf{a}_i\}}(\mu),$$
$$w_i = d_{\{\mathsf{b}_i\}}(\nu), \qquad x_i = d_{\{\mathsf{a}_i\}}(\nu)$$

for $i = 1, \ldots, s$. Finally, let

$$\begin{aligned} T &= \left\{ i \in \{1, \ldots, s\} \,\middle|\, \mathsf{a}_i \in W_1 \right\}, \\ U &= \left\{ i \in \{1, \ldots, s\} \,\middle|\, \mathsf{b}_i \in W_1 \right\}, \\ W_1' &= W_1 \cap \{ \mathsf{a}_j, \mathsf{b}_j \mid j \in J' \}. \end{aligned}$$

Corollary B.11 implies that W_1', T, and U satisfy the equality

$$d_{W_1'}(\rho) = \sum_{\kappa \in W_1'} d_{\{\kappa\}}(\rho) = \sum_{i \in T} d_{\{\mathsf{a}_i\}}(\rho) + \sum_{i \in U} d_{\{\mathsf{b}_i\}}(\rho)$$

for all monomials $\rho \in \mathcal{F}(J)$. It also implies that $d_{W_1}(\rho) = d_{W_1'}(\rho)$ for monomials $\rho \in \mathcal{F}(J')$, since $d_{W_1}(\rho) = d_{W_1'}(\rho) + d_{W_1 \setminus W_1'}(\rho)$ for all monomials ρ, $W_1 \setminus W_1' = W_1 \setminus \{ \mathsf{a}_j, \mathsf{b}_j \mid j \in J' \}$, and $d_{W_1 \setminus \{ \mathsf{a}_j, \mathsf{b}_j \mid j \in J' \}}(\gamma) = 0$ for every $\gamma \in \mathcal{F}(J')$.

With all preliminaries sorted out, it is now time to combine them into a proof of (4.3). Note that

$$\begin{aligned} d_V(\alpha\beta) &= d_{W_1}\big(t^S(\hat{\alpha}\hat{\beta})\big) = d_{W_1}(\lambda) = d_{W_1'}(\lambda) = \\ &= \sum_{i \in T} d_{\{\mathsf{a}_i\}}(\lambda) + \sum_{i \in U} d_{\{\mathsf{b}_i\}}(\lambda). \end{aligned}$$

From the explicit description of λ given in Theorem 4.2, one sees that $d_{\{\mathsf{a}_i\}}(\lambda) = v_i + x_i$ and $d_{\{\mathsf{b}_i\}}(\lambda) = u_i + w_i$. Thus

$$\begin{aligned} d_V(\alpha\beta) &= \sum_{i \in T}(v_i + x_i) + \sum_{i \in U}(u_i + w_i) = \\ &= \left(\sum_{i \in T} d_{\{\mathsf{a}_i\}}(\mu) + \sum_{i \in U} d_{\{\mathsf{b}_i\}}(\mu) \right) + \\ &\qquad + \left(\sum_{i \in T} d_{\{\mathsf{a}_i\}}(\nu) + \sum_{i \in U} d_{\{\mathsf{b}_i\}}(\nu) \right) = \\ &= d_{W_1'}(\mu) + d_{W_1'}(\nu) = d_{W_1}(\mu) + d_{W_1}(\nu) = \\ &= d_{W_1}(\hat{\alpha}) + d_{W_1}(\hat{\beta}) = d_V(\alpha) + d_V(\beta). \end{aligned}$$

□

4.2 Grading $\mathcal{H}(\mathbf{q}, J)$

Of the three basic operations in an $\mathcal{H}(\mathbf{q}, J)$ algebra (addition, multiplication by a scalar, and general multiplication), the first two are easily described in terms of the corresponding elements in $\operatorname{Irr}(S(\mathbf{q}, J))$ and the reason for this is of course that $\operatorname{Irr}(S(\mathbf{q}, J))$ is a module. As was seen in the previous section however, multiplication becomes quite a headache, since it does not produce results that are confined to $\operatorname{Irr}(S(\mathbf{q}, J))$. This is, on the other hand, just the way things must be, since the whole point of the Diamond Lemma apparatus is to allow various rules of reduction to appear in an algebra.

This does not mean that multiplication is totally out of control. A careful study of the rules of reduction in $S(\mathbf{q}, J)$ will show that they all leave some linear combinations of the basic degree functions unchanged. Back in the quotient $\mathcal{H}(\mathbf{q}, J)$, this means that some aspects of multiplication are relatively simple to predict. The basic means of this prediction are the various gradations of $\mathcal{H}(\mathbf{q}, J)$ that are defined and examined in this section.

Definition 4.2 Let $\mathbb{Z}(J)$ denote the additive group with the set

$$\left\{ \text{functions } f\colon J \longrightarrow \mathbb{Z} \,\middle|\, f^{-1}(\mathbb{Z} \setminus \{0\}) \text{ is finite} \right\}, \tag{4.4}$$

where $f^{-1}(\mathbb{Z} \setminus \{0\})$ simply means the set $\{ j \in J \mid f(j) \in \mathbb{Z} \setminus \{0\} \}$.

In other words, $\mathbb{Z}(J)$ is a free abelian group with rank $|J|$.

Definition 4.3 Let Y be the set of all monomials in $\mathcal{F}(J)$. Let

$$Y_f = \{ \mu \in Y \mid d_{\{\mathsf{b}_j\}}(\mu) - d_{\{\mathsf{a}_j\}}(\mu) = f(j) \text{ for all } j \in J \}$$
$$Y_n^j = \{ \mu \in Y \mid d_{\{\mathsf{b}_j\}}(\mu) - d_{\{\mathsf{a}_j\}}(\mu) = n \}$$

for all $f \in \mathbb{Z}(J)$, $n \in \mathbb{Z}$, and $j \in J$. Let $\phi\colon \mathcal{F}(J) \longrightarrow \mathcal{H}(\mathbf{q}, J)$ be the natural homomorphism. Set

$$K_f = \phi(Y_f) \quad \text{and} \quad K_n^j = \phi(Y_n^j)$$

for all $f \in \mathbb{Z}(J)$, $n \in \mathbb{Z}$, and $j \in J$.

The set K_f will be called the **chain*** for f. For any chain K_f, the vector space $\operatorname{Span}(K_f)$ will be called the **chain space** of K_f.

*These sets are *not* chains in the normal mathematical sense of the word, nor do they have any close connection to chains in that sense. Therefore we do not recommend using this terminology.

Theorem 4.4 *The algebra $\mathcal{H}(\mathbf{q}, J)$ decomposes as*

$$\mathcal{H}(\mathbf{q}, J) = \bigoplus_{f \in \mathbb{Z}(J)} \operatorname{Span}(K_f), \tag{4.5}$$

and because of this composition, $\mathcal{H}(\mathbf{q}, J)$ is a $\mathbb{Z}(J)$-graded $\mathcal{K}$-algebra. For each $j \in J$, the algebra $\mathcal{H}(\mathbf{q}, J)$ also decomposes as

$$\mathcal{H}(\mathbf{q}, J) = \bigoplus_{n \in \mathbb{Z}} \operatorname{Span}(K_n^j), \tag{4.6}$$

and each of these compositions makes $\mathcal{H}(\mathbf{q}, J)$ a $\mathbb{Z}$-graded $\mathcal{K}$-algebra.

Proof. Let the set Y and the sets Y_f and Y_n^j be as in Definition 4.3. Thus Y is a basis for $\mathcal{F}(J)$, and $\{Y_f\}_{f \in \mathbb{Z}(J)}$ is a partition of Y. Hence

$$\mathcal{F}(J) = \bigoplus_{f \in \mathbb{Z}(J)} \operatorname{Span}(Y_f).$$

Likewise, $\{Y_n^j\}_{n \in \mathbb{Z}}$ is a partition of Y for each $j \in J$, and hence

$$\mathcal{F}(J) = \bigoplus_{n \in \mathbb{Z}} \operatorname{Span}(Y_n^j).$$

Now let $f, g \in \mathbb{Z}(J)$ be arbitrary. Let $\mu \in Y_f$ and $\nu \in Y_g$. It then follows from Theorem B.13 that

$$\begin{aligned} d_{\{\mathsf{b}_j\}}(\mu\nu) - d_{\{\mathsf{a}_j\}}(\mu\nu) &= d_{\{\mathsf{b}_j\}}(\mu) + d_{\{\mathsf{b}_j\}}(\nu) - d_{\{\mathsf{a}_j\}}(\mu) - d_{\{\mathsf{a}_j\}}(\nu) = \\ &= f(j) + g(j) \end{aligned}$$

for all $j \in J$. Hence $\mu\nu \in Y_{f+g}$. In general this means that $Y_f \cdot Y_g \subseteq Y_{f+g}$, which in turn implies that $\operatorname{Span}(Y_f) \cdot \operatorname{Span}(Y_g) \subseteq \operatorname{Span}(Y_{f+g})$. Thus the submodules $\{\operatorname{Span}(Y_f)\}_{f \in \mathbb{Z}(J)}$ define a $\mathbb{Z}(J)$-gradation of $\mathcal{F}(J)$. That the submodules $\{\operatorname{Span}(Y_n^j)\}_{n \in \mathbb{Z}}$ define a $\mathbb{Z}$-gradation of $\mathcal{F}(J)$ for each $j \in J$ is shown similarly.

Now consider the reduction system $S(\mathbf{q}, J)$ for $\mathcal{F}(J)$. Let $1_j \colon J \longrightarrow \mathbb{Z}$ denote the function that maps j to 1 and all other elements in J to 0. This function 1_j is an element in the group $\mathbb{Z}(J)$. With this, the following observations can be made about the elements $\mu_s - \alpha_s$ for $s \in S(\mathbf{q}, J)$ that generate $\mathcal{I}\big(S(\mathbf{q}, J)\big)$:

$$\begin{aligned} \mathsf{a}_i\mathsf{a}_j - \mathsf{a}_j\mathsf{a}_i &\in Y_{-1_j-1_i} \\ \mathsf{a}_i\mathsf{b}_j - \mathsf{b}_j\mathsf{a}_i &\in Y_{1_j-1_i} \\ \mathsf{b}_i\mathsf{a}_j - \mathsf{a}_j\mathsf{b}_i &\in Y_{-1_j+1_i} \\ \mathsf{b}_i\mathsf{b}_j - \mathsf{b}_j\mathsf{b}_i &\in Y_{1_j+1_i} \\ \mathsf{a}_j\mathsf{b}_j - q_j\mathsf{b}_j\mathsf{a}_j - 1 &\in Y_0. \end{aligned}$$

Thus $\mu_s - \alpha_s \in \bigcup_{f\in\mathbb{Z}(J)} Y_f$ for all $s \in S(\mathbf{q}, J)$.

Let $\phi\colon \mathcal{F}(J) \longrightarrow \mathcal{H}(\mathbf{q}, J)$ be the natural homomorphism. Theorem B.19 then implies that

$$\Big\{\phi\big(\mathrm{Span}(Y_f)\big)\Big\}_{f\in\mathbb{Z}(J)}$$

is a gradation of $\mathcal{H}(\mathbf{q}, J)$. Now the first part of the present theorem follows from the equality

$$\phi\big(\mathrm{Span}(Y_f)\big) = \mathrm{Span}\big(\phi(Y_f)\big) = \mathrm{Span}(K_f),$$

which holds for all $f \in \mathbb{Z}(J)$.

Finally observe that since $\mathrm{Span}(K_f) \subseteq \mathrm{Span}(K_n^j)$ for each n and j such that $f(j) = n$, the gradations $\{Y_n^j\}_{n\in\mathbb{Z}}$ also fulfill the conditions of Theorem B.19. Thus the second part of this theorem holds for the same reasons as the first did. □

We will now introduce some new notations.

Definition 4.4 Let $\alpha \in \mathcal{H}(\mathbf{q}, J)$, $j \in J$, $n \in \mathbb{Z}$, and $f \in \mathbb{Z}(J)$. Then

$$\begin{aligned} \alpha \sqsubseteq K_f &\quad \text{will mean} \quad \alpha \in \mathrm{Span}(K_f), \\ \alpha \sqsubseteq K_n^j &\quad \text{will mean} \quad \alpha \in \mathrm{Span}(K_n^j). \end{aligned}$$

An α is said to be **fully homogeneous** if there is an f such that $\alpha \sqsubseteq K_f$. The α is said to be **homogeneous in** j if there is a j and an n such that $\alpha \sqsubseteq K_n^j$.

The **intersection** of α and K_f, which is denoted $\alpha \sqcap K_f$, is the projection of α into $\mathrm{Span}(K_f)$ that is associated with the decomposition (4.5). Similarly, the **intersection** of α and K_n^j, which is denoted $\alpha \sqcap K_n^j$, is the projection of α into $\mathrm{Span}(K_n^j)$ that is associated with the decomposition (4.6).

Let

$$\chi\colon \left\{ \alpha \in \mathcal{H}(\mathbf{q}, J) \,\middle|\, \alpha \text{ is nonzero and fully homogeneous} \right\} \longrightarrow \mathbb{Z}(J)$$

be the function that is defined by the fact that $\chi(\alpha)$ is the unique element in $\mathbb{Z}(J)$ that satisfies $\alpha \sqsubseteq K_{\chi(\alpha)}$. This function χ will be called the **chain function** (this is sometimes called the degree of α, but that term has another meaning in this book). Let

$$\chi_j\colon \left\{ \alpha \in \mathcal{H}(\mathbf{q}, J) \,\middle|\, \alpha \text{ is nonzero and homogeneous in } j \right\} \longrightarrow \mathbb{Z}$$

be the function that is defined by the fact that $\chi_j(\alpha)$ is the unique integer that satisfies $\alpha \sqsubseteq K^j_{\chi_j(\alpha)}$. This function χ_j will be called the **chain function in** j.

The two classes of functions $\overline{\chi}_j, \underline{\chi}_j : \mathcal{H}(\mathbf{q}, J) \setminus \{0\} \longrightarrow \mathbb{Z}$ are called the **upper chain functions** and **lower chain functions** respectively. They are defined by

$$\overline{\chi}_j(\alpha) = \max \left\{ n \in \mathbb{Z} \,\middle|\, \alpha \sqcap K^j_n \neq 0 \right\}, \tag{4.7}$$
$$\underline{\chi}_j(\alpha) = \min \left\{ n \in \mathbb{Z} \,\middle|\, \alpha \sqcap K^j_n \neq 0 \right\}. \tag{4.8}$$

Note: In the special case $\mathcal{H}(q) = \mathcal{H}(\mathbf{q}, J)$, i.e. when $|J| = 1$, the j's are dropped from the above notations. This causes no harm since the two gradations are equal in that case.

Note: Intersection is defined as a projection, hence it is a linear map. Thus for all $\alpha, \beta \in \mathcal{H}(q)$, $c \in \mathcal{K}$, and $f \in \mathbb{Z}(J)$:

$$(\alpha + \beta) \sqcap K_f = \alpha \sqcap K_f + \beta \sqcap K_f \tag{4.9}$$
$$(c\alpha) \sqcap K_f = c(\alpha \sqcap K_f). \tag{4.10}$$

Similar equalities are true for the $\mathbb{Z}$-gradations.

Note: The easiest way too look at intersection in probably to regard it as a "removal of all terms from other chains", i.e., if $\alpha \in \mathcal{H}(\mathbf{q}, J)$ is expressed as

$$\alpha = \sum_{\mu \in P} a_\mu \mu,$$

where P is a finite set of monomials in $\mathcal{H}(\mathbf{q}, J)$ and $\{a_\mu\}_{\mu\in P} \subseteq \mathcal{K}$, then

$$\alpha \sqcap K_f = \sum_{\mu\in P\cap K_f} a_\mu \mu$$

for all $f \in \mathbb{Z}(J)$. The analogous equalities hold for the K_n^j chains.

Note: If $\alpha \in \mathcal{H}(\mathbf{q}, J)$ is nonzero and homogeneous in j, then

$$\underline{\chi}_j(\alpha) = \chi_j(\alpha) = \overline{\chi}_j(\alpha).$$

Corollary 4.5 (to Theorem 4.4) *Let $j \in J$, $k \in \mathbb{N}$, and $m, n \in \mathbb{Z}$. Let $\alpha \sqsubseteq K_m^j$ and $\beta \sqsubseteq K_n^j$. Then $\alpha\beta \sqsubseteq K_{m+n}^j$ and $\alpha^k \sqsubseteq K_{km}^j$.*

Also, if $f, g \in \mathbb{Z}(J)$, $k \in \mathbb{N}$, $\alpha \sqsubseteq K_f$, and $\beta \sqsubseteq K_g$ then $\alpha\beta \sqsubseteq K_{f+g}$ and $\alpha^k \sqsubseteq K_{kf}$.

The notations introduced in Definition 4.4 can also be used to reformulate Lemma B.20. This gives the following corollary to Theorem 4.4.

Corollary 4.6 *Let $\alpha, \beta \in \mathcal{H}(\mathbf{q}, J)$. Let $j \in J$ and $n, m \in \mathbb{Z}$, also let $\gamma \sqsubseteq K_n^j$. Let $f, g \in \mathbb{Z}(J)$ and $\delta \sqsubseteq K_g$. Then*

$$(\alpha\gamma) \sqcap K_{m+n}^j = (\alpha \sqcap K_m^j)\gamma, \tag{4.11}$$

$$(\gamma\alpha) \sqcap K_{m+n}^j = \gamma(\alpha \sqcap K_m^j), \tag{4.12}$$

$$(\alpha\beta) \sqcap K_n^j = \sum_{\substack{k,l\in\mathbb{Z}\\ k+l=n}} (\alpha \sqcap K_k^j)(\beta \sqcap K_l^j), \tag{4.13}$$

$$[\alpha, \beta] \sqcap K_n^j = \sum_{\substack{k,l\in\mathbb{Z}\\ k+l=n}} [\alpha \sqcap K_k^j, \beta \sqcap K_l^j], \tag{4.14}$$

$$(\alpha\delta) \sqcap K_{f+g} = (\alpha \sqcap K_f)\delta, \tag{4.15}$$

$$(\delta\alpha) \sqcap K_{f+g} = \delta(\alpha \sqcap K_f), \tag{4.16}$$

$$(\alpha\beta) \sqcap K_f = \sum_{\substack{g,h\in\mathbb{Z}(J)\\ g+h=f}} (\alpha \sqcap K_g)(\beta \sqcap K_h), \tag{4.17}$$

$$[\alpha, \beta] \sqcap K_f = \sum_{\substack{g,h\in\mathbb{Z}(J)\\ g+h=f}} [\alpha \sqcap K_g, \beta \sqcap K_h]. \tag{4.18}$$

Theorem 4.7 *Let $j \in J$ be arbitrary. If $\alpha, \beta \in \mathcal{H}(\mathbf{q}, J)$ are nonzero, then*

$$\overline{\chi}_j(\alpha\beta) \leqslant \overline{\chi}_j(\alpha) + \overline{\chi}_j(\beta), \tag{4.19}$$

$$\underline{\chi}_j(\alpha\beta) \geqslant \underline{\chi}_j(\alpha) + \underline{\chi}_j(\beta). \tag{4.20}$$

Furthermore, if $q_j \neq 0$ for all $j \in J$ then

$$\overline{\chi}_j(\alpha\beta) = \overline{\chi}_j(\alpha) + \overline{\chi}_j(\beta), \tag{4.21}$$

$$\underline{\chi}_j(\alpha\beta) = \underline{\chi}_j(\alpha) + \underline{\chi}_j(\beta). \tag{4.22}$$

Proof. Let $\alpha_k = \alpha \sqcap K^j_k$, $\beta_l = \beta \sqcap K^j_l$, and $\gamma_i = (\alpha\beta) \sqcap K^j_i$. Then

$$\gamma_i = \sum_{\substack{k,l \in \mathbb{Z} \\ k+l=i}} \alpha_k \beta_l.$$

Let $n = \overline{\chi}_j(\alpha)$ and $m = \overline{\chi}_j(\beta)$. If $i > n + m$ and $k + l = i$ then $k > n$ or $l > m$. Thus $\gamma_i = 0$ and hence $\overline{\chi}_j(\alpha\beta) \leqslant n + m$. This has proved (4.19).

If $i = n + m$, $k + l = i$, $\alpha_k \neq 0$, and $\beta_l \neq 0$ then $k = n$ and $l = m$. Thus $\gamma_i = \alpha_n \beta_m$ and by Theorem 4.9, $\alpha_n \beta_m \neq 0$. Hence $\overline{\chi}_j(\alpha\beta) \geqslant n + m$, and this has proved (4.21).

The claims about $\underline{\chi}_j$ are proved analogously. □

Theorem 4.8 *If $\alpha \in \mathcal{H}(\mathbf{q}, J)$ is homogeneous in j, then*

$$\deg_{\mathrm{A}_j} \alpha + \deg_{\mathrm{B}_j} \alpha = \deg_j \alpha. \tag{4.23}$$

Proof. If $\alpha = 0$ then the theorem is trivial. Thus assume that $\alpha \in \mathcal{H}(\mathbf{q}, J) \setminus \{0\}$. Let $\phi \colon \mathcal{F}(J) \longrightarrow \mathcal{H}(\mathbf{q}, J)$ be the natural homomorphism, and let $\hat{\alpha} \in \operatorname{Irr}\bigl(S(\mathbf{q}, J)\bigr)$ be the unique element that satisfies $\phi(\hat{\alpha}) = \alpha$. Let $J' = \{j_1, \dots, j_s\}$ be a finite subset of J such that $\hat{\alpha} \in \mathcal{F}(J')$. Let $f_{i+1} = d_{\{\mathsf{a}_{j_i}\}} \circ \mathrm{pr}_2$ for $i = 1, \dots, s$, let $f_{i+s+1} = d_{\{\mathsf{b}_{j_i}\}} \circ \mathrm{pr}_2$ for $i = 1, \dots, s$, let $f_1 = d_{\{\mathsf{a}_j\}} \circ \mathrm{pr}_2$, and let $f_0 = d_{\{\mathsf{a}_j, \mathsf{b}_j\}} \circ \mathrm{pr}_2$. Let $\mathbf{f} = (f_1, \dots, f_{2s+1})$ and $\mathbf{f}' = (f_0, f_1, \dots, f_{2s+1})$.

Now there exist pairs (a, μ) and (b, ν) such that

$$\bigl\{(a, \mu)\bigr\} = \operatorname{Maj}\bigl(\operatorname{Mon}(\hat{\alpha}), \mathbf{f}'\bigr) \quad \text{and} \quad \bigl\{(b, \nu)\bigr\} = \operatorname{Maj}\bigl(\operatorname{Mon}(\hat{\alpha}), \mathbf{f}\bigr).$$

Clearly these satisfy that $\deg_j \alpha = d_{\{\mathsf{a}_j, \mathsf{b}_j\}}(\mu)$ and $\deg_{\mathrm{A}_j} \alpha = d_{\{\mathsf{a}_j\}}(\nu)$. It is also the case that $\deg_{\mathrm{B}_j} \alpha = d_{\{\mathsf{b}_j\}}(\nu)$, because $d_{\{\mathsf{b}_j\}}(\lambda) - d_{\{\mathsf{a}_j\}}(\lambda) = n$ for all $\lambda \in K^j_n$, including μ and ν.

Hence

$$\begin{aligned}\deg_j \alpha &= d_{\{\mathsf{a}_j,\mathsf{b}_j\}}(\mu) = d_{\{\mathsf{a}_j\}}(\mu) + d_{\{\mathsf{b}_j\}}(\mu) = 2d_{\{\mathsf{a}_j\}}(\mu) + n \leqslant \\ &= 2\deg_{\mathrm{A}_j}(\alpha) + n = 2d_{\{\mathsf{a}_j\}}(\nu) + n = d_{\{\mathsf{a}_j\}}(\nu) + d_{\{\mathsf{b}_j\}}(\nu) = \\ &= d_{\{\mathsf{a}_j,\mathsf{b}_j\}}(\nu) \leqslant \deg_j \alpha,\end{aligned}$$

from which follows that

$$\deg_j \alpha = d_{\{\mathsf{a}_j\}}(\nu) + d_{\{\mathsf{b}_j\}}(\nu) = \deg_{\mathrm{A}_j} \alpha + \deg_{\mathrm{B}_j} \alpha.$$

□

4.3 Some useful properties

The concepts of degree and chains are far from only being related to themselves. This section gives three examples of how they can be used to derive results about such mainstream mathematical concepts as factorisation and dimension of a subspace, all of which will be used in the following chapters.

Half the first theorem is really a direct corollary to Theorem 4.2, but we have preferred to use the degree function deg in our proof here, as it relieves us of having to spend most of the proof repeating the constructions in the proof of Theorem 4.3.

Theorem 4.9 *There are no $\alpha, \beta \in \mathcal{H}(\mathbf{q}, J) \setminus \{0\}$ such that $\alpha\beta = 0$ —i.e., there are no zero divisors in $\mathcal{H}(\mathbf{q}, J)$ —if and only if $q_j \neq 0$ for all $j \in J$.*

Proof. To prove the 'if' part, note that $\deg \gamma \geqslant 0$ for all $\gamma \in \mathcal{H}(\mathbf{q}, J) \setminus \{0\}$. Thus in particular, by Theorem 4.3, $\deg(\alpha\beta) = \deg \alpha + \deg \beta \geqslant 0$. As $\deg 0 = -\infty < 0$, it follows that $\alpha\beta \neq 0$.

To prove the 'only if' part, let $j \in J$ be such that $q_j = 0$. Then

$$(B_j A_j - I)B_j = B_j A_j B_j - B_j = B_j \cdot I - B_j = 0,$$

but by Theorem 3.1, neither $B_j A_j - I = 0$ nor $B_j = 0$. This proves the theorem. □

The next theorem establishes a connection between degree and dimension that will be one of the key ingredients in many of our more advanced results.

Theorem 4.10 *Let $f \in \mathbb{Z}(J)$ and $D \subseteq \mathbb{Z}(J)$. Let $\mathcal{V}$ be any subspace of* $\operatorname{Span}(K_f)$ *which satisfies*

$$\forall\, \alpha{\in}\mathcal{V}\colon\; \big(\alpha \neq 0 \implies \exists\, d{\in}D\colon\; \forall\, j{\in}J\colon\; \deg_j \alpha = d(j)\big). \tag{4.24}$$

(i) *Then*

$$\dim \mathcal{V} \leqslant |D| \tag{4.25}$$

and for every $g \in \mathbb{N}$,

$$\dim \{\, \alpha \in \mathcal{V} \mid \deg \alpha \leqslant g \,\} \leqslant \left| \left\{ d \in D \,\middle|\, \textstyle\sum_{j\in J} d(j) \leqslant g \right\} \right|. \tag{4.26}$$

(ii) *If $W \subseteq \mathcal{V}$ is such that*

$$\forall\, d{\in}D\colon\; \exists\, \beta{\in}W\colon\; \forall\, j{\in}J\colon\; \deg_j \beta = d(j), \tag{4.27}$$

then $\operatorname{Span}(W) = \mathcal{V}$.

Proof. If $D = \emptyset$ then (4.25) is trivial, because it follows that $\mathcal{V} = \{0\}$. Hence what is left is to prove (4.25) for nonempty D.

Define a vector space homomorphism

$$\psi\colon \operatorname{Span}(K_f) \longrightarrow \{h\colon D \longrightarrow \mathcal{K}\}$$

as follows. Let μ be any element of the basis (3.4) that belongs to K_f. Then

$$\psi(\mu)(d) = \begin{cases} 1 & \text{if } \deg_j \mu = d(j) \text{ for all } j \in J, \\ 0 & \text{otherwise.} \end{cases}$$

Now (4.24) implies that 0 is the only element in $\mathcal{V}$ that ψ maps to 0, which in turn is equivalent to that the restriction of ψ to $\mathcal{V}$ is injective. Hence

$$\dim \mathcal{V} \leqslant \dim\{h\colon D \longrightarrow \mathcal{K}\} = |D|,$$

which proves (4.25).

Substituting

$$\{\, \alpha \in \mathcal{V} \mid \deg \alpha \leqslant g \,\} \text{ for } \mathcal{V} \quad \text{and} \quad \left\{ d \in D \,\middle|\, \textstyle\sum_{j\in J} d(j) \leqslant g \right\} \text{ for } D$$

in (4.25) gives (4.26).

In order to prove (ii), assume that W satisfies (4.27). Let $J' \subseteq J$ be finite and let $k \in \mathbb{N}$. Define

$$\begin{aligned} D_k(J') &= \Big\{ d \in D \Bigm| \textstyle\sum_{j \in J} d(j) \leqslant k \text{ and } d(j) = 0 \text{ for all } j \in J \setminus J' \Big\}, \\ \mathcal{V}_k(J') &= \big\{ \alpha \in \mathcal{V} \bigm| \deg \alpha \leqslant k \text{ and } \deg_j \alpha = 0 \text{ for all } j \in J \setminus J' \big\}, \\ W_k(J') &= \big\{ \beta \in W \bigm| \deg \beta \leqslant k \text{ and } \deg_j \beta = 0 \text{ for all } j \in J \setminus J' \big\}. \end{aligned}$$

Since $W_k(J')$ has a subset with $|D_k(J')|$ nonzero members of different degree, it follows that

$$|D_k(J')| \leqslant \dim \operatorname{Span}\big(W_k(J')\big) \leqslant \dim \mathcal{V}_k(J') \leqslant |D_k(J')|.$$

Hence $\dim \operatorname{Span}\big(W_k(J')\big) = \dim \mathcal{V}_k(J')$. As both these spaces are finite-dimensional (J' is finite) and $\operatorname{Span}\big(W_k(J')\big) \subseteq V_k(J')$, they must be equal. Thus $\operatorname{Span}\big(W_k(J')\big) = V_k(J')$

Now let $\alpha \in \mathcal{V}$ be arbitrary, then there is a finite $J' \subseteq J$ such that $\alpha \in \mathcal{H}(\mathbf{q}, J')$, or equivalently $\deg_j \alpha = 0$ for all $j \in J \setminus J'$. Let $k = \deg \alpha$. Then

$$\alpha \in \mathcal{V}_k(J') = \operatorname{Span}\big(W_k(J')\big) \subseteq \operatorname{Span}(W).$$

Since $\alpha \in \mathcal{V}$ is arbitrary, $\operatorname{Span}(W) = \mathcal{V}$. □

Our last theorem uses a rather trivial observation to prove the existence of certain factorisations. The result as such is hardly enough to get anyone excited, but there are a couple of instances in which it comes in handy.

Theorem 4.11 *Let $n \in \mathbb{Z}$, $j \in J$, and $\alpha \sqsubseteq K_n^j$. If $n \geqslant 0$ then for each m satisfying $0 \leqslant m \leqslant n$ there are $\beta, \gamma \sqsubseteq K_{n-m}^j$ such that*

$$\alpha = B_j^m \beta = \gamma B_j^m. \tag{4.28}$$

If $n \leqslant 0$ then for each m satisfying $0 \leqslant m \leqslant -n$ there are $\beta, \gamma \sqsubseteq K_{n+m}^j$ such that

$$\alpha = A_j^m \beta = \gamma A_j^m. \tag{4.29}$$

Proof. Let $J' = J \setminus \{j\}$. Let $\prec$ be a total ordering of J such that $i \prec j$ for all $i \in J'$. This total ordering should be used in each application of Theorems 3.1 and 3.2 below.

Assume $0 \leqslant m \leqslant n$. Then, as a consequence of Theorem 3.1, there exist a $k \in \mathbb{N}$ and $\{\alpha_i\}_{i=0}^{k} \subset \mathcal{H}(\mathbf{q}, J')$ such that

$$\alpha = \sum_{i=0}^{k} \alpha_i B_j^{n+i} A_j^i.$$

But then

$$\alpha = B_j^m \underbrace{\sum_{i=0}^{k} \alpha_i B_j^{n-m+i} A_j^i}_{\beta} = B_j^m \beta,$$

where $\beta \sqsubseteq K_{n-m}^j$ as required.

If $q_j \neq 0$ then as a consequence of Theorem 3.2, there exists a $k \in \mathbb{N}$ and $\{\alpha_i\}_{i=0}^{k} \subset \mathcal{H}(\mathbf{q}, J')$ such that

$$\alpha = \sum_{i=0}^{k} \alpha_i A_j^i B_j^{n+i}$$

(just choose $D = \{j\}$). Then

$$\alpha = \underbrace{\sum_{i=0}^{k} \alpha_i A_j^i B_j^{n-m+i}}_{\gamma} B_j^m = \gamma B_j^m,$$

where $\gamma \sqsubseteq K_{n-m}^j$ as required.

If, on the other hand, $q_j = 0$ then

$$\alpha = \alpha A_j^m B_j^m = \gamma B_j^m,$$

where $\gamma \sqsubseteq K_{n-m}^j$ as required.

The case $n \leqslant 0$ is handled similarly. □

Corollary 4.12 *Let $f \in \mathbb{Z}(J)$ and $\alpha \sqsubseteq K_f$ be arbitrary. Let $\{j_i\}_{i=1}^{m} \subseteq J$ be a set which contains $\{ j \in J \mid f(j) \neq 0 \}$. Let $r_i = \max\{0, f(j_i)\}$ and $s_i = \max\{0, -f(j_i)\}$ for $1 \leqslant i \leqslant m$. Then there exists a $\beta \sqsubseteq K_0$—note that the 0 in the subscript of K_0 is the $0 \in \mathbb{Z}(J)$—such that*

$$\alpha = B_{j_m}^{r_m} \cdots B_{j_1}^{r_1} \beta A_{j_1}^{s_1} \cdots A_{j_m}^{s_m}. \tag{4.30}$$

Proof. The proof is by induction on m. The base for the induction is the (notationally perhaps somewhat awkward) case $m = 0$, in which $f = 0$ and $\beta = \alpha$.

In the induction step, assume that the corollary holds for all $m < k$, where $k > 0$, and attempt to prove it for $m = k$. Let $f' = f - r_k 1_{j_k}$ (where 1_j as usual denotes the element of $\mathbb{Z}(J)$ that is 1 in j and 0 everywhere else) and $f'' = f' + s_k 1_{j_k}$. Note that $f''(j_k) = 0$ and $f''(j) = f(j)$ for all other $j \in J$.

By Theorem 4.11, there exists an $\alpha' \sqsubseteq K_{f'}$ such that $\alpha = B_{j_k}^{r_k}\alpha'$ and an $\alpha'' \sqsubseteq K_{f''}$ such that $\alpha' = \alpha'' A_{j_k}^{s_k}$. Hence for this α'',

$$\alpha = B_{j_k}^{r_k}\alpha'' A_{j_k}^{s_k}.$$

Since there are at most $k-1$ positions in f'' which are nonzero, the induction hypothesis can be applied and hence there exists a $\beta \sqsubseteq K_0$ that satisfies

$$\begin{aligned}\alpha &= B_{j_k}^{r_k}\alpha'' A_{j_k}^{s_k} = \\ &= B_{j_k}^{r_k} B_{j_{k-1}}^{r_{k-1}} \cdots B_{j_1}^{r_1}\beta A_{j_1}^{s_1} \cdots A_{j_{k-1}}^{s_{k-1}} A_{j_k}^{s_k}\end{aligned}$$

exactly as the corollary claims. Thus by the principle of induction, the corollary holds for arbitrary $m \in \mathbb{N}$. □

Chapter 5

Centralisers of elements in $\mathcal{H}(\mathbf{q}, J)$

Never underestimate a theorem that counts something.

J. B. FRALEIGH

In this chapter, we begin our study of centralisers in the $\mathcal{H}(\mathbf{q}, J)$ algebras. In Section 5.1 the gradations of $\mathcal{H}(\mathbf{q}, J)$ introduced in Chapter 4 are used to decompose centralisers and the centre into direct sums of homogeneous parts, and to describe in Theorem 5.3 a set of commuting elements in $\mathcal{H}(\mathbf{q}, J)$. In Section 5.2 vectors of deformation parameters $\mathbf{q} = (q_j)_{j\in J}$, and hence the algebras $\mathcal{H}(\mathbf{q}, J)$, are classified depending on the properties of $\{q_j\}_{j\in J}$ as subsets of the multiplicative group of the coefficient field $\mathcal{K}$. The relevance of this classification to the description of commuting elements in $\mathcal{H}(\mathbf{q}, J)$ is seen best in Theorem 5.4 which is of key importance for most of the subsequent results. In the last three sections of this chapter we treat some of the more important classes of deformation vectors $\mathbf{q}$ and study in more detail the properties of the corresponding algebras $\mathcal{H}(\mathbf{q}, J)$.

5.1 General definitions and theorems

When studying centralisers in $\mathcal{H}(\mathbf{q}, J)$, it turns out to be fruitful to consider not only the whole centraliser, but also various subsets defined using the chains introduced in Chapter 4.

Definition 5.1 For all $\beta \in \mathcal{H}(\mathbf{q}, J)$, let $X(\beta)$ denote the set

$$\{ g \in \mathbb{Z}(J) \mid \beta \sqcap K_g \neq 0 \}.$$

Let $\alpha \in \mathcal{H}(\mathbf{q}, J)$. Let $m, n \in \mathbb{Z}$ and $j \in J$. Let $f \in \mathbb{Z}(J)$, $F \subseteq \mathbb{Z}(J)$, and $d \in \mathbb{N}$. Then define

$$\begin{aligned}
\mathrm{Cen}(\alpha) &= \{ \beta \in \mathcal{H}(\mathbf{q}, J) \mid [\alpha, \beta] = 0 \}, \\
\mathrm{Cen}(f, \alpha) &= \{ \beta \sqsubseteq K_f \mid [\alpha, \beta] = 0 \}, \\
\mathrm{Cen}(d, f, \alpha) &= \{ \beta \sqsubseteq K_f \mid [\alpha, \beta] = 0 \text{ and } \deg \beta \leqslant d \}, \\
\mathrm{Cen}(n{:}j, \alpha) &= \{ \beta \sqsubseteq K_n^j \mid [\alpha, \beta] = 0 \}, \\
\mathrm{Cen}(F, \alpha) &= \{ \beta \in \mathcal{H}(\mathbf{q}, J) \mid [\alpha, \beta] = 0 \text{ and } X(\beta) \subseteq F \}, \\
\mathrm{Cen}\big([m, n]{:}j, \alpha\big) &= \Big\{ \beta \in \mathrm{Cen}(\alpha) \Bigm| \beta = 0, \text{ or } m \leqslant \underline{\chi}_j(\beta) \text{ and } \overline{\chi}_j(\beta) \leqslant n \Big\}.
\end{aligned}$$

The set $\mathrm{Cen}(\alpha)$ is called the **centraliser** of α. All of the sets defined above are submodules of $\mathcal{H}(\mathbf{q}, J)$.

Theorem 5.1 *Let $n, m \in \mathbb{Z}$, $j \in J$, and $\alpha \sqsubseteq K_n^j$. Let $f, g \in \mathbb{Z}(J)$ and $\beta \sqsubseteq K_f$. Then*

$$\mathrm{Cen}(m{:}j, \alpha) = \{ \gamma \sqcap K_m^j \mid \gamma \in \mathrm{Cen}(\alpha) \}, \tag{5.1}$$

$$\mathrm{Cen}(g, \beta) = \{ \gamma \sqcap K_g \mid \gamma \in \mathrm{Cen}(\beta) \}. \tag{5.2}$$

Therefore

$$\mathrm{Cen}(\alpha) = \bigoplus_{m \in \mathbb{Z}} \mathrm{Cen}(m{:}j, \alpha), \tag{5.3}$$

$$\mathrm{Cen}(\beta) = \bigoplus_{g \in \mathbb{Z}(J)} \mathrm{Cen}(g, \beta). \tag{5.4}$$

Proof. The proof of (5.1) is by showing that both sides are subsets of the other side, as shown below. The proof of (5.2) is completely analogous.

($\subseteq$): Let $\gamma \in \mathrm{Cen}(m{:}j, \alpha)$ be arbitrary. Since $\gamma \sqsubseteq K_m^j$, it follows that $\gamma \sqcap K_m^j = \gamma$. Furthermore $\gamma \in \mathrm{Cen}(\alpha)$, and hence

$$\mathrm{Cen}(m{:}j, \alpha) \subseteq \left\{ \gamma \sqcap K_m^j \,\middle|\, \gamma \in \mathrm{Cen}(\alpha) \right\}.$$

($\supseteq$): Let $\gamma \in \mathrm{Cen}(\alpha)$ be arbitrary. Then by Corollary 4.6,

$$\alpha(\gamma \sqcap K_m^j) = (\alpha\gamma) \sqcap K_{n+m}^j = (\gamma\alpha) \sqcap K_{m+n}^j = (\gamma \sqcap K_m^j)\alpha$$

and thus $\gamma \sqcap K_m^j \in \mathrm{Cen}(m{:}j, \alpha)$. It follows that

$$\left\{ \gamma \sqcap K_m^j \,\middle|\, \gamma \in \mathrm{Cen}(\alpha) \right\} \subseteq \mathrm{Cen}(m, \alpha).$$

The proofs of (5.3) and (5.4) are likewise analogous, hence only the proof of (5.3) is shown. Let $\gamma \in \mathrm{Cen}(\alpha)$ be arbitrary, then

$$\gamma = \sum_{m \in \mathbb{Z}} \gamma \sqcap K_m^j \in \sum_{m \in \mathbb{Z}} \mathrm{Cen}(m{:}j, \alpha) \subseteq \mathrm{Cen}(\alpha).$$

As γ is arbitrary, it follows that the $\subseteq$ is in fact an equality. Thus

$$\mathrm{Cen}(\alpha) = \sum_{m \in \mathbb{Z}} \mathrm{Cen}(m{:}j, \alpha).$$

Proving that the sum is in fact direct is simpler, it suffices to notice that

$$\mathrm{Cen}(m{:}j, \alpha) \subseteq \mathrm{Span}(K_m^j)$$

and that the sum of $\mathrm{Span}(K_m^j)$ for all $m \in \mathbb{Z}$ is direct by Theorem 4.4. □

In view of the above, the next theorem should not come as a surprise.

Theorem 5.2 *Let $\mathcal{C}$ be the centre of $\mathcal{H}(\mathbf{q}, J)$. For each $f \in \mathbb{Z}(J)$, let*

$$\mathcal{C}_f = \left\{ \alpha \sqsubseteq K_f \,\middle|\, [\alpha, A_j] = 0 = [\alpha, B_j] \text{ for all } j \in J \right\}.$$

Then

$$\mathcal{C}_f = \{ \alpha \sqcap K_f \mid \alpha \in \mathcal{C} \} \tag{5.5}$$

and

$$\mathcal{C} = \bigoplus_{f \in \mathbb{Z}(J)} \mathcal{C}_f. \tag{5.6}$$

Proof. Obviously $\mathcal{C}_f \subseteq \mathcal{C}$, and hence $\mathcal{C}_f \subseteq \{ \alpha \sqcap K_f \mid \alpha \in \mathcal{C} \}$. To prove the reverse inclusion, let $\beta \in \mathcal{C}$ be arbitrary. Then

$$\begin{aligned} A_j(\beta \sqcap K_f) &= (A_j\beta) \sqcap K_{f-1_j} = (\beta A_j) \sqcap K_{f-1_j} = (\beta \sqcap K_f)A_j, \\ B_j(\beta \sqcap K_f) &= (B_j\beta) \sqcap K_{f+1_j} = (\beta B_j) \sqcap K_{f+1_j} = (\beta \sqcap K_f)B_j \end{aligned}$$

for all $j \in J$, where 1_j denotes the function in $\mathbb{Z}(J)$ which maps j to 1 and everything else to 0, just as it did in the proof of Theorem 4.4. This shows that $\beta \sqcap K_f \in \mathcal{C}_f$.

The decomposition (5.6) can be proved as follows. Let $\beta \in \mathcal{C}$ be arbitrary. Then

$$\beta = \sum_{f \in \mathbb{Z}(J)} \beta \sqcap K_f \in \sum_{f \in \mathbb{Z}(J)} \mathcal{C}_f \subseteq \mathcal{C}.$$

As β is arbitrary, it follows that $\mathcal{C} \subseteq \sum_{f \in \mathbb{Z}(J)} \mathcal{C}_f \subseteq \mathcal{C}$, and hence these sets are equal. Proving that the sum is in fact direct is simpler, it suffices to notice that $\mathcal{C}_f \subseteq \operatorname{Span}(K_f)$ and that the sum of all $\operatorname{Span}(K_f)$ for $f \in \mathbb{Z}(J)$ is direct by Theorem 4.4. □

Theorem 5.3 *If $q_j \neq 0$ for all $j \in J$ then $\alpha\beta = \beta\alpha$ for all $\alpha, \beta \sqsubseteq K_0$, where the* 0 *in the subscript of K_0 is the $0 \in \mathbb{Z}(J)$.*

Proof. By Theorem 3.3, every $\gamma \in \mathcal{H}(\mathbf{q}, J)$ can be written as a linear combination of monomials $\prod_{i=1}^{n} P(j_i, k_i, l_i)$, where

$$P(j,k,l) = \begin{cases} C_j^k B_j^l & \text{if } l > 0 \\ C_j^k & \text{if } l = 0 \\ C_j^k A_j^{-l} & \text{if } l < 0 \end{cases}.$$

It is easily seen that $\chi_j\bigl(P(j,k,l)\bigr) = l$. Hence every $\gamma \sqsubseteq K_0$ can be written as a linear combination of monomials $\prod_{i=1}^{n} C_{j_i}^{k_i} = C_{j_1}^{k_1} C_{j_2}^{k_2} \cdots C_{j_n}^{k_n}$.

It is however the case that $C_iC_j = C_jC_i$ for all $i, j \in J$. This implies that both α and β belong to a *commutative* subalgebra of $\mathcal{H}(\mathbf{q}, J)$. Thus in particular $\alpha\beta = \beta\alpha$. □

Theorem 5.3 could alternatively be proved by using Theorem 2.8 and the basis (3.4). A special case of it was obtained within another context in [373].

5.2 Classification of $\mathcal{H}(\mathbf{q}, J)$

The following sections, and in particular the next chapter, will show that the nature of $\mathcal{H}(\mathbf{q}, J)$ is very dependent on the properties of $\{q_j\}_{j \in J}$ seen as a subset of the group $\mathcal{K}^*$. This group is the commutative group with set $\mathcal{K} \setminus \{0\}$ and whose group operation is the multiplication operation of the *field* $\mathcal{K}$.

Definition 5.2 Let $q_j \in \mathcal{K}^*$ be given. Then

- q_j is said to be of **torsion type** if $q_j \neq 1$ and there is a positive integer solution p to $q_j^p = 1$. In that case the least such p is called the **order** of q_j.
- $q_j \neq 1$ is said to be of **free type** if the only integer solution p to $q_j^p = 1$ is $p = 0$. In this case the **order** of q_j is defined to be the integer 0.
- $q_j = 1$, finally, is said to be of **free type** if the field $\mathcal{K}$ has characteristic 0 and of **torsion type** otherwise. The **order** of $q_j = 1$ is defined to be the characteristic of $\mathcal{K}$.

Please note that the order of a q_j of torsion type defined above is not quite the order of q_j as element in the group $\mathcal{K}^*$, although in many cases the two concepts are equal.

We use the notation $\langle q \rangle$ for the subgroup of $\mathcal{K}^*$ generated by $q \in \mathcal{K}^*$. More generally, we write $\langle q_j \rangle_{j \in J}$ for the subgroup of $\mathcal{K}^*$ that is generated by the elements of the set $\{q_j\}_{j \in J}$.

Let $\mathbf{q} \in (\mathcal{K}^*)^{\square \supseteq J}$. Then

- $\mathbf{q}$ is said to be of **direct type** on J, or simply of direct type in the case when $\mathbf{q} \in (\mathcal{K}^*)^J$, if

$$\langle q_j \rangle_{j \in J} = \bigoplus_{j \in J} \langle q_j \rangle. \tag{5.7}$$

  (Please note that since this is a direct sum of subgroups of the group $\mathcal{K}^*$, the "addition" in this sum is the same thing as the multiplication operation in the field $\mathcal{K}$.)
- $\mathbf{q}$ is said to be of **strictly direct type** on J, or simply of strictly direct type in the case when $\mathbf{q} \in (\mathcal{K}^*)^J$, if $\mathbf{q}$ is of direct type on J and $q_j \neq 1$ for all $j \in J$.

- $\mathbf{q}$ is said to be of **direct free type** on J, or simply of direct free type in the case when $\mathbf{q} \in (\mathcal{K}^*)^J$, if $\mathbf{q}$ is of direct type on J and all q_j for $j \in J$ are of free type.
- $\mathbf{q}$ is said to be of **direct torsion type** on J, or simply of direct torsion type in the case when $\mathbf{q} \in (\mathcal{K}^*)^J$, if $\mathbf{q}$ is of direct type on J and all q_j for $j \in J$ are of torsion type.

Example 5.1 What will the above concepts mean in the case that $\mathcal{K} = \mathbb{C}$? The q's of torsion type will be the roots of 1, with the exception of 1 itself, so all torsion type elements can be written on the form

$$q = e^{2\pi i k/l}, \tag{5.8}$$

where i is the imaginary unit, k and l are integers, $l \neq 0$, and $\frac{k}{l} \notin \mathbb{Z}$. The order of one of these q's will be the smallest positive integer l for which there exists an integer k that solves (5.8). This number can also be expressed as $|l|/\gcd(k, l)$.

The element 1 is of free type whilst 0 is neither of free nor of torsion type since $0 \notin \mathcal{K}^*$. All other complex numbers are of free type.

A $\mathbf{q} \in (\mathcal{K}^*)^{\square\supseteq J}$ is of strictly direct type on J if the only solutions to

$$1 = \prod_{j \in J'} q_j^{n_j},$$

where J' is an arbitrary nonempty finite subset of J and $n_j \in \mathbb{Z}$ for all $j \in J'$, are such that $n_j = 0$ for all q_j of free type and the order of q_j divides n_j for all q_j that are of torsion type. In particular these conditions say that $q_j \neq 1$ for all $j \in J$, but in the case of a direct type $\mathbf{q}$, the requirement is merely that what remains of the $\mathbf{q}$ after all elements q_j that are 1 have been removed must be of strictly direct type.

Thus $\mathbf{q} = (2, 1, -1)$ is of direct type, but not of strictly direct type. The vector $\mathbf{q} = (5, -2, e^i)$ is of strictly direct type, and even of strictly direct free type, since none of the elements is a root of 1.

For the moment some of these classifications might seem peculiar, but what follows will show that they are relevant. Most of the classifications are motivated by the following important theorem, on which almost everything that follows is based.

Theorem 5.4 *Let $j \in J$ be given and let $\mathbf{q} \in (\mathcal{K}^*)^{\square \supseteq J}$ be such that $q_j \neq 1$ and*

$$\langle q_i \rangle_{i \in J} = \langle q_j \rangle \oplus \langle q_i \rangle_{i \in J \setminus \{j\}}. \tag{5.9}$$

Also let $\alpha, \beta \in \mathcal{H}(\mathbf{q}, J)$ be nonzero elements homogeneous in j such that $[\alpha, \beta] = 0$.

If q_j is of free type then

$$\chi_j(\alpha) \deg_j \beta = \chi_j(\beta) \deg_j \alpha. \tag{5.10}$$

If instead q_j is of torsion type then

$$\chi_j(\alpha) \deg_j \beta \equiv \chi_j(\beta) \deg_j \alpha \pmod{2p}, \tag{5.11}$$

where p is the order of q_j.

Proof. Let $\phi \colon \mathcal{F}(J) \longrightarrow \mathcal{H}(\mathbf{q}, J)$ be the natural homomorphism, and let $\hat{\alpha}, \hat{\beta} \in \operatorname{Irr}\bigl(S(\mathbf{q}, J)\bigr)$ be the unique elements that satisfy $\alpha = \phi(\hat{\alpha})$ and $\beta = \phi(\hat{\beta})$.

Let J' be a finite subset of J such that $\hat{\alpha}, \hat{\beta} \in \mathcal{F}(J')$ and $j \in J'$. Let $s = |J'|$ and $\{j_1, \dots, j_s\} = J'$ where, in particular, $j_1 = j$. Let

$$W_1 = \{\mathsf{a}_j, \mathsf{b}_j\} \qquad W_{1+i} = \{\mathsf{a}_{j_i}\} \qquad W_{s+1+i} = \{\mathsf{b}_{j_i}\}$$

for $i = 1, \dots, s$. Finally let $f_i = d_{W_i} \circ \mathrm{pr}_2$ for $i = 1, \dots, 2s+1$ and $\mathbf{f} = (f_1, \dots, f_{2s+1})$.

Now it follows from Theorem 4.2 that there exist pairs (a, μ) and (b, ν) such that

$$\bigl\{(a, \mu)\bigr\} = \operatorname{Maj}\bigl(\operatorname{Mon}(\hat{\alpha}), \mathbf{f}\bigr) \quad \text{and} \quad \bigl\{(b, \nu)\bigr\} = \operatorname{Maj}\bigl(\operatorname{Mon}(\hat{\beta}), \mathbf{f}\bigr).$$

That same theorem also implies that

$$\begin{aligned} \operatorname{Maj}\Bigl(\operatorname{Mon}\bigl(t^{S(\mathbf{q},J)}(\hat{\alpha}\hat{\beta})\bigr), \mathbf{f}\Bigr) &= \bigl\{(c, \lambda)\bigr\} \\ \operatorname{Maj}\Bigl(\operatorname{Mon}\bigl(t^{S(\mathbf{q},J)}(\hat{\beta}\hat{\alpha})\bigr), \mathbf{f}\Bigr) &= \bigl\{(d, \lambda)\bigr\}, \end{aligned}$$

where

$$\begin{aligned} \lambda &= \mathsf{b}_{j_1}^{k_1} \mathsf{a}_{j_1}^{l_1} \cdots \mathsf{b}_{j_s}^{k_s} \mathsf{a}_{j_s}^{l_s}, \\ c &= ab \prod_{i=1}^{s} q_{j_i}^{m_i}, \end{aligned}$$

$$\begin{aligned} d &= ab\prod_{i=1}^{s} q_{j_i}^{n_i}, \\ k_i &= d_{\{\mathsf{b}_{j_i}\}}(\mu) + d_{\{\mathsf{b}_{j_i}\}}(\nu), \\ l_i &= d_{\{\mathsf{a}_{j_i}\}}(\mu) + d_{\{\mathsf{a}_{j_i}\}}(\nu), \\ m_i &= d_{\{\mathsf{a}_{j_i}\}}(\mu)\, d_{\{\mathsf{b}_{j_i}\}}(\nu), \\ n_i &= d_{\{\mathsf{a}_{j_i}\}}(\nu)\, d_{\{\mathsf{b}_{j_i}\}}(\mu). \end{aligned}$$

But since $\alpha\beta = \beta\alpha$, one must have that $t^{S(\mathsf{q},J)}(\hat{\alpha}\hat{\beta}) = t^{S(\mathsf{q},J)}(\hat{\beta}\hat{\alpha})$. Hence $(c, \lambda) = (d, \lambda)$, and it follows that

$$\prod_{i=1}^{s} q_{j_i}^{m_i} = \prod_{i=1}^{s} q_{j_i}^{n_i}. \tag{5.12}$$

By (5.9), this implies that

$$q_j^{m_1} = q_j^{n_1}.$$

This is equivalent to $m_1 - n_1 = pr$, where r is some integer and p is the order of q_j. Thus

$$d_{\{\mathsf{a}_j\}}(\mu)\, d_{\{\mathsf{b}_j\}}(\nu) - d_{\{\mathsf{a}_j\}}(\nu)\, d_{\{\mathsf{b}_j\}}(\mu) = pr. \tag{5.13}$$

Let

$$\begin{aligned} u &= \deg_{\mathrm{A}} \alpha, & v &= \deg_{\mathrm{B}} \alpha, \\ w &= \deg_{\mathrm{A}} \beta, & x &= \deg_{\mathrm{B}} \beta. \end{aligned}$$

From the homogeneity in j of α and β it now follows that

$$\begin{aligned} d_{\{\mathsf{b}_j\}}(\mu) - d_{\{\mathsf{a}_j\}}(\mu) &= v - u = \chi_j(\alpha), \\ d_{\{\mathsf{b}_j\}}(\nu) - d_{\{\mathsf{a}_j\}}(\nu) &= x - w = \chi_j(\beta), \end{aligned}$$

and from Corollary B.11 it follows that

$$\begin{aligned} d_{\{\mathsf{b}_j\}}(\mu) + d_{\{\mathsf{a}_j\}}(\mu) &= d_{W_1}(\mu) = \deg_j \alpha = v + u, \\ d_{\{\mathsf{b}_j\}}(\nu) + d_{\{\mathsf{a}_j\}}(\nu) &= d_{W_1}(\nu) = \deg_j \beta = x + w. \end{aligned}$$

This implies

$$\begin{aligned} \chi_j(\alpha)\deg_j\beta - \chi_j(\beta)\deg_j\alpha &= \\ = (v-u)(x+w) - (x-w)(v+u) &= \\ = 2vw - 2ux = 2(n_1 - m_1) &= 2pr, \end{aligned}$$

which is equivalent to (5.10) or (5.11), depending on the type of q_j. □

It is only to be expected that similar relations can be shown even if q_j cannot be separated as above. If for example

$$\langle q_i \rangle_{i\in J} = \langle q_j, q_{j'} \rangle \oplus \langle q_i \rangle_{i \in J\setminus\{j,j'\}}$$

then it seems probable that some slightly weaker integer equation can be deduced, involving not only $\deg_j$ and χ_j but also $\deg_{j'}$ and $\chi_{j'}$. One might for example get one modulo equation and one normal, even if q_j and $q_{j'}$ are both of free type.

A similar theorem exists for $\mathcal{H}(1)$. That theorem appears in this book as Theorem 6.1 and is the reason for the classification of $q_j = 1$.

5.3 The case q_j of torsion type for some $j \in J$

This section treats $\mathcal{H}(\mathbf{q}, J)$ where there is a q_j that belongs to the torsion submodule of $\mathcal{K}^*$. In non-algebraic terms, this would be that one of the q_j is a root of unity.

For the rest of this section, let a q_j of torsion type be given and let p be the order of q_j. Then the central observation is that $\{p\}_{q_j} = 0$. If $q_j \neq 1$ then it follows from the formula for a geometric sum that

$$\{p\}_{q_j} = \sum_{i=0}^{p-1} q_j^i = \frac{q_j^p - 1}{q_j - 1} = 0.$$

If instead $q_j = 1$ then this is simply the defining property of the characteristic of a field. An immediate consequence of this equality is the following lemma.

Lemma 5.5 *The generators A_j and B_j satisfy the relations*

$$A_j^p A_i = A_i A_j^p \qquad A_j^p B_i = B_i A_j^p \tag{5.14}$$

$$B_j^p A_i = A_i B_j^p \qquad B_j^p B_i = B_i B_j^p \tag{5.15}$$

for all $i \in J$.

Proof. Most of the relations follow directly from the definition of a generalised q-deformed Heisenberg algebra, but two of them are not true in general. These are the second and third of the above equations for $i = j$. In that case however,

$$A_j B_j^p = q_j^p B_j^p A_j + \{p\}_{q_j} B^{p-1} = B_j^p A_j$$
$$A_j^p B_j = q_j^p B_j A_j^p + \{p\}_{q_j} A^{p-1} = B_j A_j^p$$

by Theorem 2.2. □

Theorem 5.6 *Every* $\alpha \in \mathcal{H}(\mathbf{q}, J)$ *satisfies*

$$\alpha A_j^p = A_j^p \alpha \tag{5.16}$$
$$\alpha B_j^p = B_j^p \alpha \tag{5.17}$$

Proof. Let $X = \{\, A_i, B_i \mid i \in J \,\}$. Then Lemma 5.5 says that $\lambda A_j^p = A_j^p \lambda$ and $\lambda B_j^p = B_j^p \lambda$ for every $\lambda \in X$. This implies that $\mu A_j^p = A_j^p \mu$ and $\mu B_j^p = B_j^p \mu$ for every monomial $\mu \in \mathcal{H}(\mathbf{q}, J)$ too, since every monomial $\mu = \lambda_1 \cdots \lambda_m$ for some $\{\lambda_i\}_{i=1}^m \subseteq X$.

It is furthermore true that every $\alpha \in \mathcal{H}(\mathbf{q}, J)$ can be written as

$$\alpha = \sum_{i=1}^{n} a_i \mu_i$$

for some $n \in \mathbb{N}$, $\{a_i\}_{i=1}^n \subseteq \mathcal{K}$, and monomials $\{\mu_i\}_{i=1}^n \subset \mathcal{H}(\mathbf{q}, J)$. Thus

$$\alpha A_j^p = \sum_{i=1}^{n} a_i \mu_i A_j^p = \sum_{i=1}^{n} A_j^p a_i \mu_i = A_j^p \alpha$$
$$\alpha B_j^p = \sum_{i=1}^{n} a_i \mu_i B_j^p = \sum_{i=1}^{n} B_j^p a_i \mu_i = B_j^p \alpha$$

which proves the theorem. □

Theorem 5.7 *The centre of* $\mathcal{H}(\mathbf{q}, J)$ *is nontrivial—it contains at least the subalgebra that is generated by* A_j^p *and* B_j^p. *In other words, all* $\alpha \in \mathcal{H}(\mathbf{q}, J)$ *and* $\beta \in \mathrm{Alg}_{\mathcal{K}}\langle A^p, B^p \rangle$ *satisfy* $[\beta, \alpha] = 0$.

Proof. By Theorem 5.6, both A_j^p and B_j^p belong to the centre. By Theorem 3.1, neither is a scalar multiple of I. Hence the centre is nontrivial.

The centre of an algebra is always a subalgebra of the algebra. Hence the centre of $\mathcal{H}(\mathbf{q}, J)$ must at least contain the $\mathcal{K}$-subalgebra of $\mathcal{H}(\mathbf{q}, J)$ that is generated by A_j^p and B_j^p. Definition 3.1 introduced the notation $\mathrm{Alg}_{\mathcal{K}}\langle A^p, B^p \rangle$ for this subalgebra. □

The above theorem is of course important for purely theoretical reasons, but there are other, more applied, consequences that also can be of interest. This result does for example lead to some computational simplifications, but we have deferred our discussion of that to Section 6.3, where the advantages of these simplifications are more apparent.

Theorem 5.8 *If $\alpha, \beta \in \mathcal{H}(\mathbf{q}, J)$ are such that $[\alpha, \beta] = 0$ then*

$$\left[B_j^{pn_1} \alpha A_j^{pn_2}, B_j^{pm_1} \beta A_j^{pm_2}\right] = 0 \tag{5.18}$$

for all $n_1, n_2, m_1, m_2 \in \mathbb{N}$.

Proof. By Theorem 5.7,

$$\begin{aligned}
&\left[B_j^{pn_1} \alpha A_j^{pn_2}, B_j^{pm_1} \beta A_j^{pm_2}\right] = \\
&\qquad = B_j^{pn_1} \alpha A_j^{pn_2} B_j^{pm_1} \beta A_j^{pm_2} - B_j^{pm_1} \beta A_j^{pm_2} B_j^{pn_1} \alpha A_j^{pn_2} = \\
&\qquad = B_j^{p(n_1+m_1)} \alpha A_j^{pn_2} \beta A_j^{pm_2} - B_j^{p(n_1+m_1)} \beta A_j^{pm_2} \alpha A_j^{pn_2} = \\
&\qquad = B_j^{p(n_1+m_1)} \alpha \beta A_j^{p(n_2+m_2)} - B_j^{p(n_1+m_1)} \beta \alpha A_j^{p(n_2+m_2)} = \\
&\qquad = B_j^{p(n_1+m_1)} [\alpha, \beta] A_j^{p(n_2+m_2)} = 0.
\end{aligned}$$

□

This theorem displays the following important property of commuting elements: *Given α and β that do commute, there will exist an infinity of α and β "look-alikes", different from the two in that the degrees in A_j and B_j in every single term are shifted some multiple of p steps.*

Although this is a "nice" property inasmuch as it means that there are plenty of commuting elements, it will also make it harder to classify them. The reason for this is of course that everything becomes periodic, and since the period p is an integer its factorisation into primes will be something that every such classification must take into account. This is most apparent in Theorem 6.22.

5.4 The centre of $\mathcal{H}(\mathbf{q}, J)$ when $\mathbf{q}$ is of strictly direct type on J

This section treats the important case when $\mathbf{q}$ is of strictly direct type.

Lemma 5.9 *Let $\mathbf{q} \in (\mathcal{K}^*)^{\square \supseteq J}$ be arbitrary and let*

$$J' = \{ j \in J \mid q_j \text{ is of torsion type} \}.$$

Let p_j be the order of q_j for every $j \in J'$ and let $p_j = 0$ for every $j \in J \setminus J'$. Assume that for every $j \in J$ and all fully homogeneous nonzero $\alpha, \beta \in \mathcal{H}(\mathbf{q}, J)$ such that $[\alpha, \beta] = 0$, there is an $n \in \mathbb{Z}$ such that

$$\chi_j(\alpha) \deg_j \beta = \chi_j(\beta) \deg_j \alpha + 2np_j. \tag{5.19}$$

Then the centre of $\mathcal{H}(\mathbf{q}, J)$ is

$$\mathrm{Alg}_{\mathcal{K}}\langle \{A_j^{p_j}, B_j^{p_j}\}_{j \in J'} \rangle. \tag{5.20}$$

Proof. Let $\mathcal{C}$ be the centre of $\mathcal{H}(\mathbf{q}, J)$ and let

$$\mathcal{C}_f = \mathcal{C} \cap \mathrm{Span}(K_f)$$

for all $f \in \mathbb{Z}(J)$. Recall that $\mathcal{C} = \bigoplus_{f \in \mathbb{Z}(J)} \mathcal{C}_f$ by Theorem 5.2. Hence to determine $\mathcal{C}$, it is sufficient to determine $\mathcal{C}_f$ for each $f \in \mathbb{Z}(J)$.

Fix an arbitrary $f \in \mathbb{Z}(J)$ such that $\mathcal{C}_f \neq \{0\}$ and let $\alpha \in \mathcal{C}_f$ be nonzero. Then in particular, $[\alpha, A_j] = [\alpha, B_j] = 0$ for all $j \in J$. All A_j and B_j are fully homogeneous, and so is α. Hence by the condition in the lemma,

$$\begin{aligned} f(j) &= -\deg_j \alpha + 2p_j r_j \\ f(j) &= \deg_j \alpha + 2p_j s_j \end{aligned}$$

for all $j \in J$ and some $\{r_j\}_{j \in J}, \{s_j\}_{j \in J} \subseteq \mathbb{Z}$. From adding these two equations one gets that $f(j)$ is an integer multiple of p_j for all $j \in J$. Then one easily deduces that

$$\deg_j \alpha = |f(j)| + 2p_j t_j$$

for some $\{t_j\}_{j \in J} \subseteq \mathbb{N}$.

Thus if f is such that $\mathcal{C}_f \neq \{0\}$ and $\alpha \in \mathcal{C}_f$ is nonzero, then $f(j) = \deg_j \alpha = 0$ for all j such that q_j is of free type. There are however infinitely many possibilities for j such that q_j is of torsion type, i.e., for $j \in J'$.

What will now be shown is that every combination of degrees not deemed impossible by the above actually occurs.

To do this, let $f_0 \in \mathbb{Z}(J)$ be arbitrary and define f by setting $f(j) = p_j f_0(j)$ for all $j \in J$. The elements of $\mathbb{Z}(J)$ which can not be expressed as such an f correspond to the chain spaces whose only element common to $\mathcal{C}$ is 0. Furthermore define $f', f'' \in \mathbb{Z}(J)$ by setting $f'(j) = \max\{0, f(j)\}$ and $f''(j) = \max\{0, -f(j)\}$ for each $j \in J$. Note that this makes $f = f' - f''$. Finally let

$$N = \{\, g \in \mathbb{Z}(J) \mid g(j) \geqslant 0 \text{ for all } j \in J \,\}.$$

Now consider the sets

$$D = \{\, h \in N \mid \exists\, g{\in}N\colon\ \forall\, j{\in}J\colon\ h(j) = |f(j)| + 2p_j g(j) \,\},$$

$$W = \left\{ \prod_{j\in J'} B_j^{f'(j)+p_j g(j)} A_j^{f''(j)+p_j g(j)} \,\middle|\, g \in N \right\}.$$

By the above, $\alpha \in \mathcal{C}_f$ implies that $\deg_j \alpha = h(j)$ for all $j \in J$, for some $h \in D$. By Theorem 5.7, it follows that $W \subseteq \mathcal{C}_f$. Finally since

$$\deg_j\left(\prod_{j\in J'} B_j^{f'(j)+p_j g(j)} A_j^{f''(j)+p_j g(j)} \right) = |f(j)| + 2p_j g(j)$$

for all $j \in J$ and $g \in N$, the conditions in part (ii) of Theorem 4.10 are fulfilled. Hence $\mathcal{C}_f = \operatorname{Span}(W)$.

This can be summarised as follows: Every element in every $\mathcal{C}_f$ can be expressed as a polynomial expression in elements of the set $\{A_j^{p_j}, B_j^{p_j}\}_{j\in J'}$. Hence this is true for the centre as a whole as well. Furthermore, as a consequence of Theorem 5.7, every element that can be expressed this way is an element of the centre. Therefore the centre of $\mathcal{H}(\mathbf{q}, J)$ is exactly the set of those elements, i.e., the set $\operatorname{Alg}_{\mathcal{K}}\langle \{A_j^{p_j}, B_j^{p_j}\}_{j\in J'} \rangle$. □

Theorem 5.10 *Let $\mathbf{q}$ be of strictly direct type on J. Let*

$$J' = \{\, j \in J \mid q_j \text{ is of torsion type} \,\}$$

and for each $j \in J'$ let p_j be the order of q_j. Then the centre of $\mathcal{H}(\mathbf{q}, J)$ is

$$\operatorname{Alg}_{\mathcal{K}}\langle \{A_j^{p_j}, B_j^{p_j}\}_{j\in J'} \rangle. \tag{5.21}$$

Proof. By Theorem 5.4, equation (5.19) is fulfilled for all α, β, and j that Lemma 5.9 assumes it is. Therefore all conditions of that lemma are fulfilled, and its conclusion follows. $\square$

5.5 $\mathcal{H}(\mathbf{q}, J)$ for $\mathbf{q} \in \mathcal{Q}(J, \mathcal{K})$

At this point, it is suitable to introduce a class $\mathcal{Q}(J,\mathcal{K}) \subseteq (\mathcal{K}^*)^{\square \supseteq J}$ of deformation vectors $\mathbf{q}$. It will be an immediate observation that this class contains the class of all $\mathbf{q}$ of strictly direct free type on J, but some results in Chapter 6 will allow us to find more elements from this class.

Definition 5.3 Let a set J be given. Let $j \in J$, let $\mathbf{q} \in (\mathcal{K}^*)^{\square \supseteq J}$, and consider the property that

$$\left.\begin{array}{c} [\alpha, \beta] = 0 \implies \chi_j(\alpha) \deg_j \beta = \chi_j(\beta) \deg_j \alpha \\ \text{for all fully homogeneous } \alpha, \beta \in \mathcal{H}(\mathbf{q}, J) \setminus \{0\}. \end{array}\right\} \tag{5.22}$$

If $J = \{j\}$ then $\mathcal{Q}(J,\mathcal{K})$ is simply the set of all $\mathbf{q} \in (\mathcal{K}^*)^J$ that satisfy (5.22).

For finite J such that $|J| > 1$, the set $\mathcal{Q}(J,\mathcal{K})$ consists of all $\mathbf{q} \in (\mathcal{K}^*)^{\square \supseteq J}$ such that there exists a $j \in J$ for which (5.22) is satisfied and for which $\mathbf{q} \in \mathcal{Q}(J \setminus \{j\}, \mathcal{K})$. Finally for infinite J,

$$\mathcal{Q}(J,\mathcal{K}) = \left\{ \mathbf{q} \in (\mathcal{K}^*)^{\square \supseteq J} \,\middle|\, \mathbf{q} \in \mathcal{Q}(J', \mathcal{K}) \text{ for all finite nonempty } J' \subset J \right\}.$$

It might at this point seem that the above definition is unnecessarily complicated. Why not simply require that (5.22) holds for all $j \in J$? The answer is that this would make $\mathcal{Q}(J,\mathcal{K})$ smaller than we want it to be.

Theorem 5.11 *If $\mathbf{q} \in (\mathcal{K}^*)^{\square \supseteq J}$ is of strictly direct free type on J, then $\mathbf{q} \in \mathcal{Q}(J,\mathcal{K})$.*

Proof. For finite J this is proved by induction on $|J|$. If $J = \{j\}$ then (5.22) is fulfilled by Theorem 5.4. This has laid the base for the induction.

Now consider the case that $|J| = k > 1$ and assume that the theorem holds for all J such that $|J| = k - 1$. Choose a $j \in J$. Then

$$\langle q_i \rangle_{i \in J} = \langle q_j \rangle \oplus \langle q_i \rangle_{i \in J \setminus \{j\}}$$

since $\mathbf{q}$ is of strictly direct type on J. Hence (5.22) is fulfilled by Theorem 5.4. Furthermore $\mathbf{q}$ satisfy the conditions in the theorem for the set

$J \setminus \{j\}$ instead of J, and hence $\mathbf{q} \in \mathcal{Q}(J \setminus \{j\}, \mathcal{K})$ by the induction hypothesis. Thus $\mathbf{q} \in \mathcal{Q}(J, \mathcal{K})$ and the induction step is completed. By the principle of induction, it follows that the theorem holds for all finite J.

For infinite J, it suffices to observe that $\mathbf{q}$ is of strictly direct type on J' for all finite $J' \subseteq J$. This implies that $\mathbf{q} \in \mathcal{Q}(J', \mathcal{K})$ for all such J' and hence by definition, $\mathbf{q} \in \mathcal{Q}(J, \mathcal{K})$. □

Lemma 5.12 *Let $j \in J$ and $\mathbf{q} \in (\mathcal{K}^*)^{\square \supseteq J}$. Let $\alpha, \beta \in \mathcal{H}(\mathbf{q}, J)$ be homogeneous in j and nonzero. Then the following hold.*

(i) *There exists $\{\alpha_i\}_{i=0}^n, \{\beta_i\}_{i=0}^m \in \mathcal{H}(\mathbf{q}, J \setminus \{j\})$ such that*

$$\alpha = \sum_{i=0}^{n} \alpha_i B_j^{k-i} A_j^{l-i} \qquad\qquad \beta = \sum_{i=0}^{m} \beta_i B_j^{r-i} A_j^{s-i},$$

where $k = \deg_{\mathrm{B}_j} \alpha$, $l = \deg_{\mathrm{A}_j} \alpha$, $n = \min\{k, l\}$, $r = \deg_{\mathrm{B}_j} \beta$, $s = \deg_{\mathrm{A}_j} \beta$, and $m = \min\{r, s\}$.

(ii) *If $[\alpha, \beta] = 0$ and*

$$\chi_j(\alpha) \deg_j \beta = \chi_j(\beta) \deg_j \alpha + 2ph$$

for some $h \in \mathbb{Z}$, where p is the order of q_j, then $[\alpha_0, \beta_0] = 0$.

Proof. Existence of $\{\alpha_i\}_{i=0}^n$ and $\{\beta_i\}_{i=0}^m$ with the required properties is easy. First write α and β in normal form using an ordering of J where j is maximal, then collect terms having the same j-degree.

In proving the second part, it is convenient to rewrite $[\alpha, \beta]$ in the same way. Therefore let $\{\gamma_i\}_{i=0}^w \subseteq \mathcal{H}(\mathbf{q}, J \setminus \{j\})$ satisfy

$$[\alpha, \beta] = \sum_{i=0}^{w} \gamma_i B_j^{u-i} A_j^{v-i},$$

where $u = k + r$, $v = l + s$, and $w = \min\{u, v\}$.

The next step is to express γ_0 in terms of $\{\alpha_i\}_{i=0}^n$ and $\{\beta_i\}_{i=0}^m$. Since

$$\begin{aligned}\deg_j\left(\left(\sum_{i=1}^{n} \alpha_i B_j^{k-i} A_j^{l-i}\right)\left(\sum_{t=0}^{m} \beta_t B_j^{r-t} A_j^{s-t}\right)\right) = \\ = \deg_j\left(\sum_{i=1}^{n} \alpha_i B_j^{k-i} A_j^{l-i}\right) + \deg_j\left(\sum_{t=0}^{m} \beta_t B_j^{r-t} A_j^{s-t}\right) < \\ < \deg_j \alpha + \deg_j \beta = u + v\end{aligned}$$

and

$$\begin{aligned}\deg_j\left(\left(\sum_{i=0}^{n} \alpha_i B_j^{k-i} A_j^{l-i}\right)\left(\sum_{t=1}^{m} \beta_t B_j^{r-t} A_j^{s-t}\right)\right) &= \\ = \deg_j\left(\sum_{i=0}^{n} \alpha_i B_j^{k-i} A_j^{l-i}\right) + \deg_j\left(\sum_{t=1}^{m} \beta_t B_j^{r-t} A_j^{s-t}\right) &< \\ < \deg_j \alpha + \deg_j \beta = u + v,\end{aligned}$$

it follows that the only term of $\alpha\beta$ that contributes to γ_0 is $\alpha_0 B_j^k A_j^l \cdot \beta_0 B_j^r A_j^s$. Similarly, the only term of $\beta\alpha$ that contributes to γ_0 is $\beta_0 B_j^r A_j^s \cdot \alpha_0 B_j^k A_j^l$. Application of (2.4) to rewrite these products gives that

$$\gamma_0 = \alpha_0 \beta_0 q_j^{lr} - \beta_0 \alpha_0 q_j^{ks}.$$

If $q_j = 1$ then of course $q_j^{lr} = 1 = q_j^{ks}$. If $q_j \neq 1$ then observe that

$$\begin{aligned}2ph &= \chi_j(\alpha) \deg_j \beta - \chi_j(\beta) \deg_j \alpha = \\ &= (\deg_{\mathrm{B}_j} \alpha - \deg_{\mathrm{A}_j} \alpha)(\deg_{\mathrm{B}_j} \beta + \deg_{\mathrm{A}_j} \beta) + \\ &\quad - (\deg_{\mathrm{B}_j} \beta - \deg_{\mathrm{A}_j} \beta)(\deg_{\mathrm{B}_j} \alpha + \deg_{\mathrm{A}_j} \alpha) = \\ &= 2(\deg_{\mathrm{A}_j} \alpha \deg_{\mathrm{B}_j} \beta - \deg_{\mathrm{A}_j} \beta \deg_{\mathrm{B}_j} \alpha) = 2(lr - sk)\end{aligned}$$

for some $h \in \mathbb{Z}$ and p being the order of q_j. Hence $q_j^{lr} = q_j^{ks} \neq 0$ in this case too. Thus

$$\gamma_0 = q_j^{lr}(\alpha_0 \beta_0 - \beta_0 \alpha_0) = q_j^{lr}[\alpha_0, \beta_0].$$

Now since $[\alpha, \beta] = 0$, it follows that $\gamma_0 = 0$, and hence $[\alpha_0, \beta_0] = 0$. □

Theorem 5.13 *Consider an $\mathcal{H}(\mathbf{q}, J)$ for which $\mathbf{q} \in \mathcal{Q}(J, \mathcal{K})$. Let $J' \subseteq J$ be finite. Let $f, g \in \mathbb{Z}(J')$ and let $\alpha \sqsubseteq K_f$ be nonzero. If $f(j) \neq 0$ for all $j \in J'$ then*

$$\dim \operatorname{Cen}(g, \alpha) \leqslant 1. \tag{5.23}$$

Proof. The proof is by induction on $|J'|$. The case $J' = \emptyset$ is trivial since $\dim \mathcal{H}(\mathbf{q}, \emptyset) = \dim \{ aI \mid a \in \mathcal{K} \} = 1$.

To verify the inductive step, let $k > 0$ and assume that the theorem holds for all sets J' for which $|J'| < k$. Now let some J', f, α, and g be given that satisfy $|J'| = k$ and the conditions in the theorem. It shall be proved that the claim holds for these as well.

Let $\beta_1, \beta_2 \in \mathrm{Cen}(g, \alpha)$ be arbitrary nonzero elements. It needs only be verified that β_1 and β_2 are linearly dependent. Let $j \in J$ be such that (5.22) is fulfilled and such that $J'' = J' \setminus \{j\}$ satisfies $\mathbf{q} \in \mathcal{Q}(J'', \mathcal{K})$. It follows from Lemma 5.12 that there are $\alpha', \beta_1', \beta_2' \in \mathcal{H}(\mathbf{q}, J'')$ such that

$$\deg_j \bigl(\alpha - \alpha' B_j^{\deg_{B_j} \alpha} A_j^{\deg_{A_j} \alpha}\bigr) < \deg_j \alpha$$

$$\deg_j \bigl(\beta_1 - \beta_1' B_j^{\deg_{B_j} \beta_1} A_j^{\deg_{A_j} \beta_1}\bigr) < \deg_j \beta_1$$

$$\deg_j \bigl(\beta_2 - \beta_2' B_j^{\deg_{B_j} \beta_2} A_j^{\deg_{A_j} \beta_2}\bigr) < \deg_j \beta_2.$$

Lemma 5.12 also implies that $[\alpha', \beta_1'] = [\alpha', \beta_2'] = 0$, which means that $\beta_1', \beta_2' \in \mathrm{Cen}(\alpha')$. The elements α', β_1', and β_2' are just as fully homogeneous as their non-primed counterparts; define $f', g' \in \mathbb{Z}(J'')$ by $f'(i) = f(i)$ and $g'(i) = g(i)$ for all $i \in J''$, then $\alpha' \sqsubseteq K_{f'}$ and $\beta_1', \beta_2' \sqsubseteq K_{g'}$ respectively.

This means that the inductive hypothesis can be applied to J'', f', α', and g'. Hence $\dim \mathrm{Cen}(g', \alpha') \leqslant 1$, and since $\beta_1', \beta_2' \in \mathrm{Cen}(g', \alpha')$, there exists some $b \in \mathcal{K}$ that satisfies $b\beta_1' = \beta_2'$.

It follows from the definition of $\mathcal{Q}(J', \mathcal{K})$ that

$$\frac{g(j) \deg_j \alpha}{f(j)} = \deg_j \beta_1 = \deg_j \beta_2.$$

This implies that $\deg_j(\beta_2 - b\beta_1) < \deg_j \beta_2$ since the terms in β_2 and $b\beta_1$ with highest j-degree are equal. As a result of this, $\beta_2 - b\beta_1 = 0$ because if it was not so, then (5.22) would have to hold for α and $\beta_2 - b\beta_1$ despite the fact that these two obviously do not satisfy it.

The conclusion of this is that any two $\beta_1, \beta_2 \in \mathrm{Cen}(g, \alpha)$ are linearly dependent and hence $\dim \mathrm{Cen}(g, \alpha) \leqslant 1$. This finishes the inductive step and thus by induction the theorem holds for all finite J'. □

The condition on f above is necessary to ensure the conclusion. An easy way to see this is that if j is such that $f(j) = 0$ then every $\gamma \sqsubseteq K_0$ of $\mathcal{H}(q_j)$ will commute with α. Hence one can multiply any nonzero $\beta \in \mathrm{Cen}(g, \alpha)$ by some γ to get a new element of $\mathrm{Cen}(g, \alpha)$ which is linearly independent of β.

The above theorem does not exhaust the information about $\mathrm{Cen}(g, \alpha)$ that can be deduced from (5.22), but the conditions tend to become rather messy when $|J| > 1$. A more complete specification of $\dim \mathrm{Cen}(g, \alpha)$ in the special case $\mathcal{H}(q)$ is given by Theorem 6.6. Analogous results for $\mathcal{H}(\mathbf{q}, 2)$, $\mathcal{H}(\mathbf{q}, 3)$, or other $\mathcal{H}(\mathbf{q}, n)$ with n small should however be possible to get, without the number of cases growing unreasonably large, simply by continuing as in the proof of Theorem 6.6.

Instead we go on to the case $\operatorname{Cen}(F, \alpha)$, where α need not be fully homogeneous. In order to do this, the following definition comes in handy.

Definition 5.4 Let $F \subset \mathbb{Z}(J)$ be finite and nonempty. F is said to be **weakly nonzero** if, for some $f \in F$, $f(j) \neq 0$ for all $j \in J$. F is said to be **strongly nonzero** if there exists a semigroup total ordering $\prec$ on $\mathbb{Z}(J)$ such that $f = \max_{\prec} F$ satisfies $f(j) \neq 0$ for all $j \in J$.

Admittedly, the concepts of being strongly and weakly nonzero respectively are abstract, but there is a more geometrical interpretation. Think of $\mathbb{Z}(J)$ as $\mathbb{Z}^n$ for some n and let

$$G = \left\{ g \in F \,\middle|\, g(j) = 0 \text{ for some } j \in J \right\},$$

i.e., G is the set of elements of F that lies on some of the coordinate hyperplanes. Then F is weakly nonzero if and only if G is not the whole of F, whereas F is strongly nonzero if and only if there is an $f \in F$ that is not included in the convex hull of G.

As it happens, strongly nonzero is the property we will assume in the next theorem, but weakly nonzero is what we suspect may be sufficient. We will discuss this some more after the proof of the theorem, but before we can start with that, we need a little lemma.

Lemma 5.14 *Let $\mathcal{U}$ be an arbitrary vector space with decomposition*

$$\mathcal{U} = \bigoplus_{i=1}^{n} \mathcal{U}_i$$

and let π_i denote the projection of $\mathcal{U}$ onto $\mathcal{U}_i$ that is associated with that decomposition.

Let $\{d_i\}_{i=1}^n \subset \mathbb{N}$ and $\mathcal{V}$ be any subspace of $\mathcal{U}$. Now if

$$\dim \left\{ \pi_i(\alpha) \,\middle|\, \alpha \in \mathcal{V}, \text{ and } \pi_j(\alpha) = 0 \text{ for all } j > i \right\} \leqslant d_i \tag{5.24}$$

for all $i = 1, \ldots, n$ then

$$\dim \mathcal{V} \leqslant \sum_{i=1}^{n} d_i. \tag{5.25}$$

Proof. The proof is by induction on n. The case $n = 1$ is trivial since π_1 is the identity map on $\mathcal{U}$.

Let $k \geqslant 1$ and assume that the lemma holds for $n = k$. It will now be shown that this implies that the lemma also holds for $n = k + 1$. Thus, $\mathcal{V}$ is a subspace of $\mathcal{U} = \bigoplus_{i=1}^{k+1} \mathcal{U}_i$ and

$$\dim \{ \pi_i(\alpha) \mid \alpha \in \mathcal{V}, \text{ and } \pi_j(\alpha) = 0 \text{ for all } j > i \} \leqslant d_i$$

for all $i = 1, \ldots, k + 1$.

Let $\{\beta_i\}_{i=1}^{l}$ be a basis for $\mathcal{V}$. Then $\{\pi_{k+1}(\beta_i)\}_{i=1}^{l}$ will be a spanning set for $\pi_{k+1}(\mathcal{V})$. Let $I = \{1, \ldots, l\}$ be a set of indices, and let $J \subseteq I$ be a set of indices such that $\{\pi_{k+1}(\beta_i)\}_{i \in J}$ is a basis for $\pi_{k+1}(\mathcal{V})$. By the conditions of the lemma, $|J| \leqslant d_{k+1}$.

For all $i \in I \setminus J$, let $\gamma_i = \beta_i + \sum_{j \in J} c_{ji}\beta_j$, where the scalars c_{ji} have been chosen so that $-\pi_{k+1}(\beta_i) = \sum_{j \in J} c_{ji}\pi_{k+1}(\beta_j)$. Then $\pi_{k+1}(\gamma_i) = 0$ for all $i \in I \setminus J$. Furthermore $\{\beta_j\}_{j \in J} \cup \{\gamma_i\}_{i \in I \setminus J}$ is also a basis for $\mathcal{V}$.

Now $\mathcal{V}' = \operatorname{Span}(\{\gamma_i\}_{i \in I \setminus J})$ is a subspace of $\bigoplus_{i=1}^{k} \mathcal{U}_i$. Hence the induction hypothesis can be applied, and thus $|I \setminus J| = \dim \mathcal{V}' \leqslant \sum_{i=1}^{k} d_i$. This implies that

$$\dim \mathcal{V} = |I \setminus J| + |J| \leqslant \sum_{i=1}^{k+1} d_i,$$

which proves (5.25) for $n = k+1$. The lemma then follows from the principle of induction. □

Theorem 5.15 *Consider an* $\mathcal{H}(\mathbf{q}, J)$ *such that* $\mathbf{q} \in \mathcal{Q}(J, \mathcal{K})$. *Let* $\alpha \in \mathcal{H}(\mathbf{q}, J)$ *be such that*

$$F = \{ f \in \mathbb{Z}(J) \mid \alpha \sqcap K_f \neq 0 \}$$

is strongly nonzero. Let $G \subset \mathbb{Z}(J)$ *be finite. Then*

$$\dim \operatorname{Cen}(G, \alpha) \leqslant |G|. \tag{5.26}$$

Proof. Let $\prec$ be a semigroup total ordering of $\mathbb{Z}(J)$ such that $f = \max_{\prec} F$ satisfies $f(j) \neq 0$ for all $j \in J$. Let $\{g_1, \ldots, g_n\} = G$ be ordered such that $g_i \prec g_j$ if and only if $i < j$. Let $k \in \mathbb{Z}$ satisfy $1 \leqslant k \leqslant n$, and let β be an arbitrary element of $\operatorname{Cen}(G, \alpha)$ which satisfies $\beta \sqcap K_{g_i} = 0$ for all

$i > k$. Then, by (4.18),

$$0 = [\alpha, \beta] \sqcap K_{f+g_k} = \sum_{i=1}^{n} [\alpha \sqcap K_{f+g_k-g_i}, \beta \sqcap K_{g_i}] \overset{(*)}{=}$$

$$\overset{(*)}{=} \sum_{i=1}^{k} [\alpha \sqcap K_{f+g_k-g_i}, \beta \sqcap K_{g_i}] \overset{(**)}{=} [\alpha \sqcap K_f, \beta \sqcap K_{g_k}], \quad (5.27)$$

where $(*)$ is because of the conditions on β and $(**)$ is because of the definition of f. Thus $\beta \sqcap K_{g_k} \in \mathrm{Cen}(g_k, \alpha \sqcap K_f)$. From Theorem 5.13 it follows that

$$\dim \left\{ \beta \sqcap K_{g_k} \,\middle|\, \beta \in \mathrm{Cen}(G, \alpha) \text{ and } \beta \sqcap K_{g_i} = 0 \text{ for all } i > k \right\} \leqslant 1.$$

Now the conditions of Lemma 5.14 are fulfilled—just let $\mathcal{U}_i = \mathrm{Span}(K_{g_i})$ and $\mathcal{V} = \mathrm{Cen}(G, \alpha)$—and since the upper bound on dimension is 1 in each of the $n = |G|$ terms in the decomposition, it follows that

$$\dim \mathrm{Cen}(G, \alpha) \leqslant |G|.$$

□

There is no doubt that more information about $\mathrm{Cen}(G, \alpha)$ can be obtained in a similar way, and the class of α for which this is possible may also be enlarged. We hope for example that the above conclusion will hold, not only for strongly nonzero F, but also for weakly nonzero F.

The same argument as above, but with an f that is not necessarily a $\max_{\prec} F$, gives sums similar to (5.27), but which cannot simply be reduced to one term. On the other hand this need not be of vital importance, since the increase in the complexity of the situation is perhaps best compared to what one gets when one faces a linear equation system in which the matrix is not triangular, after having seen only such systems with triangular matrices.

Another path of research would be to further examine $\mathrm{Cen}(g, \alpha')$, where $\alpha' \sqsubseteq K_f$ and $f(j) = 0$ for some $j \in J$. This should also give more information about $\mathrm{Cen}(G, \alpha)$ when F is not strongly nonzero.

Finally, it is possible to get similar results where the 'strongly nonzero' condition is weakened by considering not $\mathrm{Cen}(F, \alpha)$, but the set

$$\bigcap_{i=1}^{k} \mathrm{Cen}(F, \alpha_i)$$

for some commuting $\{\alpha_i\}_{i=1}^n$. That the α_i commute with each other makes it possible to weaken the condition to that every $j \in J$ has some i such that $f_i(j) \neq 0$, where

$$f_i = \max\nolimits_{\prec} \{ f \in \mathbb{Z}(J) \mid \alpha_i \sqcap K_f \neq 0 \}$$

for $1 \leqslant i \leqslant n$.

Chapter 6

Centralisers of elements in $\mathcal{H}(q)$

The method which the Centre Party wants to use in community work is decentralization. Decentralization means spreading influence and power to give the individual a greater opportunity to influence developments. Decentralization is a useful method in all situations.

— from *The Centre Party's system of ideas*,
a Web page published by THE SWEDISH CENTRE PARTY

If a q-deformed Heisenberg algebra $\mathcal{H}(q)$ is considered instead of a generalised q-deformed Heisenberg algebra $\mathcal{H}(\mathbf{q}, J)$, more precise theorems can be given. It is likely that many of the theorems in this chapter can be generalised to $\mathcal{H}(\mathbf{q}, J)$, but we have chosen not to take this step and instead concentrate on some less obvious directions of investigation.

6.1 Classification of $\mathcal{H}(1)$

This section studies the algebra $\mathcal{H}(1)$, which has a somewhat unexpected structure. One might expect it to be like the algebras where q is of torsion type, but this is true only when 1 is of torsion type. The base of it all is the following theorem.

Theorem 6.1 *Let $\alpha, \beta \in \mathcal{H}(1)$ be nonzero, homogeneous, and commute with each other.*

If 1 is of free type, i.e., if $\mathcal{K}$ has characteristic 0, then

$$\chi(\alpha)\deg\beta = \chi(\beta)\deg\alpha. \tag{6.1}$$

If 1 is of torsion type, i.e., $\mathcal{K}$ has nonzero characteristic, then

$$\chi(\alpha)\deg\beta \equiv \chi(\beta)\deg\alpha \pmod{2p}, \tag{6.2}$$

where p is the order of 1, i.e., p is the characteristic of $\mathcal{K}$.

Proof. By Theorem 3.1, there exist numbers $r, s, t, u, v, w \in \mathbb{N}$ and coefficients $\{a_i\}_{i=0}^t, \{b_i\}_{i=0}^w \subseteq \mathcal{K}$ such that

$$\alpha = \sum_{i=0}^{t} a_i B^{r-i} A^{s-i}, \qquad \beta = \sum_{i=0}^{w} a_i B^{u-i} A^{v-i}.$$

It is possible to choose $r = \deg_{\mathrm{B}}\alpha$, $s = \deg_{\mathrm{A}}\alpha$, $t = \min\{r, s\}$, $u = \deg_{\mathrm{B}}\beta$, $v = \deg_{\mathrm{A}}\beta$, and $w = \min\{u, v\}$.

For the numbers $k = r + u$, $l = s + v$, and $m = \min\{k, l\}$, there are also coefficients $\{c_i\}_{i=0}^m \subseteq \mathcal{K}$ such that

$$[\alpha, \beta] = \sum_{i=0}^{m} c_i B^{k-i} A^{l-i}.$$

The next step is to express c_0 and c_1 in a_i and b_i. It turns out that there are only a few terms in the expansion of

$$\left[\sum_{i=0}^{t} a_i B^{r-i} A^{s-i}, \sum_{j=0}^{w} b_j B^{u-j} A^{v-j}\right] \tag{6.3}$$

that can contribute to these coefficients, because only a few terms have degrees $k + l$ and $k + l - 2$ respectively (no terms have degree higher than $k + l$).

By Corollary 2.4,

$$A^i B^j = \sum_{k=0}^{\min(i,j)} \left\{ k! \binom{i}{k} \binom{j}{k} \right\}_1 B^{j-k} A^{i-k}.$$

If this is applied to (6.3), one gets

$$c_0 = \{1\}_1 a_0 b_0 - \{1\}_1 b_0 a_0 = 0,$$

because $a_0 B^r A^s \cdot b_0 B^u A^v$ is the only term of $\alpha\beta$ and $b_0 B^u A^v \cdot a_0 B^r A^s$ is the only term of $\beta\alpha$ which contribute to c_0.

Now assume that $m \geqslant 1$ (if $m = 0$ then c_1 does not make sense). There are three terms of $\alpha\beta$ that can contribute to c_1: $a_0 B^r A^s \cdot b_1 B^{u-1} A^{v-1}$, $a_1 B^{r-1} A^{s-1} \cdot b_0 B^u A^v$, and $a_0 B^r A^s \cdot b_0 B^u A^v$. The total contribution is

$$\{1\}_1 a_0 b_1 + \{1\}_1 a_1 b_0 + \{su\}_1 a_0 b_0 = a_0 b_1 + a_1 b_0 + \{su\}_1 a_0 b_0.$$

A similar situation holds for the $-\beta\alpha$ term, from which the total contribution is $-a_0 b_1 - a_1 b_0 - \{rv\}_1 a_0 b_0$. Thus

$$\begin{aligned} c_1 &= a_0 b_1 + a_1 b_0 + \{su\}_1 a_0 b_0 - b_0 a_1 - b_1 a_0 - \{rv\}_1 a_0 b_0 = \\ &= \{su - rv\}_1 a_0 b_0 \end{aligned}$$

If $[\alpha, \beta] = 0$ then in particular $c_0 = 0 = c_1$. The first of these equalities holds anyway, but $c_1 = 0$ is equivalent to

$$\deg_{\mathrm{A}} \alpha \deg_{\mathrm{B}} \beta - \deg_{\mathrm{A}} \beta \deg_{\mathrm{B}} \alpha = pn \tag{6.4}$$

for some $n \in \mathbb{Z}$, where p is the order of 1.

If $m = 0$ then things are a bit different. As

$$m = \min\{\deg_{\mathrm{B}} \alpha + \deg_{\mathrm{B}} \beta, \deg_{\mathrm{A}} \alpha + \deg_{\mathrm{A}} \beta\}$$

and all these degrees are nonnegative, it must be the case that $\deg_{\mathrm{A}} \alpha = \deg_{\mathrm{A}} \beta = 0$ or $\deg_{\mathrm{B}} \alpha = \deg_{\mathrm{B}} \beta = 0$. This means that the left hand side of (6.4) is 0 if $m = 0$, so the equality is indeed fulfilled.

Finally, by Theorem 4.8 and using (6.4),

$$\begin{aligned} \chi(\alpha) \deg \beta - \chi(\beta) \deg \alpha &= (\deg_{\mathrm{B}} \alpha - \deg_{\mathrm{A}} \alpha)(\deg_{\mathrm{B}} \beta + \deg_{\mathrm{A}} \beta) \\ &\quad - (\deg_{\mathrm{B}} \beta - \deg_{\mathrm{A}} \beta)(\deg_{\mathrm{B}} \alpha + \deg_{\mathrm{A}} \alpha) = \\ &= 2(\deg_{\mathrm{B}} \alpha \deg_{\mathrm{A}} \beta - \deg_{\mathrm{A}} \alpha \deg_{\mathrm{B}} \beta) = -2pn, \end{aligned}$$

where n is some integer and p is the order of 1. This proves equations (6.1) and (6.2). □

Theorem 6.1 demonstrates why the case $q_j = 1$ has been classified as it is in Section 5.2. With this classification, it simply depends on the type of q whether the quantity

$$\chi(\alpha) \deg \beta - \chi(\beta) \deg \alpha$$

is known to be zero (free case) or simply a multiple of some $2p$ (torsion case). This similarity raises the question of whether the theorems based on Theorem 5.4 extend to $\mathcal{H}(1)$ as well, and indeed they do. The counterpart to Theorem 5.10 in $\mathcal{H}(1)$ is the following corollary.

Corollary 6.2 *Let $\mathcal{H}(1)$ be given and let p be the order of $1 \in \mathcal{K}$. Then the centre of $\mathcal{H}(1)$ is $\{\, aI \mid a \in \mathcal{K} \,\}$ if $p = 0$ and $\mathrm{Alg}_{\mathcal{K}}\langle A^p, B^p \rangle$ otherwise.*

Proof. Note that, by Theorem 6.1, the conditions in Lemma 5.9 are fulfilled. The corollary follows. □

The really interesting result however, is Theorem 5.11, which can be slightly extended.

Theorem 6.3 *If $\mathbf{q} \in (\mathcal{K}^*)^{\square \supseteq J}$ is of direct type on J, at most one $j \in J$ satisfy $q_j = 1$, and q_j is of free type for all $j \in J$ then $\mathbf{q} \in \mathcal{Q}(J, \mathcal{K})$.*

Proof. The proof is almost identical to that of Theorem 5.11, so only the new points about it will be mentioned. If $\mathcal{K}$ has characteristic zero, $J = \{j\}$, and $q_j = 1$ then $\mathcal{H}(\mathbf{q}, J) = \mathcal{H}(1)$ and hence by Theorem 6.1,

$$\chi(\alpha) \deg \beta = \chi(\beta) \deg \alpha$$

for all homogeneous nonzero $\alpha, \beta \in \mathcal{H}(\mathbf{q}, J)$ such that $[\alpha, \beta] = 0$. This is the definition of $\mathbf{q} \in \mathcal{Q}(J, \mathcal{K})$ for this particular J.

For all other J, the proof of Theorem 5.11 works as it stands but proves the slightly stronger claim in this theorem, since $\mathcal{Q}(J', \mathcal{K})$ for $J' \subset J$ is known to be slightly larger. □

A consequence of this is that Theorems 5.13 and 5.15 holds for all $\mathcal{H}(q)$ that have q of free type. These results will be stated in the next section.

Is one $q_j = 1$ the best one can get or is there space for more in the $\mathbf{q}$'s that are in $\mathcal{Q}(J, \mathcal{K})$? This is indeed a question to which it would be of interest to have an answer.

If one tries to perform the calculations in the proof of Theorem 6.1 in $\mathcal{H}(\mathbf{q}, 2)$ with $\mathbf{q} = (1, 1)$, it leads to

$$0 = \gamma_0 = [\alpha_0, \beta_0],$$
$$0 = \gamma_1 = [\alpha_0, \beta_1] + [\alpha_1, \beta_0] + \{su - rv\}_1 \alpha_0 \beta_0$$

(a_i, b_i, and c_i become α_i, β_i, and γ_i since they must be elements of $\mathcal{H}(1)$, not simply of $\mathcal{K}$), and that's it! The commutators no longer have to be zero and this opens up for the possibility that $\{su - rv\}_1 \neq 0$.

If additional constraints are added, the conclusion is once again what we want. If $\deg_1 \alpha_1 \leqslant \deg_1 \alpha_0$ and $\deg_1 \beta_1 \leqslant \deg_1 \beta_0$ (r, s, u, and v are degrees in B_2 and A_2) then $\{su - rv\}_1 = 0$ must hold, but additional constraints make the rest of the theory weaker and much less useful.

We believe that it is possible to prove a generalisation of Theorem 6.1 for the above $\mathcal{H}(\mathbf{q}, 2)$ too, but that would seem to require better methods than the ones we have applied for $\mathcal{H}(1)$.

6.2 $\mathcal{H}(q)$ when q is of free type

At this point, there are a few results proved earlier in this book that are valid in all $\mathcal{H}(q)$ algebras for which q is of free type. As these appear in Sections 5.2, 5.4, and 6.1, but are stated in such a way that they do not include all the $\mathcal{H}(q)$ algebras for which q is of free type, it seems best to restate the two most important results here.

Corollary 6.4 (to Theorems 5.4 and 6.1) *Let $q \in \mathcal{K}$ be of free type. If $\alpha, \beta \in \mathcal{H}(q)$ are nonzero, homogeneous, and commute then*

$$\chi(\alpha) \deg \beta = \chi(\beta) \deg \alpha. \tag{6.5}$$

Corollary 6.5 (to Corollary 6.2 and Theorem 5.10) *Let $q \in \mathcal{K}$ be of free type. Then the centre of $\mathcal{H}(q)$ is the set $\{ aI \mid a \in \mathcal{K} \}$.*

Part of the next theorem has previously been stated as a special case of Theorem 5.13. The next theorem is however more detailed in that it treats some cases which Theorem 5.13 does not treat.

Theorem 6.6 *Let $q \in \mathcal{K}$ be of free type and consider the algebra $\mathcal{H}(q)$. Let $n, m \in \mathbb{Z}$, and let $\alpha \sqsubseteq K_m$ be nonzero.*

- *If $m = n = 0$ then $\operatorname{Cen}(n, \alpha) = \operatorname{Span}(K_0)$ and $\dim \operatorname{Cen}(n, \alpha) = \infty$.*
- *If $m = 0 = \deg \alpha$ then $\alpha = aI$ for some $a \in \mathcal{K}$ and $\dim \operatorname{Cen}(n, \alpha) = \infty$.*
- *If $m \neq 0 = n$ then $\dim \operatorname{Cen}(n, \alpha) = 1$.*
- *If $nm > 0$ and $m \mid n \deg \alpha$ then $\dim \operatorname{Cen}(n, \alpha) \leqslant 1$.*
- *Otherwise $\dim \operatorname{Cen}(n, \alpha) = 0$.*

Proof. Assume that $m = 0 = n$. According to Theorem 5.3 any $\beta \sqsubseteq K_n = K_0$ commutes with α, and hence $\operatorname{Cen}(0, \alpha) = \operatorname{Span}(K_0)$, which is infinite-dimensional.

Assume that $m = 0 = \deg \alpha$. This means that α is of the form aI, where $a \in \mathcal{K}$. Thus α is an element in the centre of $\mathcal{H}(q)$ and $\mathrm{Cen}(n, \alpha) = \mathrm{Span}(K_n)$, which is infinite-dimensional.

Assume that $m = 0$, but $n \neq 0$ and $\deg \alpha \neq 0$. Then for all nonzero $\beta \sqsubseteq K_n$ it follows that $n \deg \alpha \neq 0 = m \deg \beta$. Since such β do not satisfy (6.5), they cannot commute with α. Hence $\mathrm{Cen}(n, \alpha) = \{0\}$ and $\dim \mathrm{Cen}(n, \alpha) = 0$.

Assume that $m \neq 0 = n$. It follows from (6.5) that every nonzero $\beta \in \mathrm{Cen}(0, \alpha)$ must satisfy $m \deg \beta = 0$, i.e., it must be of the form bI, where $b \in \mathcal{K}$. Thus $\mathrm{Cen}(0, \alpha) = \mathrm{Span}\big(\{I\}\big)$, which is one-dimensional.

Assume that $mn > 0$ and $m \mid n \deg \alpha$. Let $D = \{\frac{n \deg \alpha}{m}\}$. Then every $\beta \in \mathrm{Cen}(n, \alpha) \setminus \{0\}$ satisfies $\deg \beta \in D$ and so by Theorem 4.10, $\dim \mathrm{Cen}(n, \alpha) \leqslant 1$.

Assume that $mn > 0$ but $m \nmid n \deg \alpha$, or that $mn < 0$. Then no $\beta \in \mathrm{Cen}(n, \alpha) \setminus \{0\}$ can solve (6.5). Thus $\mathrm{Cen}(n, \alpha) = \{0\}$, and consequently $\dim \mathrm{Cen}(n, \alpha) = 0$. □

Corollary 6.7 *Let $q \in \mathcal{K}$ be of free type and consider the algebra $\mathcal{H}(q)$. Let $m \in \mathbb{Z}$ be nonzero and let $k \in \mathbb{Z}^+$. If $\alpha \sqsubseteq K_m$ also is nonzero then*

(i) $\dim \mathrm{Cen}(km, \alpha) = 1$
(ii) $\mathrm{Cen}(km, \alpha) = \{\, a\alpha^k \mid a \in \mathcal{K} \,\}$.

Proof. By Theorem 6.6, $\dim \mathrm{Cen}(km, \alpha) \leqslant 1$. By Corollary 4.5, $\alpha^k \sqsubseteq K_{km}$. This implies that $\dim \mathrm{Cen}(km, \alpha) \geqslant 1$, which shows the claim about the dimension of $\mathrm{Cen}(km, \alpha)$. The identification of $\mathrm{Cen}(km, \alpha)$ follows from that the single element $\alpha^k \in \mathrm{Cen}(km, \alpha)$ will span that one-dimensional space. □

Theorem 6.6 and Corollary 6.7 give an almost complete description of $\mathrm{Cen}(n, \alpha)$ for a homogeneous $\alpha \in \mathcal{H}(q)$. The case that is not completely specified is when n and $\chi(\alpha)$ have the same sign, $\chi(\alpha) \mid n \deg \alpha$, and there is no β which satisfies both $\beta^k = \alpha$ and $\beta^l \in \mathrm{Cen}(n, \alpha)$ for some integers k and l. Are there any such α and n for which $\dim \mathrm{Cen}(n, \alpha) = 1$? At the moment, we do not know.

In [46], Burchnall and Chaundy give the following example of a similar situation: If $\mathcal{K}$ is $\mathbb{C}$,

$$\begin{aligned} \alpha &= A^2 - 2B^{-2} = B^{-2}(BA + I)(BA - 2I), \quad \text{and} \\ \beta &= A^3 - 3B^{-2}A + 3B^{-3} = B^{-3}(BA + I)(BA - I)(BA - 3I) \end{aligned} \tag{6.6}$$

then $[\alpha, \beta] = 0$ in a Heisenberg algebra where $AB - BA = I$. Unfortunately, this example does not apply here. Burchnall and Chaundy studied an algebra where B can have negative exponents, while we have confined ourselves to nonnegative powers of A and B, a context in which there are no examples of this that we know of.

The situation becomes even more complex if one considers formal power series of the form $\sum_{i=-\infty}^{0} a_i B^{k+i} A^{l+i}$; these can be viewed as homogeneous elements in some extension of $\mathcal{H}(q)$. It seems like Theorem 6.6 would hold for these things, furthermore it appears as $\dim \text{Cen}(n, \alpha) = 1$ for all n for which Theorem 6.6 allows this! At the first glance, one would believe that this generates plenty of examples like (6.6), but what happens is quite the contrary: Every pair of homogeneous elements α and β that commute must be scalar multiples of some powers of a common element γ. The reason for this is the following corollary to Theorem 6.6.

Corollary 6.8 *Let* $q \in \mathcal{K}$ *be of free type and consider the algebra* $\mathcal{H}(q)$. *Let* $m, n \in \mathbb{Z} \setminus \{0\}$. *Let* $\alpha \sqsubseteq K_m$ *and* $\beta \sqsubseteq K_n$ *be nonzero and satisfy* $[\alpha, \beta] = 0$. *If there exists a common divisor* d *of* m *and* n *such that* $\dim \text{Cen}(d, \alpha) = 1$ *then there exists a* $\gamma \in \text{Cen}(d, \alpha)$ *and* $a, b \in \mathcal{K}$ *such that* $a\gamma^r = \alpha$ *and* $b\gamma^s = \beta$, *where* $r = m/d$ *and* $s = n/d$.

Proof. Obviously there is a nonzero $\gamma \in \text{Cen}(n, \alpha)$, hence $\gamma^r \in \text{Cen}(m, \gamma)$ and $\gamma^s \in \text{Cen}(n, \gamma)$. The definition of γ implies $\alpha \in \text{Cen}(m, \gamma)$, hence α must be a scalar multiple of γ^r, because $\dim \text{Cen}(rd, \gamma) = 1$ by Corollary 6.7. This also implies that

$$\text{Cen}(n, \alpha) = \text{Cen}(n, \gamma^r) \supseteq \text{Cen}(n, \gamma),$$

but all these spaces are one-dimensional and therefore coincide. In particular, $\beta \in \text{Cen}(n, \gamma)$. Thus β is a scalar multiple of γ^s. □

Of course, this corollary need not say anything about what really happens for formal power series—there are lots of basic machinery that would have to be established before one could claim an analogous result for them—but we do have the general impression that "things seem to work pretty much as in $\mathcal{H}(q)$".

Leaving hypothetical arguments about power series aside however, one should note that there is an obvious application of the above corollary—finding all radicals of a homogeneous element in $\mathcal{H}(q)$. The corollary says that simply by searching for any nonzero element (in the chain space of interest) that *commutes* with the given (which can be done by solving a homogeneous linear equation system), one finds something that is a scalar multiple of the radical. Thus the non-commutative part of finding a radical can be done by solving linear equations, the only non-linear equation one ever needs to solve will be for an element in the field of scalars $\mathcal{K}$, where one only has to deal with problems in commutative algebra.

Corollary 6.9 (of Theorem 5.15) *Let $q \in \mathcal{K}$ be of free type and consider the algebra $\mathcal{H}(q)$. Let $m, n \in \mathbb{Z}$ be such that $m \leqslant n$, and let $\alpha \in \mathcal{H}(q)$ satisfy $\alpha \not\sqsubseteq K_0$. Then*

$$\dim \mathrm{Cen}\bigl([m,n],\alpha\bigr) \leqslant n - m + 1 \tag{6.7}$$

Proof. Theorem 5.15 says that this is true for α such that the set

$$F = \{\, l \in \mathbb{Z} \mid \alpha \sqcap K_l \neq 0 \,\}$$

is strongly nonzero (just take $G = \{\, i \in \mathbb{Z} \mid m \leqslant i \leqslant n \,\}$). What needs to be proved is that $\alpha \not\sqsubseteq K_0$ implies that F is strongly nonzero.

If $\overline{\chi}(\alpha) > 0$ then F is strongly nonzero—just choose the usual 'greater than' as the total ordering of $\mathbb{Z}$. If $\underline{\chi}(\alpha) < 0$ then F is also strongly nonzero—just choose the usual 'less than' as the total ordering of $\mathbb{Z}$. Finally, if neither of these are fulfilled then $\underline{\chi}(\alpha) = 0 = \overline{\chi}(\alpha)$ and $\alpha \sqsubseteq K_0$, contrary to the conditions stated above. □

Theorem 6.10 *Let q be of free type. Let $\alpha, \beta \in \mathcal{H}(q)$ be nonzero and satisfy $[\alpha, \beta] = 0$. Then the following hold:*

- *If $\overline{\chi}(\alpha) > 0$ then either $\overline{\chi}(\beta) > 0$ or $\overline{\chi}(\beta) = 0$, but in the second case β belongs to the centre of $\mathcal{H}(q)$.*
- *If $\overline{\chi}(\alpha) = 0$ then $\overline{\chi}(\beta) = 0$ or $\alpha \sqcap K_0 = cI$ for some $c \in \mathcal{K}$.*
- *If $\overline{\chi}(\alpha) < 0$ then either $\overline{\chi}(\beta) < 0$ or $\overline{\chi}(\beta) = 0$, but in the second case $\beta \sqcap K_0 = cI$ for some $c \in \mathcal{K}$.*
- *If $\underline{\chi}(\alpha) > 0$ then either $\underline{\chi}(\beta) > 0$ or $\underline{\chi}(\beta) = 0$, but in the second case $\beta \sqcap K_0 = cI$ for some $c \in \mathcal{K}$.*
- *If $\underline{\chi}(\alpha) = 0$ then $\underline{\chi}(\beta) = 0$ or $\alpha \sqcap K_0 = cI$ for some $c \in \mathcal{K}$.*
- *If $\underline{\chi}(\alpha) < 0$ then either $\underline{\chi}(\beta) < 0$ or $\underline{\chi}(\beta) = 0$, but in the second case β belongs to the centre of $\mathcal{H}(q)$.*

Proof. For each of the cases above regarding $\overline{\chi}$ corresponds a case regarding $\underline{\chi}$ as follows

$$\begin{aligned} \overline{\chi}(\alpha) < 0 &\quad\longleftrightarrow\quad \underline{\chi}(\alpha) > 0 \\ \overline{\chi}(\alpha) = 0 &\quad\longleftrightarrow\quad \underline{\chi}(\alpha) = 0 \\ \overline{\chi}(\alpha) > 0 &\quad\longleftrightarrow\quad \underline{\chi}(\alpha) < 0. \end{aligned}$$

Therefore it is enough to consider the $\underline{\chi}$ cases.

For the first case, assume that $\underline{\chi}(\alpha) > 0$. Let $a = \underline{\chi}(\alpha)$ and $b = \underline{\chi}(\beta)$. Then

$$0 = [\alpha, \beta] \sqcap K_{a+b} = [\alpha \sqcap K_a, \beta \sqcap K_b].$$

Thus Theorem 6.6 can be applied to $\alpha \sqcap K_a$ and $\beta \sqcap K_b$, which yields the two possibilities (i) $b > 0$ and (ii) $b = 0$ and $\beta \sqcap K_0 = cI$ for some $c \in \mathcal{K}$. This proves the first case.

For the second case, assume that $\underline{\chi}(\alpha) = 0$. Let $b = \underline{\chi}(\beta)$. Then

$$0 = [\alpha, \beta] \sqcap K_b = [\alpha \sqcap K_0, \beta \sqcap K_b].$$

Thus Theorem 6.6 can be applied to $\alpha \sqcap K_0$ and $\beta \sqcap K_b$, which yields the two possibilities (i) $b = 0$ and (ii) $\alpha \sqcap K_0 = cI$ for some $c \in \mathcal{K}$. This proves the second case.

And for the last case, assume that $\underline{\chi}(\alpha) < 0$. Let $a = \underline{\chi}(\alpha)$ and $b = \underline{\chi}(\beta)$. Then

$$0 = [\alpha, \beta] \sqcap K_{a+b} = [\alpha \sqcap K_a, \beta \sqcap K_b].$$

Thus Theorem 6.6 can be applied to $\alpha \sqcap K_a$ and $\beta \sqcap K_b$, which yields the two possibilities (i) $b < 0$ and (ii) $b = 0$ and $\beta \sqcap K_0 = cI$ for some $c \in \mathcal{K}$. In case (i) the proof is finished. Therefore assume (ii).

Let $\beta' = \beta - cI$, where $cI = \beta \sqcap K_0$. It follows that

$$0 = [\alpha, \beta] = [\alpha, cI] + [\alpha, \beta'] = [\alpha, \beta']$$

and hence $\beta' = 0$, because if $\beta' \neq 0$ then $\underline{\chi}(\beta') > 0$ and the first case could be applied to α and β', but that yields a contradiction with $\underline{\chi}(\alpha) < 0$. Thus $\beta = \beta \sqcap K_0 = cI$, which belongs to the centre of $\mathcal{H}(q)$. This proves the last case. □

Example 6.1 Corollary 6.9 and Theorem 6.10 can be used to determine $\operatorname{Cen}(\alpha)$ in some important cases. Let $P \in \mathcal{K}[x] \setminus \mathcal{K}$ and $\alpha = P(A)$. Then we claim that every $\beta \in \operatorname{Cen}(\alpha)$ satisfies $\beta = Q(A)$ for some $Q \in \mathcal{K}[x]$.

For $\beta = 0$, this is trivial. Hence let $\beta \in \operatorname{Cen}(\alpha)$ be nonzero and let $n = \underline{\chi}(\beta)$. Note that $\deg_{\mathrm{B}} \alpha = 0$, therefore $\overline{\chi}(\alpha) \leqslant 0$. This implies that $\overline{\chi}(\beta)$ cannot be positive, because if it was then Theorem 6.10 would imply either $\overline{\chi}(\alpha) > 0$ or that α is in the centre of $\mathcal{H}(q)$, neither of which are possible. Hence $\overline{\chi}(\beta) \leqslant 0$ as well and $\beta \in \operatorname{Cen}\bigl([n,0],\alpha\bigr)$.

By Corollary 6.9, $\dim \operatorname{Cen}\bigl([n,0],\alpha\bigr) \leqslant 1-n$. On the other hand, it is clear that

$$\sum_{i=0}^{|n|} b_i A^i \in \operatorname{Cen}\bigl([n,0],\alpha\bigr)$$

for every $\{b_i\}_{i=0}^{|n|} \subset \mathcal{K}$. Since $\{I, A, \dots, A^{|n|}\}$ spans a $(1-n)$-dimensional subspace of $\operatorname{Cen}\bigl([n,0],\alpha\bigr)$, that space is the whole of $\operatorname{Cen}\bigl([n,0],\alpha\bigr)$. Thus $\beta = \sum_{i=0}^{|n|} b_i A^i$ for some $\{b_i\}_{i=0}^{|n|} \subset \mathcal{K}$, which means that β can be expressed as $Q(A)$ for some $Q \in \mathcal{K}[x]$.

There is nothing special about A in the example; A can just as well be replaced by any nonzero $\alpha \sqsubseteq K_{-1}$ or $\alpha \sqsubseteq K_1$. It is in fact possible to repeat the same argument for all homogeneous and nonzero α for which $\deg \alpha$ and $\chi(\alpha)$ are coprime, but then Corollary 6.9 needs to be replaced by something slightly stronger that makes use of the fact that $\dim \operatorname{Cen}(n,\alpha)$ will be zero for some n (Theorem 5.15 only uses that the dimension is $\leqslant 1$).

6.3 $\mathcal{H}(q)$ when q is of torsion type

Like the case of $\mathcal{H}(q)$ when q is of free type, there are a few important results about $\mathcal{H}(q)$ when q is of torsion type that appear in various places in Chapter 5 and Section 6.1. In the same way as there, the two most important will be repeated here for easy reference. For the rest of this section, let $q \in \mathcal{K}^*$ be of torsion type and p be the order of q. Then the following holds.

Corollary 6.11 (to Theorems 5.4 and 6.1) *If $\alpha, \beta \in \mathcal{H}(q)$ are nonzero, homogeneous, and commute then*

$$\chi(\beta) \deg \alpha \equiv \chi(\alpha) \deg \beta \pmod{2p}. \tag{6.8}$$

Corollary 6.12 (to Corollary 6.2 and Theorem 5.10) *The centre of* $\mathcal{H}(q)$ *is the subalgebra of* $\mathcal{H}(q)$ *that is spanned by* A^p *and* B^p.

An alternative way to express this latter result is that the centre of $\mathcal{H}(q)$ is the set $\{P(A^p, B^p)\}_{P \in \mathcal{K}[x,y]}$.

Since the respective centres of the $\mathcal{H}(q)$ algebras that are considered here are nontrivial and completely known, one might use this to simplify computations even more than is possible for $\mathcal{H}(\mathbf{q}, J)$ algebras in general. Every associative ring can be considered as an algebra over its centre, thus $\mathcal{H}(q)$ can be considered as an $\mathrm{Alg}_{\mathcal{K}}\langle A^p, B^p\rangle$-algebra. As a such, $\mathcal{H}(q)$ is no longer infinite-dimensional; instead it has the finite dimension p^2 and one basis for the algebra is

$$\left\{ B^i A^j \,\middle|\, 0 \leqslant i < p \text{ and } 0 \leqslant j < p \right\}.$$

We should perhaps remark at this point that because the centre is not a field (it is in fact, just like $\mathcal{K}[x, y]$, not even a principal ideal domain), the definitions of basis and dimension in general are not as trivial as for vector spaces. The above set does however possess all the usual properties of a basis: it is linearly independent and spans the algebra. This justifies saying that the dimension of the algebra over $\mathrm{Alg}_{\mathcal{K}}\langle A^p, B^p\rangle$ is p^2.

The centre $\mathrm{Alg}_{\mathcal{K}}\langle A^p, B^p\rangle$ is isomorphic to $\mathcal{K}[x, y]$, and one isomorphism is $P \mapsto P(A^p, B^p)$. Computations in the centre should not be a problem, since multiplication there is commutative. The algebra $\mathcal{H}(q)$ is of course still noncommutative, but the fact that it is p^2-dimensional means that all the noncommutative properties of this algebra can be captured in a $p^2 \times p^2$ multiplication table. If one is to write a program that makes computations in these algebras, one might therefore be able to avoid implementing the q-combinatorial functions introduced earlier altogether, and do just fine with a precomputed table.

Similar things can of course be done for all $\mathcal{H}(\mathbf{q}, J)$ algebras where some $j \in J$ is of torsion type, but the presentation is more straight on if the centre is known. Otherwise $\mathcal{H}(\mathbf{q}, J)$ must be considered as an $\mathcal{R}$-algebra, where $\mathcal{R}$ is merely known to be a subring of the centre.

The rest of this section is devoted to the problem of finding a theorem that corresponds to Theorem 6.6, but in the present case where q is of torsion type. This goal is only partially achieved, the strongest result is

Theorem 6.21 which covers approximately half the cases covered by Theorem 6.6. We believe though, that the other half should be possible to obtain using similar techniques.

The next lemma is in a way Theorem 5.8 taken the other way around, being focused more on reducing problems than generating new examples from old.

Lemma 6.13 *Let α be an arbitrary member of $\mathcal{H}(q)$. Let $n \in \mathbb{Z}$. Let $\beta \sqsubseteq K_n$ be nonzero. Then there exist an $m \in \mathbb{Z}$ and a $\gamma \sqsubseteq K_m$ such that*

(i) $[\alpha, \beta] = 0$ *if and only if* $[\alpha, \gamma] = 0$,
(ii) $-p < m < p$,
(iii) $n \equiv m \pmod{p}$, *and*
(iv) $|n - m| = \deg \beta - \deg \gamma$.

Proof. There are three cases to consider: $n \leqslant -p$, $-p < n < p$, and $p \leqslant n$. The middle case is obvious since it is possible to choose $\gamma = \beta$.

The last case is proved as follows. There exists a positive integer r such that $pr \leqslant n < p(r+1)$. Let $m = n - pr$. By Theorem 4.11, there exists a $\gamma \sqsubseteq K_m$ such that $\beta = B^{pr}\gamma$. It follows that

- $[\alpha, \beta] = [\alpha, B^{pr}\gamma] = B^{pr}[\alpha, \gamma]$, and since by Theorem 4.9 there are no zero divisors in $\mathcal{H}(q)$, it follows that $[\alpha, \beta] = 0$ if and only if $[\alpha, \gamma] = 0$.
- $-p < 0 = pr - pr \leqslant n - pr = m = n - pr < p(r+1) - pr = p$.
- $p \mid pr = n - m$ and thus $n \equiv m \pmod{p}$.
- $\deg \beta - \deg \gamma = \deg(B^{pr}\gamma) - \deg \gamma = \deg B^{pr} = pr = n - m = |n - m|$.

The first case is considered in exactly the same way, with the only difference that β is rewritten as γA^{pr} instead of as $B^{pr}\gamma$. □

The next theorem is a strengthening of Theorem 5.3 that follows immediately from the periodic nature of these $\mathcal{H}(q)$ algebras.

Theorem 6.14 *If $\alpha \sqsubseteq K_n$, $\beta \sqsubseteq K_m$, and $n \equiv 0 \equiv m \pmod{p}$, then*

$$[\alpha, \beta] = 0.$$

Proof. By Lemma 6.13, there exists a $\gamma \sqsubseteq K_0$ such that $[\alpha, \beta] = 0$ if and only if $[\alpha, \gamma] = 0$. The theorem also says that there exists a $\delta \sqsubseteq K_0$ such that $[\alpha, \gamma] = 0$ if and only if $[\delta, \gamma] = 0$. By Theorem 5.3, γ and δ do commute, hence so do α and β. □

Theorem 6.15 *Let* $n \in \mathbb{N}$ *and* $\alpha \sqsubseteq K_0$. *Let* $\beta \sqsubseteq K_n$ *and* $\gamma \sqsubseteq K_{-n}$ *both be nonzero. Then*

- $[\alpha, \beta] = 0$ *if and only if* $[\alpha, B^n] = 0$.
- $[\alpha, \gamma] = 0$ *if and only if* $[\alpha, A^n] = 0$.

Proof. Let $\delta \in \mathcal{H}(q)$ and $\varepsilon \sqsubseteq K_0$. Then

$$[\alpha, \delta\varepsilon] = \alpha\delta\varepsilon - \delta\varepsilon\alpha = \alpha\delta\varepsilon - \delta\alpha\varepsilon + \delta\alpha\varepsilon - \delta\varepsilon\alpha = [\alpha, \delta]\varepsilon + \delta[\alpha, \varepsilon] = [\alpha, \delta]\varepsilon.$$

By Theorem 4.11, $\beta = B^n\varepsilon$ for some $\varepsilon \sqsubseteq K_0$. Hence $[\alpha, \beta] = [\alpha, B^n]\varepsilon$. Since $\varepsilon \neq 0$ and by Theorem 4.9 there are no zero divisors in $\mathcal{H}(q)$, it follows that $[\alpha, \beta] = 0$ if and only if $[\alpha, B^n] = 0$.

Similarly, by the same theorems, $\gamma = A^n\varepsilon$ for some $\varepsilon \sqsubseteq K_0$. Hence $[\alpha, \gamma] = [\alpha, A^n]\varepsilon$. Since $\varepsilon \neq 0$, it follows that $[\alpha, \gamma] = 0$ if and only if $[\alpha, A^n] = 0$. □

Corollary 6.16 *Let* $\alpha \sqsubseteq K_0$ *and* $n, d \in \mathbb{N}$.

- *If* $[\alpha, B^n] \neq 0$ *then* $\dim \operatorname{Cen}(d, n, \alpha) = 0$.
- *If* $[\alpha, A^n] \neq 0$ *then* $\dim \operatorname{Cen}(d, -n, \alpha) = 0$.
- *If* $[\alpha, B^n] = 0$ *then* $\dim \operatorname{Cen}(2d + n, n, \alpha) \leqslant d + 1$.
- *If* $[\alpha, A^n] = 0$ *then* $\dim \operatorname{Cen}(2d + n, -n, \alpha) \leqslant d + 1$.

Proof. The first two points above are immediate consequences of the theorem. The third point is a trivial consequence of

$$\begin{aligned}\dim \{\, \beta \sqsubseteq K_n \mid \deg \beta \leqslant 2d + n \,\} &= \\ &= \dim \operatorname{Span}\Big(\{\, B^{n+i}A^i \,\big|\, n + 2i \leqslant 2d + n \,\}\Big) = \\ &= \dim \operatorname{Span}\Big(\{\, B^{n+i}A^i \,\big|\, i \leqslant d \,\}\Big) = d + 1\end{aligned}$$

and the fourth follows similarly. □

Definition 6.1 The function $r \colon \mathbb{Z} \longrightarrow \mathbb{Z}^+$ is defined by

$$r(n) = \frac{p}{\gcd(n, p)}. \tag{6.9}$$

This function r is used in this section to simplify some expressions. It comes in handy when one wants to describe how dimensions of various parts of centralisers vary with chain number.

Lemma 6.17 *Let $n \in \mathbb{Z}$, $\alpha \sqsubseteq K_n$, and $\beta \in \mathrm{Cen}(0,\alpha) \setminus \{0\}$. Then $\deg\beta = 2jr(n)$ for some $j \in \mathbb{N}$.*

Proof. If $n = 0$ then $r(n) = 1$, and hence the lemma simply claims that $\deg\beta$ is even, which is trivial to see. Therefore assume that $n \neq 0$. By (6.8), it follows that $n \deg\beta = 2ip$ for some $i \in \mathbb{Z}$. Let $r = r(n)$ and $s = n/\gcd(n,p)$. Then

$$\deg\beta = \frac{2ip}{n} = \frac{2ir}{s}.$$

Which are the possible values of i? Since r and s are coprime, it must be the case that $s \mid 2i$. Now there are two cases.

If s is odd then $i = sj$ for some $j \in \mathbb{Z}$, which implies

$$\deg\beta = \frac{2ip}{n} = \frac{2spj}{n} = \frac{2pj}{\gcd(n,p)} = 2rj.$$

If s is even then $i = \frac{1}{2}sk$ for some $k \in \mathbb{Z}$, which implies

$$\deg\beta = \frac{2ip}{n} = \frac{skp}{n} = \frac{kp}{\gcd(n,p)} = rk.$$

Since r and s are coprime, r must be odd. On the other hand every $\gamma \sqsubseteq K_0$ has even degree, and hence $\deg\beta$ is even. Thus $k = 2j$ for some $j \in \mathbb{Z}$.

In any case, $\deg\beta = 2rj$ for some $j \in \mathbb{Z}$. Finally, since $\deg\beta$ cannot be negative, it follows that $j \in \mathbb{N}$. □

Lemma 6.18 *Let $n \in \mathbb{Z}$, $d \in \mathbb{N}$, and $\alpha \sqsubseteq K_n$. Then*

$$\dim \mathrm{Cen}(2d,0,\alpha) \leqslant \left\lfloor \frac{d}{r(n)} \right\rfloor + 1. \tag{6.10}$$

Proof. Let $D = \{\, 2ir(n) \mid i \in \mathbb{N} \,\}$ and $D_d = \{m \in D \mid m \leqslant 2d\}$. By Lemma 6.17, all nonzero $\beta \in \mathrm{Cen}(0,\alpha)$ satisfy $\deg\beta \in D$. Hence Theorem 4.10 applies, and it follows that

$$\dim \mathrm{Cen}(2d,0,\alpha) \leqslant |D_d|.$$

Since $D_d \setminus D_{d-1}$ can be nonempty only for those d that satisfy $r(n) \mid d$, and since $D_0 = \{0\}$, a trivial induction yields

$$|D_d| \leqslant \left\lfloor \frac{d}{r(n)} \right\rfloor + 1.$$

The lemma follows. □

Lemma 6.19 *Let $n, s \in \mathbb{N}$ and $\{a_i\}_{i=0}^s \subseteq \mathcal{K}$. Let $C = BA$ and $\alpha = \sum_{i=0}^s a_i C^i$. Then the following properties are all equivalent:*

$$[\alpha, A^n] = 0, \tag{6.11}$$

$$[\alpha, B^n] = 0, \tag{6.12}$$

$$a_i = \sum_{j=i}^{s} \binom{j}{i} \left(\{n\}_q\right)^{j-i} (q^n)^i a_j \quad \textit{for all } i = 0, \ldots, s. \tag{6.13}$$

Proof. By Theorem 2.7,

$$\begin{aligned} A^n \alpha &= \sum_{j=0}^{s} a_j A^n C^j = \sum_{j=0}^{s} a_j \left(q^n C + \{n\}_q I\right)^j A^n = \\ &= \sum_{j=0}^{s} a_j \sum_{i=0}^{j} \binom{j}{i} \left(\{n\}_q\right)^{j-i} (q^n)^i C^i A^n = \\ &= \sum_{0 \leqslant i \leqslant j \leqslant s} a_j \binom{j}{i} \left(\{n\}_q\right)^{j-i} (q^n)^i C^i A^n = \\ &= \sum_{i=0}^{s} \left(\sum_{j=i}^{s} \binom{j}{i} \left(\{n\}_q\right)^{j-i} (q^n)^i a_j \right) C^i A^n. \end{aligned}$$

Since $\alpha A^n = \sum_{i=0}^s a_i C^i A^n$, and since $\left\{ C^i, C^i A^j, C^i B^j \mid i \in \mathbb{N}, j \in \mathbb{Z}^+ \right\}$ by Theorem 3.3 is a basis for $\mathcal{H}(q)$, it follows that (6.11) and (6.13) are equivalent.

Next note that by Theorem 2.7,

$$B^n C^i = q^{-ni} \left(C - \{n\}_q I\right)^i B^n = q^{-ni} \sum_{j=0}^{i} \left(-\{n\}_q\right)^{i-j} C^j B^n$$

and hence $\{B^n C^i\}_{i=0}^s$ is a linearly independent subset of $\mathcal{H}(q)$. Thus it is a basis for the subspace of $\mathcal{H}(q)$ that it spans. Theorem 2.7 also implies that

$$\begin{aligned} \alpha B^n &= \sum_{j=0}^{s} a_j C^j B^n = \sum_{j=0}^{s} a_j B^n \left(q^n C + \{n\}_q I\right)^j = \\ &= \sum_{j=0}^{s} a_j B^n \sum_{i=0}^{j} \binom{j}{i} \left(\{n\}_q\right)^{j-i} (q^n)^i C^i = \end{aligned}$$

$$= \sum_{0 \leqslant i \leqslant j \leqslant s} a_j \tbinom{j}{i} \left(\{n\}_q\right)^{j-i} (q^n)^i a_j B^n C^i =$$

$$= \sum_{i=0}^{s} \left(\sum_{j=i}^{s} \tbinom{j}{i} \left(\{n\}_q\right)^{j-i} (q^n)^i a_j \right) B^n C^i,$$

whilst as usual $B^n \alpha = \sum_{i=0}^{s} a_i B^n C^i$. Thus $\alpha B^n = B^n \alpha$ if and only if (6.13) holds. This means that (6.12) and (6.13) are equivalent. Hence the lemma is proved. □

The unexpected result here is that (6.11) and (6.12) are equivalent for $\alpha \sqsubseteq K_0$. It is only to be expected that there are equivalent systems of linear equations, but that a single system of linear equations is equivalent to both of these is quite a surprise. This is yet another example of how similar A and B behave, quite contrary to the different nature of A and B in some representations.

Lemma 6.20 *Let $n \in \mathbb{N}$ and set $s = r(n)$. Let $\{a_i\}_{i=0}^{s} \subseteq \mathcal{K}$. Then the system*

$$a_i = \sum_{j=i}^{s} \tbinom{j}{i} \left(\{n\}_q\right)^{j-i} (q^n)^i a_j \quad \textit{for all } i = 0, \ldots, s \tag{6.14}$$

of linear equations has a solution with $a_s = 1$.

Proof. Set $x = q^n$, $y = \{n\}_q$, and

$$c_{ij} = \begin{cases} 0 & \text{if } i > j, \\ x^i - 1 & \text{if } i = j, \\ \binom{j}{i} x^i y^{j-i} & \text{if } i < j. \end{cases}$$

Then the matrix equation $\left[c_{ij}\right]_{i,j=0}^{s} \cdot \left[a_j\right]_{j=0}^{s} = \left[0\right]_{i=0}^{s}$ is equivalent to (6.14). Next note that $c_{i0} = 0$ for all i. Hence a_0 can be chosen arbitrarily and the other unknowns must satisfy

$$\left[c_{ij}\right]_{i=0;j=1}^{s;s} \cdot \left[a_j\right]_{j=1}^{s} = \left[0\right]_{i=0}^{s}.$$

If $q = 1$ then $c_{is} = 0$ for all i too. In such a case a trivial solution of the wanted kind is to set $a_s = 1$ and all other a_i to 0, so in that case the lemma has been proved. Therefore assume that $q \neq 1$.

Next set $z = 1 - q$ and

$$d_{ki} = \begin{cases} 0 & \text{if } k < i, \\ z^{k-i} & \text{if } k \geqslant i. \end{cases}$$

Note that $\Big[d_{ki}\Big]_{k,i=0}^{s}$ is an invertible matrix. Hence

$$\Big[d_{ki}\Big]_{k,i=0}^{s} \cdot \Big[c_{ij}\Big]_{i=0;j=1}^{s;s} \cdot \Big[a_j\Big]_{j=1}^{s} = \Big[0\Big]_{i=0}^{s} \tag{6.15}$$

has a solution of the desired kind if and only if (6.14) has one. Then set

$$\Big[f_{kj}\Big]_{k=0;j=1}^{s;s} = \Big[d_{ki}\Big]_{k,i=0}^{s} \cdot \Big[c_{ij}\Big]_{i=0;j=1}^{s;s}$$

and consider f_{kj} for $k \geqslant j$. It follows that

$$\begin{aligned} f_{kj} &= \sum_{i=0}^{s} d_{ki}c_{ij} = \sum_{i=0}^{j} z^{k-i}c_{ij} = z^{k-j}\Big(\sum_{i=0}^{j-1} \tbinom{j}{i} z^{j-i}x^i y^{j-i} + x^j - 1\Big) = \\ &= z^{k-j}\Big(\sum_{i=0}^{j} \tbinom{j}{i} (yz)^{j-i}x^i - 1\Big) = z^{k-j}\big((yz+x)^j - 1\big) = \\ &= z^{k-j}\Big(\big(\{n\}_q(1-q) + q^n\big)^j - 1\Big) = z^{k-j}\Big(\big(1 - q^n + q^n\big)^j - 1\Big) = \\ &= z^{k-j}\big(1^j - 1\big) = 0, \end{aligned}$$

hence $\Big[f_{kj}\Big]_{k=0;j=1}^{s;s}$ is an upper triangular matrix (pay attention to that $k = j - 1$ in the main diagonal of this matrix, not $k = j$). Also consider f_{kj} for $k = j - 1$. In this case it follows that

$$\begin{aligned} f_{kj} &= \sum_{i=0}^{s} d_{j-1,i}\, c_{ij} = \sum_{i=0}^{j-1} \tbinom{j}{i} z^{j-1-i}x^i y^{j-i} = \\ &= z^{-1}\Big(\sum_{i=0}^{j-1} \tbinom{j}{i} z^{j-i}x^i y^{j-i}\Big) = z^{-1}\Big(\sum_{i=0}^{j} \tbinom{j}{i} (yz)^{j-i}x^i - x^j\Big) = \\ &= z^{-1}\big((yz+x)^j - x^j\big) = z^{-1}\big(1 - x^j\big) = \{nj\}_q. \end{aligned}$$

Thus

$$\left[f_{kj}\right]_{k=0;j=1}^{s;s} = \begin{pmatrix} \{n\}_q & \cdot & \cdot & \dots & \cdot & \cdot \\ 0 & \{2n\}_q & \cdot & \dots & \cdot & \cdot \\ 0 & 0 & \{3n\}_q & \dots & \cdot & \cdot \\ \vdots & \vdots & \vdots & \ddots & \vdots & \vdots \\ 0 & 0 & 0 & \dots & \{(s-1)n\}_q & \cdot \\ 0 & 0 & 0 & \dots & 0 & \{sn\}_q \\ 0 & 0 & 0 & \dots & 0 & 0 \end{pmatrix}$$

Note that the entries $f_{j-1,j} = \{jn\}_q$ that form the main diagonal of $\left[f_{kj}\right]_{k=0;j=1}^{s;s}$ are nonzero for $j = 1, 2, \dots, s-1$, but $\{sn\}_q = 0$. Since $\left[f_{kj}\right]_{k=0;j=1}^{s;s}$ is an upper triangular matrix, it follows that a_s can be taken as a parameter of the solution set. Hence (6.14) has a solution with $a_s = 1$, just as the lemma claims. □

The next theorem combines the results in the previous lemmas. It approximately covers half the cases covered by Theorem 6.6.

Theorem 6.21 *Let $n, d \in \mathbb{N}$. Let $\alpha \sqsubseteq K_{-n}$ and $\beta \sqsubseteq K_n$ be nonzero. Then the following equations hold:*

$$\mathrm{Cen}(0, \alpha) = \mathrm{Cen}(0, \beta) \tag{6.16}$$

$$\dim \mathrm{Cen}(2d, 0, \alpha) = \dim \mathrm{Cen}(2d, 0, \beta) = \left\lfloor \frac{d}{r(n)} \right\rfloor + 1 \tag{6.17}$$

Proof. Let $\gamma \sqsubseteq K_0$ be nonzero. By Theorem 6.15, $[\gamma, \alpha] = 0$ if and only if $[\gamma, A^n] = 0$. This theorem also implies that $[\gamma, \beta] = 0$ if and only if $[\gamma, B^n] = 0$. By Lemma 6.19, $[\gamma, A^n] = 0$ if and only if $[\gamma, B^n] = 0$. Hence $[\gamma, \alpha] = 0$ if and only if $[\gamma, \beta] = 0$. This proves (6.16).

By Lemmas 6.19 and 6.20, there exists a $\delta \in \mathrm{Cen}(0, A^n)$ such that $\deg \delta = 2r(n)$. Set $s = \lfloor d/r(n) \rfloor + 1$. Then $\{\delta^i\}_{i=0}^{s-1}$ is a linearly independent set with s elements which all clearly belong to $\mathrm{Cen}(0, A^n)$ and have degree less than or equal to $2d$. Hence $\dim \mathrm{Cen}(2d, 0, A^n) \geqslant s$.

By Lemma 6.18, $\dim \mathrm{Cen}(2d, 0, A^n) \leqslant s$. By the above,

$$\mathrm{Cen}(0, \alpha) = \mathrm{Cen}(0, A^n) = \mathrm{Cen}(0, \beta).$$

This proves (6.17). □

Theorem 6.21 suggests an interesting question. If $\gcd(n,p) = \gcd(m,p)$ then $\dim \text{Cen}(k,0,B^n)$ and $\dim \text{Cen}(k,0,B^m)$ are equal for all values of k, i.e., elements that are linearly independent of all elements of lower degree appear at the same degrees in both $\text{Cen}(0,B^n)$ and $\text{Cen}(0,B^m)$. Might this be simply because these sets are equal? That question is answered positively by the next theorem, but there is much more that relates the $\text{Cen}(0,B^n)$ subalgebras of $\text{Span}(K_0)$ than that.

Theorem 6.22 *Let*

$$H = \{\, \text{Cen}(0,B^n) \mid n \in \mathbb{N} \,\} \qquad \textit{and} \qquad G = \{\, n \in \mathbb{N} \mid n \mid p \,\}.$$

Then $(H,\subseteq)$ *is a lattice that is isomorphic to the lattice* $(G,|)$*, where the bar in* $(G,|)$ *is the relation 'divides'. An isomorphism* $\psi\colon H \longrightarrow G$ *is defined by letting* $\psi(\text{Cen}(0,B^n)) = \gcd(n,p)$ *for all* $n \in \mathbb{N}$.

Proof. To prove that ψ is well-defined, it suffices to prove for arbitrary $n,m \in \mathbb{N}$ that if $\gcd(n,p) \neq \gcd(m,p)$ then $\text{Cen}(0,B^n) \neq \text{Cen}(0,B^m)$. Therefore, assume these conditions.

By Theorem 6.21, $\dim \text{Cen}(2p,0,B^n) = \lfloor p/r(n) \rfloor + 1 = \lfloor \gcd(n,p) \rfloor + 1 = \gcd(n,p)+1$. Similarly, $\dim \text{Cen}(2p,0,B^m) = \gcd(m,p)+1$. It follows from this that $\dim \text{Cen}(2p,0,B^m) \neq \dim \text{Cen}(2p,0,B^n)$, hence $\text{Cen}(2p,0,B^m) \neq \text{Cen}(2p,0,B^n)$, thus $\text{Cen}(0,B^m) \neq \text{Cen}(0,B^n)$ as required.

Once it is established that ψ is well-defined, it is a trivial matter to prove that ψ is surjective. Since ψ maps $\text{Cen}(0,B^n)$ to n for all $n \in G$, it follows that the image of $\{\, \text{Cen}(0,B^n) \mid n \in G \,\}$ is the whole of G.

Let $m \in \mathbb{N}$ be arbitrary and let $d = \gcd(m,p)$. To prove that ψ is injective, it will be enough to prove that $\text{Cen}(0,B^m) = \text{Cen}(0,B^d)$. In addition, let $D_n = B^n$ if $n \geqslant 0$ and let $D_n = A^{-n}$ if $n < 0$. Then $D_n \sqsubseteq K_n$ is nonzero.

Clearly, there is an $i \in \mathbb{N}$ such that $m = id$. Thus it follows that

$$\text{Cen}(0,B^d) \subseteq \text{Cen}(0,B^{id}) = \text{Cen}(0,B^m).$$

There are also $r,s \in \mathbb{Z}$ such that $d = rm + sp$. Therefore

$$\begin{aligned} \text{Cen}(0,B^m) &= \text{Cen}(0,B^m) \cap \text{Cen}(0,B^p) && \text{(i)} \\ &\subseteq \text{Cen}(0,B^{|rm|}) \cap \text{Cen}(0,B^{|sp|}) && \text{(ii)} \\ &= \text{Cen}(0,D_{rm}) \cap \text{Cen}(0,D_{sp}) && \text{(by Theorem 6.21)} \\ &\subseteq \text{Cen}(0,D_{rm}D_{sp}) && \text{(iii)} \end{aligned}$$

$$= \mathrm{Cen}(0, B^d) \qquad \text{(by Theorem 6.15)}.$$

Here (i) is since $\mathrm{Cen}(0, B^p) = \mathrm{Span}(K_0)$, (ii) is because $[\alpha, \beta] = 0$ implies $[\alpha^n, \beta] = 0$, and (iii) is due to that if $[\alpha, \beta] = [\alpha, \gamma] = 0$ then $[\alpha, \beta\gamma] = 0$.

Combining these inequalities gives $\mathrm{Cen}(0, B^d) = \mathrm{Cen}(0, B^m)$, which is the desired equality.

By the above, ψ is a bijection. To prove that ψ is in fact an isomorphism of partially ordered sets, one needs to prove that $\mathrm{Cen}(0, B^n) \subseteq \mathrm{Cen}(0, B^m)$ if and only if $\gcd(n, p) \mid \gcd(m, p)$.

To prove the 'if' part, assume that $\gcd(n, p) \mid \gcd(m, p)$ and let $d = \gcd(n, p)$. By the above, $\mathrm{Cen}(0, B^n) = \mathrm{Cen}(0, B^d)$, and since $d \mid m$ it follows that

$$\mathrm{Cen}(0, B^n) = \mathrm{Cen}(0, B^d) \subseteq \mathrm{Cen}(0, B^{\frac{m}{d}d}) = \mathrm{Cen}(0, B^m).$$

To prove the 'only if' part, assume that $\mathrm{Cen}(0, B^n) \subseteq \mathrm{Cen}(0, B^m)$. By Theorem 6.21, there exists an $\alpha \in \mathrm{Cen}(0, B^n)$ with degree $2r(n)$. By Lemma 6.17, every $\alpha \in \mathrm{Cen}(0, B^m)$ must satisfy $\deg \alpha = 2jr(m)$ for some $j \in \mathbb{N}$. Therefore

$$\frac{2p}{\gcd(n, p)} = 2r(n) = \deg \alpha = 2jr(m) = \frac{2jp}{\gcd(m, p)},$$

which simplifies to $j \gcd(n, p) = \gcd(m, p)$, which in turn implies the desired $\gcd(n, p) \mid \gcd(m, p)$.

The above sums up to that $(H, \subseteq)$ and $\big(G, \mid\big)$ are isomorphic partially ordered sets. Finally, since $\big(G, \mid\big)$ is a lattice, it follows that $(H, \subseteq)$ must be a lattice as well. □

A nice consequence of this is the following corollary, which gives a method for *constructing* commuting elements.

Corollary 6.23 *Let $n \in \mathbb{Z}$ and $\alpha \sqsubseteq K_n$ be arbitrary. Let $m \in \mathbb{Z}$ be such that $r(m) \bigm| r(n)$ and let $\beta \sqsubseteq K_m$ be arbitrary. Then*

$$\left[\alpha^{r(n)}, \beta\right] = 0. \tag{6.18}$$

Proof. If $\alpha = 0$ or $\beta = 0$ then the above equality is trivial, therefore assume that α and β are nonzero.

By Theorem 4.11, $\alpha^{r(n)}$ can be factorised as $\gamma\delta$, where $\delta \sqsubseteq K_0$ and

$$\gamma = \begin{cases} B^{nr(n)} & \text{if } n \geqslant 0, \\ A^{-nr(n)} & \text{if } n < 0. \end{cases}$$

Since

$$nr(n) = \frac{np}{\gcd(n,p)} = p\frac{n}{\gcd(n,p)},$$

$p \mid nr(n)$ and γ belongs to the centre of $\mathcal{H}(q)$. Therefore $\left[\alpha^{r(n)}, \beta\right] = 0$ if and only if $[\delta, \beta] = 0$.

It is the case that

$$0 = \left[\alpha^{r(n)}, \alpha\right] = [\gamma\delta, \alpha] = \gamma[\delta, \alpha],$$

which by Theorem 4.9 implies that $0 = [\delta, \alpha]$. Hence $\delta \in \operatorname{Cen}(0, \alpha)$ which, by Theorem 6.21, is the same set as $\operatorname{Cen}\left(0, B^{|n|}\right)$.

The condition $r(m) \mid r(n)$ means that there is a k such that $kr(m) = r(n)$, which is equivalent to $k\gcd(n,p) = \gcd(m,p)$, or simply $\gcd(n,p) \mid \gcd(m,p)$. By the isomorphism of lattices of Theorem 6.22, is this equivalent to

$$\operatorname{Cen}\left(0, B^{|n|}\right) \subseteq \operatorname{Cen}\left(0, B^{|m|}\right).$$

Another application of Theorem 6.21 yields $\delta \in \operatorname{Cen}(0, \beta)$, which by the above is equivalent to (6.18). □

The obvious next step to take in the exploration of the centralisers of $\mathcal{H}(q)$ would be to complete the covering of the cases in Theorem 6.6 by determining $\dim \operatorname{Cen}(k, n, \alpha)$ in the cases where $\alpha \sqsubseteq K_m$ and neither n nor m are divisible by p. We expect that this will turn out to be at least as complex as the other cases.

One mechanism that complicates things is that $\operatorname{Cen}(0, \alpha)$ is nontrivial. This means that given a $\beta \in \operatorname{Cen}(n, \alpha)$, one can choose any $\gamma \in \operatorname{Cen}(0, \alpha)$ and find that $\beta\gamma \in \operatorname{Cen}(n, \alpha)$ too. This suggests that a (probably in some way deformed) copy of the lattice $(H, \subseteq)$ from Theorem 6.22 exists in each $\operatorname{Span}(K_n)$. It seems likely that a thorough understanding of the lattice is a key to the next step in the exploration of the centralisers.

Chapter 7

Algebraic dependence of commuting elements in $\mathcal{H}(q)$ and $\mathcal{H}(\mathbf{q}, n)$

Men bra mycke poänga	Many points
skräpa vi ihop	we scraped up
tack vare taktiken	thanks to the tactics
du lärd oss	that you taught us
då dä stog 0–0:	when we held them to a draw:
"Spärk en åt hälvet!"	"Kick the ball to hell!"

— from *Homage to The Coach* (Lagledar'n)
by N.-E. SJÖDIN,
translation from *röåmål* (a Swedish dialect) based on that
by J. DALE, P. DALE, and G. HÄLLSTRÖM

The final property of commuting elements in $\mathcal{H}(\mathbf{q}, n)$ that we will treat is the question of whether the fact that all elements in a set commute means that they also satisfy some *commutative* polynomial relation. This question is at least as old as [46], in which it was proved that if α and β are two commuting (complex) differential operators of a certain class then there must exist a nonzero polynomial $P \in \mathbb{C}[x, y]$ such that $P(\alpha, \beta) = 0$. It is shown in Chapter 8 that a subalgebra of that class of differential operators is a faithful representation of $\mathcal{H}(1, \mathbb{C})$, and therefore it follows that the same thing is true in $\mathcal{H}(1, \mathbb{C})$ itself: *If* $\alpha, \beta \in \mathcal{H}(1, \mathbb{C})$ *satisfy* $[\alpha, \beta] = 0$ *then there exists some* $P \in \mathbb{C}[x, y] \setminus \{0\}$ *such that* $P(\alpha, \beta) = 0$.

The proof of this in [46] is by analysis—it is based indirectly on Picard's theorem—which is why it has not been modified into a proof of the same thing in other $\mathcal{H}(\mathbf{q}, n)$ algebras. Nevertheless there are some generalisations of that result to most of them, and the purpose of this section is to present these. Before presenting the proofs, which by the way are pretty combinatorial, we shall develop some algebra.

Definition 7.1 Let $k \in \mathbb{Z}^+$ and let $\mathbf{q} \in (\mathcal{K}^*)^k$. Let $\alpha_0, \alpha_1, \dots, \alpha_k \in \mathcal{H}(\mathbf{q}, k)$ commute, i.e., satisfy $[\alpha_i, \alpha_j] = 0$ for all $1 \leqslant i, j \leqslant k$. Then define

$$\mathcal{P}(\alpha_0, \alpha_1, \dots, \alpha_k) = \left\{ P \in \mathcal{K}[x_0, x_1, \dots, x_k] \,\middle|\, P(\alpha_0, \alpha_1, \dots, \alpha_k) = 0 \right\}.$$

It is not hard to see that every set $\mathcal{P}(\alpha_0, \alpha_1, \dots, \alpha_k)$ is actually an ideal in $\mathcal{K}[x_0, x_1, \dots, x_k]$. If one defines $\psi\colon \mathcal{K}[x_0, x_1, \dots, x_k] \longrightarrow \mathcal{H}(\mathbf{q}, k)$ through letting $\psi(P) = P(\alpha_0, \alpha_1, \dots, \alpha_k)$ then ψ is an algebra homomorphism whose kernel is the set $\mathcal{P}(\alpha_0, \alpha_1, \dots, \alpha_k)$.

Lemma 7.1 *Let $k \in \mathbb{Z}^+$ and let $\mathbf{q} \in (\mathcal{K}^*)^k$. Let $\alpha_0, \alpha_1, \dots, \alpha_k \in \mathcal{H}(\mathbf{q}, k)$ commute. Then $\mathcal{P}(\alpha_0, \alpha_1, \dots, \alpha_k)$ is a prime ideal.*

Proof. Let $\psi\colon \mathcal{K}[x_0, x_1, \dots, x_k] \longrightarrow \mathcal{H}(\mathbf{q}, k)$ be defined by

$$\psi(P) = P(\alpha_0, \alpha_1, \dots, \alpha_k).$$

Then the image of ψ is a commutative subalgebra of $\mathcal{H}(\mathbf{q}, k)$. Furthermore, since there are no zero divisors in $\mathcal{H}(\mathbf{q}, k)$ by Theorem 4.9, the image of ψ is an integral domain. It is a well known theorem that the kernel of a ring homomorphism that maps into an integral domain is a prime ideal. This completes the proof. □

Theorem 7.2 *Let $k \in \mathbb{Z}^+$ and let $\mathbf{q} \in \mathcal{K}^k$ satisfy $q_j \neq 0$ for all $1 \leqslant j \leqslant k$. Let $\alpha_0, \alpha_1, \dots, \alpha_k \in \mathcal{H}(\mathbf{q}, k)$ commute. If $\mathcal{P}(\alpha_0, \alpha_1, \dots, \alpha_k) \neq \{0\}$ then there is an irreducible element in $\mathcal{P}(\alpha_0, \alpha_1, \dots, \alpha_k)$.*

Proof. Denote $\mathcal{P}(\alpha_0, \alpha_1, \dots, \alpha_k)$ by $\mathcal{P}$ for short. Let

$$\psi\colon \mathcal{K}[x_0, x_1, \dots, x_k] \longrightarrow \mathcal{H}(\mathbf{q}, k)$$

be defined by $\psi(P) = P(\alpha_0, \alpha_1, \dots, \alpha_k)$. Let $P_0 \in \mathcal{P}$ be a nonzero element of minimal degree. If $\deg P_0 = 0$ then $P_0 \in \mathcal{K}$, and hence

$$\mathcal{P}(\alpha_0, \alpha_1, \dots, \alpha_k) = \mathcal{K}[x_0, x_1, \dots, x_k]$$

which of course contains irreducible elements.

Therefore consider the case that $\deg P_0 > 0$. Let $P_1 P_2 = P_0$ be an arbitrary factorisation of P_0 such that $\deg P_1 < \deg P_0$. Then $P_1 \notin \mathcal{P}$ and hence $P_2 \in \mathcal{P}$ because $\mathcal{P}$ is a prime ideal. Since the degree of P_0 is minimal among the nonzero elements of $\mathcal{P}$, it follows that $\deg P_0 \leqslant \deg P_2$. Degree is additive in $\mathcal{K}[x_0, x_1, \dots, x_k]$, so one has the equality $\deg P_0 =$

$\deg P_1 + \deg P_2$, which implies that $\deg P_1 \leqslant 0$. As P_1 is nonzero, its degree cannot be negative, and hence the degree is exactly 0.

All elements in $\mathcal{K}[x_0, x_1, \ldots, x_k]$ that have degree 0 are invertible. Thus the factorisation $P_0 = P_1 P_2$ is trivial, and since it was allowed to be arbitrary it follows that all factorisations of P_0 are trivial. Hence P_0 is irreducible. $\square$

The next theorem gives sufficient (although not necessary) conditions for the claim $\mathcal{P}(\alpha_0, \alpha_1, \ldots, \alpha_k) \neq \{0\}$.

Theorem 7.3 *Let $k \in \mathbb{Z}^+$ and let $J = \{1, \ldots, k\}$. Let $\mathbf{q} \in \mathcal{Q}(J, \mathcal{K})$ and consider the algebra $\mathcal{H}(\mathbf{q}, k)$.*

Let $\alpha_0, \alpha_1, \ldots, \alpha_k \in \mathcal{H}(\mathbf{q}, k)$ commute, i.e., satisfy $[\alpha_i, \alpha_j] = 0$ for all $i, j \in \{0, 1, \ldots, k\}$. Now, if there exists a $\beta \in \mathcal{H}(\mathbf{q}, k)$ such that the set

$$\left\{ f \in \mathbb{Z}(J) \,\middle|\, \beta \sqcap K_f \neq 0 \right\}$$

is strongly nonzero and $[\alpha_i, \beta] = 0$ for all $i \in \{0, 1, \ldots, k\}$, then there exists a polynomial $P \in \mathcal{K}[x_0, \ldots, x_k] \setminus \{0\}$ such that

$$P(\alpha_0, \alpha_1, \ldots, \alpha_k) = 0. \tag{7.1}$$

Proof. Observe that if any of the $\{\alpha_i\}_{i=0}^k$ is zero, then the claim of the theorem is trivially true. Therefore assume that all α_i for $0 \leqslant i \leqslant k$ are nonzero. Then one can let

$$\begin{aligned} r_j &= \min\{0, \underline{\chi}_j(\alpha_0), \underline{\chi}_j(\alpha_1), \ldots, \underline{\chi}_j(\alpha_k)\}, \\ s_j &= \max\{0, \overline{\chi}_j(\alpha_0), \overline{\chi}_j(\alpha_1), \ldots, \overline{\chi}_j(\alpha_k)\} \end{aligned}$$

for all $j \in J$. Observe that, for every $(i_0, i_1, \ldots, i_k) = \mathbf{i} \in \mathbb{N}^{k+1}$ and $j \in J$,

$$\begin{aligned} \overline{\chi}_j(\alpha_0^{i_0}\alpha_1^{i_1}\cdots\alpha_k^{i_k}) &= \sum_{l=0}^{k} \overline{\chi}_j(\alpha_l^{i_l}) = \sum_{l=0}^{k} i_l \overline{\chi}_j(\alpha_l) \leqslant \\ &\leqslant \sum_{l=0}^{k} i_l s_j = (i_0 + i_1 + \cdots + i_k) s_j = |\mathbf{i}| s_j, \\ \underline{\chi}_j(\alpha_0^{i_0}\alpha_1^{i_1}\cdots\alpha_k^{i_k}) &= \sum_{l=0}^{k} \underline{\chi}_j(\alpha_l^{i_l}) = \sum_{l=0}^{k} i_l \underline{\chi}_j(\alpha_l) \geqslant \\ &\geqslant \sum_{l=0}^{k} i_l r_j = (i_0 + i_1 + \cdots + i_k) r_j = |\mathbf{i}| r_j. \end{aligned}$$

We refer readers unfamiliar with the notation for rate of growth used below to Appendix D. Let

$$S(k,n) = \{\, (i_0, i_1, \ldots, i_k) \in \mathbb{N}^{k+1} \,|\, i_0 + i_1 + \cdots + i_k \leqslant n \,\}.$$

Observe that

$$\begin{aligned} |S(k,n)| &= \left|\left\{\, (i_0, i_1, \ldots, i_k, i_{k+1}) \in \mathbb{N}^{k+2} \,\middle|\, \textstyle\sum_{l=0}^{k+1} i_l = n \,\right\}\right| = \\ &= \text{the number of ways to distribute } n \text{ identical} \\ &\quad\ \text{items in } k+2 \text{ boxes} = \\ &= \binom{n+k+1}{k+1} \end{aligned}$$

for all $n \in \mathbb{N}$. Since

$$\binom{n+k+1}{k+1} = \frac{1}{(k+1)!} \prod_{i=1}^{k+1} (n+i) = \Theta(n^{k+1}),$$

it follows that $|S(k,n)| = \Theta(n^{k+1})$ as $n \to \infty$.

Let

$$G_n = \{\, f \in \mathbb{Z}(J) \,|\, nr_j \leqslant f(j) \leqslant ns_j \text{ for all } j \in J \,\}$$

for all $n \in \mathbb{N}$. By Theorem 5.15 it follows that

$$\dim \operatorname{Cen}(G_n, \beta) \leqslant |G_n| = \prod_{j=1}^{k} (ns_j - nr_j + 1). \tag{7.2}$$

This means that $\dim \operatorname{Cen}(G_n, \beta) = O(n^k)$ as $n \to \infty$.

Clearly $\alpha_0^{i_0} \alpha_1^{i_1} \cdots \alpha_k^{i_k} \in \operatorname{Cen}(\beta)$, but the bounds on $\overline{\chi}_j$ and $\underline{\chi}_j$ established above actually imply that

$$\alpha_0^{i_0} \alpha_1^{i_1} \cdots \alpha_k^{i_k} \in \operatorname{Cen}(G_n, \beta)$$

for all $(i_0, i_1, \ldots, i_k) = \mathbf{i} \in \mathbb{N}^{k+1}$ such that $i_0 + i_1 + \cdots + i_k \leqslant n$, i.e., for all $\mathbf{i} \in S(k,n)$. This means that

$$\sum_{\mathbf{i} \in S(k,n)} a_{\mathbf{i}} \alpha_0^{i_0} \cdots \alpha_k^{i_k} \in \operatorname{Cen}(G_n, \beta)$$

for all $n \in \mathbb{N}$ and $\{a_{\mathbf{i}}\}_{\mathbf{i} \in S(k,n)} \subseteq \mathcal{K}$.

This can be used to construct a family of linear mappings. Let $V(k,n)$ be the vector space $\{f\colon S(k,n) \longrightarrow \mathcal{K}\}$ over $\mathcal{K}$. Then for each $n \in \mathbb{N}$ there is a linear map

$$\psi_n\colon V(k,n) \longrightarrow \mathrm{Cen}(G_n,\beta),$$
$$f \longmapsto \sum_{\mathbf{i}\in S(k,n)} f(\mathbf{i}) \prod_{l=0}^{k} \alpha_l^{i_l}.$$

Next, compare the dimensions of the domains and codomains of these maps. It was shown above that $|S(k,n)|$, which is the dimension of $V(k,n)$, grows $\Theta(n^{k+1})$ as $n \to \infty$. It was also shown that the dimension of $\mathrm{Cen}(G_n,\beta)$ grows $O(n^k)$ as $n \to \infty$. This means that for some n, the domain of ψ_n will have more dimensions than the codomain of ψ_n. This in turn means that for such an n, the kernel of ψ_n will be nontrivial.

Let n_0 be such an n, and let f_0 be a nonzero element in the kernel of ψ_{n_0}. Consider the polynomial P defined by

$$P(x_0,x_1,\ldots,x_k) = \sum_{\mathbf{i}\in S(k,n_0)} f_0(\mathbf{i}) \prod_{l=0}^{k} x_l^{i_l}.$$

Since f_0 is nonzero, P will be nonzero. Since f_0 is in the kernel of ψ_{n_0}, it follows that $P(\alpha_0,\alpha_1,\ldots,\alpha_k) = 0$. This completes the proof. □

In many cases, one can use α_0, α_1, ..., or α_k as β. If none of these satisfy the requirements for β then there is often some polynomial expression in the α_i's that does. A good starting point is to examine expressions of the type $\alpha_{j_1}^{i_1}\alpha_{j_2}^{i_2}$ and if that doesn't turn up anything useful then the next step is to try to remove troublesome terms through some suitable linear combination of monomials in α_0, α_1, ..., and α_k.

When $k = 1$ in the above theorem, we have an even stronger result.

Theorem 7.4 *Let $q \in \mathcal{K}^*$ be of free type. If $\alpha,\beta \in \mathcal{H}(q)$ satisfy $[\alpha,\beta] = 0$ then there exists a nonzero $P \in \mathcal{K}[x,y]$ such that*

$$P(\alpha,\beta) = 0. \tag{7.3}$$

Proof. There are three cases to consider. In the simplest case, when α is of the form cI for some $c \in \mathcal{K}$, the polynomial $P(x,y) = x - c$ satisfies (7.3) since $\alpha - cI = 0$.

In the second case assume that $\alpha, \beta \sqsubseteq K_0$ and that there is no $c \in \mathcal{K}$ such that $\alpha = cI$. Let $a = \deg \alpha > 0$ and $b = \deg \beta$. A general expression for $P(\alpha, \beta)$, where P has at most degree b in the first variable and at most degree a in the second is

$$\sum_{\substack{0 \leqslant i \leqslant b \\ 0 \leqslant j \leqslant a}} p_{ij} \alpha^i \beta^j. \tag{7.4}$$

This sum contains $(a+1)(b+1)$ terms, so it might be looked at as a linear combination of the $(a+1)(b+1)$ vectors $\{\alpha^i \beta^j\}_{i=0;j=0}^{b;a}$. All these vectors belong to the vector space $\mathrm{Cen}\big(ab, 0, \alpha\big)$, which is $(\frac{1}{2}ab + 1)$-dimensional since that is exactly the number of elements from the basis (3.5) that are $\sqsubseteq K_0$ and have degree at most ab. As

$$\tfrac{1}{2}ab + 1 < (\tfrac{1}{2}ab + 1) + (\tfrac{1}{2}ab + a + b) = (a+1)(b+1),$$

the set of vectors $\{\alpha^i \beta^j\}_{i=0;j=0}^{b;a}$ is linearly dependent. Hence there exist numbers $\{p_{ij}\}_{i,j}$, not all zero, which make (7.4) zero. Thus there exists a P as required.

Next to take care of is the case when $\alpha \not\sqsubseteq K_0$ or $\beta \not\sqsubseteq K_0$. It was largely dealt with in Theorem 7.3, since by letting $k = 1$ and $\mathbf{q} = (q)$ one can make sure that the conditions about k and $\mathbf{q}$ in that theorem are satisfied. By letting $\alpha_0 = \alpha$ and $\alpha_1 = \beta$, the conclusion (7.1) is exactly the wanted (7.3). The β of Theorem 7.3, finally, can be chosen as β if $\beta \not\sqsubseteq K_0$ and as α otherwise. This is due to that in $\mathcal{H}(q)$, as it was shown in the proof of Corollary 6.9, $\beta \not\sqsubseteq K_0$ implies that $\{\, n \in \mathbb{Z} \mid \alpha_0 \sqcap K_n \,\}$ is strongly nonzero. Hence all conditions in Theorem 7.3 are fulfilled and there exists a polynomial $P \in \mathcal{K}[x, y] \setminus \{0\}$ such that $P(\alpha, \beta) = 0$ as required. □

The idea of the proofs of the two last theorems is to show that the number of coefficients in the polynomial P grows an order of magnitude faster than the dimension of the vector space $\mathrm{Cen}(G_n, \alpha_0)$, or $\mathrm{Cen}\big([rn, sn], \alpha_0\big)$ in $\mathcal{H}(q)$, to which $P(\alpha_0, \ldots, \alpha_k)$ must be confined (given that the degree of P grows $\Theta(n)$ and r and s are suitably chosen). Thus sooner or later (i.e., for some n large enough) one will have that there are more coefficients in P than there are dimensions in the vector space to which P maps $(\alpha_0, \ldots, \alpha_k)$.

This can be translated to a system of linear equations with more unknowns (coefficients of P) than independent equations (dimensions of the vector space). This system is furthermore homogeneous because one wants $P(\alpha, \beta)$ to be 0 and this corresponds to that the right hand side of the system consists of all zeros. Hence the system must have a nontrivial solution, which corresponds to a set of coefficients for P, not all zero, such that $P(\alpha, \beta) = 0$.

The proof does not give a formula for P, but it is rather easy to devise a method of computing a P directly from the proof of Theorem 7.3. The problematic part is to find an n and a nonzero $f_0 \in \ker \psi_n$, but this can be done and here we give some bounds on how much work it will take.

The first part is to find the n. Clearly, any n so large that

$$\big|S(k,n)\big| > \dim \mathrm{Cen}(G_n, \alpha_0) \tag{7.5}$$

will be sufficient. The right hand side of this inequality is too hard to compute, but it can be replaced by its upper bound $|G_n|$. The left hand side is known, but it might be hard to use when searching for an n so some simplification might be in place.

$$\big|S(k,n)\big| = \binom{n+k+1}{k+1} = \frac{1}{(k+1)!}\prod_{i=1}^{k+1}(n+i) > \frac{1}{(k+1)!}n^{k+1}$$

Trying to put the right hand side in a similar form yields

$$\dim \mathrm{Cen}(G_n, \alpha_0) \leqslant |G_n| = \prod_{j=1}^{k}\big((s_j - r_j)n + 1\big) \leqslant \prod_{j=1}^{k}\big((s_j - r_j + 1)n\big)$$

Thus (7.5) is fulfilled for all $n \geqslant (k+1)!\prod_{j=1}^{k}(s_j - r_j + 1)$, even though this is probably not the best possible n.

Once n is fixed, it gives rise to the equation $0 = \psi_n(f_0)$, which can be expanded to

$$0 = \sum_{\mathbf{i}\in S(k,n)} p_{\mathbf{i}}\alpha_0^{i_0}\cdots\alpha_k^{i_k}, \tag{7.6}$$

where $p_{\mathbf{i}} = f_0(\mathbf{i})$ for all $\mathbf{i} \in S(k,n)$. By using one of the bases from Chapter 3, all elements $\alpha_0^{i_0}\cdots\alpha_k^{i_k}$ can be expressed as linear combinations of the elements in that basis. We will assume that the BA-basis is used. Then (7.6) transforms into the system of linear equations

$$0 = \sum_{\mathbf{i}\in S(k,n)} p_{\mathbf{i}}a_{\mathbf{ji}} \qquad \text{for all } \mathbf{j} \in E, \tag{7.7}$$

where $E \subset \mathbb{N}^{2k}$ and the coefficients $\{a_{\mathbf{ji}}\}_{\mathbf{i} \in S(k,n);\, \mathbf{j} \in E}$ are defined by that the equation

$$\alpha_0^{i_0} \cdots \alpha_k^{i_k} = \sum_{\mathbf{j} \in E} a_{\mathbf{ji}} B_1^{j_1} A_1^{j_2} B_2^{j_3} A_2^{j_4} \cdots B_k^{j_{2k-1}} A_k^{j_{2k}} \tag{7.8}$$

holds for every $\mathbf{i} \in S(k,n)$. Clearly, this is a situation in which it is necessary to be able to make reasonably fast computations in $\mathcal{H}(\mathbf{q}, k)$.

It is obvious that E can always be chosen as a finite set, but there is actually an upper bound on the size of E that is polynomial in n. Simply let $d = \max_{0 \leqslant l \leqslant k} \deg \alpha_l$. Then

$$\deg(\alpha_0^{i_0} \cdots \alpha_k^{i_k}) = \sum_{l=0}^{k} \deg \alpha_l^{i_l} = \sum_{l=0}^{k} i_l \deg \alpha_l \leqslant d \sum_{l=0}^{k} i_l = d|\mathbf{i}| \leqslant dn,$$

and hence $|\mathbf{j}| \leqslant dn$ for all $\mathbf{j} \in E$. It follows from this, in analogy with the argument that gave the size of $S(k,n)$, that $|E| = O(n^{2k})$.

At first glance this seems to be bad news, as this means that the matrix $M = \left[a_{\mathbf{ji}}\right]_{\mathbf{j} \in E;\, \mathbf{i} \in S(k,n)}$ can have far more rows (asymptotically $O(n^{2k})$) than columns (asymptotically $O(n^{k+1})$). Very few linear systems $N\mathbf{x} = \mathbf{0}$, with N being such a matrix, have any solutions apart from the trivial one. According to the proof of Theorem 7.3 however, the system formed from M is one of these few.

The reason for this is that even if the number of rows in M grows very fast, the rank of M will grow slower. As the rank of a matrix equals the dimension of its column space, and the column space of M is simply the coordinate view on the image of ψ_n, it follows that the rank of M will actually be less than the number of columns (by the choice of n), and hence M has a nontrivial nullspace. The nullspace of M corresponds to the kernel of ψ_n, so in order to find a f_0, one only has to locate a nonzero element in the nullspace of M. This element will then give the coefficients of the polynomial P.

Finding nonzero elements in the nullspace of a matrix is a common computational operation for which there are lots of methods in use; the only problem is to choose one. This is, however, something which must be done with care, as the shape of the matrix M is far from being quadratic. Note that even small rounding errors can effectively increase the rank of M enough to make its nullspace contain only the zero vector, hence exact representation of numbers is in general preferable to floating point, but specific situations may of course call for other considerations as well. As the problem is far from being a novelty, the reader should have no problem handling it once it is observed.

A final remark is that the bounds on n deduced above are probably not optimal. As the time it takes to find f_0 is likely to be in the order $O(n^{4k+2})$, one does not want to use larger n than necessary. The reader should also observe that no matter what n is used for constructing the matrix M, the nullspace of M will correspond to the subset

$$\{ P \in \mathcal{P}(\alpha_0, \alpha_1, \dots, \alpha_k) \mid \deg P \leqslant n \} .$$

This means that the matrix M can be used to compute arbitrarily large portions of $\mathcal{P}(\alpha_0, \alpha_1, \dots, \alpha_k)$, which should be useful for anyone who wants to perform numerical experiments regarding the structure of that ideal.

As a conclusion, we will treat the topic of possible generalisations of Theorems 7.3 and 7.4 to other $\mathcal{H}(\mathbf{q}, J)$ algebras. In some cases such a generalisation seems to be plausible, in others we will show that the theorems does not hold.

To start with Theorem 7.3, a natural question is whether the condition that β is strongly nonzero can be dropped (this would mean that one can always choose $\beta = \alpha_0$), so that it will be more like Theorem 7.4. We believe it can, since strongly nonzero is a property we introduced in order to make a previous proof for $\mathcal{H}(q)$ work in the more general context of $\mathcal{H}(\mathbf{q}, k)$. We have no doubt that (5.10) does imply (5.26) even if the conditions on β is weakened or completely removed, but we are uncertain as to how difficult it will be to prove.

In any case, we would like to encourage those who have explicit examples of $\alpha_0, \dots, \alpha_k \in \mathcal{H}(\mathbf{q}, k)$ that fulfill all the conditions in the theorem, except the one about strongly nonzero, and which are not all $\sqsubseteq K_0$, to go ahead and try to compute P as described above. The method does not only give an f_0 if it exists, it can also be used as a test for the nonexistence of such an f_0, as it can be used to determine the dimension of $\ker \psi_n$. If it works then you have got the P you wanted; if it does not, then you have discovered a case in which the generalisation of the theorem does not hold, a discovery of theoretical interest. (Actually, it is not necessarily the generalisation of Theorem 7.3 that does not hold, it could merely be the generalisation of Theorem 5.15 that fails. In either case, it *is* interesting.)

Another question is whether it might suffice with fewer than $k+1$ commuting elements in order to ensure the existence of a polynomial P as in the theorem. The answer to this is that it does not. The easiest counterexample is to take $\alpha_l = A_l$ for $l = 1, \dots, k$—these commute but do not satisfy any nontrivial commutative polynomial relation. Thus in this aspect, the theorem gives a sharp bound.

Next we consider whether Theorem 7.4 holds for q of torsion type. The answer is once again negative, since if p is taken to be the order of q then $\alpha = A^p$ and $\beta = B^p$ commute but do not satisfy any commutative polynomial relation. This argument can be generalised to $\mathcal{H}(\mathbf{q}, k)$ as well. As soon as q_1 is of torsion type and p is its order then $\alpha_0 = A_1^p$, $\alpha_1 = B_1^p$, and $\alpha_j = B_j$ for $2 \leqslant j \leqslant k$ will not satisfy $P(\alpha_0, \alpha_1, \dots, \alpha_k) = 0$ for any nonzero $P \in \mathcal{K}[x_0, \dots, x_k]$.

The claim of Theorem 7.4 is therefore not true in these $\mathcal{H}(q)$ algebras, but might it be true for α, β that satisfies some additional conditions? It has been suggested that if

$$\gcd(\deg \alpha, p) = \gcd(\deg \beta, p) = 1$$

then the claim of Theorem 7.4 would be true. This extra condition does indeed seem to exclude all counterexamples to the claim of the theorem that we know of, but it is far from being a necessary condition. Neither does this condition have an obvious connection to the methods used to prove Theorem 7.4, so if it finally proves to be sufficient, then it is still not very likely that the proof is an adaptation of the proof of Theorem 7.4.

There is however a slightly different formulation of Theorem 7.4 that does generalise to these $\mathcal{H}(q)$ algebras.

Theorem 7.5 *Let $q \in \mathcal{K}^*$ and let $\mathcal{C}$ be the centre of $\mathcal{H}(q)$. Let $\alpha, \beta \in \mathcal{H}(q)$ satisfy $[\alpha, \beta] = 0$. Then there exists a polynomial $P \in \mathcal{C}[x, y] \setminus \{0\}$ such that*

$$P(\alpha, \beta) = 0. \tag{7.9}$$

Proof. For q of free type, $\mathcal{C}$ is isomorphic to $\mathcal{K}$ and hence the theorem follows from Theorem 7.4. Thus assume that q is of torsion type and that p is its order.

Observe that $\mathcal{H}(q)$, seen as a $\mathcal{C}$-module, contains a spanning set of p^2 elements, namely $\{B^i A^j\}_{0 \leqslant i,j < p}$. Also observe that $\mathcal{C}$, by Theorem 4.9 and Theorem 5.7, is an integral domain. Hence any subset of $\mathcal{H}(q)$ that contains more elements than a spanning set will be linearly dependent. Consider the polynomial

$$P(x,y) = \sum_{i=0}^{p} \sum_{j=0}^{p} c_{ij} x^i y^j$$

where $c_{ij} \in \mathcal{C}$. Clearly $P(\alpha, \beta)$ will be a linear combination of $(p+1)^2 > p^2$ elements in $\mathcal{H}(q)$ and these elements are linearly dependent. Thus there are coefficients $\{c_{ij}\}_{0 \leqslant i,j \leqslant p} \subset \mathcal{C}$, not all zero, such that $P(\alpha, \beta) = 0$. This is exactly what the theorem claims. $\square$

Chapter 8

Representations of $\mathcal{H}(\mathbf{q}, J)$ by q-difference operators

'It's usual to say how-do-you-do when you come in,' said Miss Black. 'And I'm not interested in who is best in the class. It won't be you, at all events.'

'Oh yes, but that's exactly who it is,' said Karlson. Then he stopped and seemed to be thinking things over.

'At least, I'm the best at arithmetic,' he said darkly, when he had finished pondering. Then he shrugged his shoulders.

'Oh well, that's a worldly matter,' he said, beginning to jump cheerfully round the kitchen.

— from *The world's best Karlson* by A. LINDGREN,
translation from the Swedish based
upon that by P. CRAMPTON,
original title: *Karlsson på taket smyger igen*

This chapter is devoted to the study of representations of the $\mathcal{H}(\mathbf{q}, J)$ algebras by q-difference operators.

Let $\mathfrak{X}$ be a (finite or infinite) set of commuting indeterminates, and let

$$P[\mathfrak{X}] = P[\mathfrak{X}, \mathcal{K}] = \\ = \left\{ f(x_1, \dots, x_n) = \sum_{k=0}^{\deg f} \sum_{\substack{\mathbf{t} \in \mathbb{N}^n \\ |\mathbf{t}|=k}} c_{\mathbf{t}} x_1^{t_1} \cdots x_n^{t_n} \;\middle|\; \begin{array}{l} c_{\mathbf{t}} \in \mathcal{K} \text{ for all } \mathbf{t}, \\ \deg f \in \mathbb{N} \cup \{\infty\}, \\ x_1, \dots, x_n \in \mathfrak{X} \end{array} \right\} \tag{8.1}$$

be the linear space of all polynomial-like expressions constructed using finite subsets of indeterminates from $\mathfrak{X}$. We will call the elements of $P[\mathfrak{X}]$ **generalised polynomials**. The notation $\mathcal{K}[\mathfrak{X}]$ will be used as usual for the linear space of all proper polynomials with coefficients from $\mathcal{K}$ and with indeterminates from a commutative set $\mathfrak{X}$. Consider the following linear operators on $P[\mathfrak{X}]$:

$$[M_{x_j} f](x_1, \ldots, x_n) = x_j f(x_1, \ldots, x_n) \tag{8.2}$$

$$[T_{x_j,q} f](x_1, \ldots, x_n) = f(x_1, \ldots, qx_j, \ldots, x_n) \tag{8.3}$$

$$[\partial_{x_j,q} f](x_1, \ldots, x_n) = \sum_{k=0}^{\deg f} \sum_{\substack{\mathbf{t}\in\mathbb{N}^n \\ |\mathbf{t}|=k \\ t_j>0}} c_{\mathbf{t}} \{t_j\}_q \prod_{i=1}^{n} x_i^{t_i - \delta_{ij}} \tag{8.4}$$

where δ_{ij} is the Kronecker delta and $c_{\mathbf{t}}$ is related to f as in (8.1). In particular, the linear operators M_{x_j}, $T_{x_j,q}$, and $\partial_{x_j,q}$ act on monomials by the formulae

$$M_{x_j}(x_1^{t_1} \cdots x_j^{t_j} \cdots x_n^{t_n}) = x_1^{t_1} \cdots x_j^{t_j+1} \cdots x_n^{t_n}.$$

$$T_{x_j,q}(x_1^{t_1} \cdots x_j^{t_j} \cdots x_n^{t_n}) = q^{t_j} x_1^{t_1} \cdots x_j^{t_j} \cdots x_n^{t_n}.$$

$$\partial_{x_j,q}(x_1^{t_1} \cdots x_j^{t_j} \cdots x_n^{t_n}) = \begin{cases} \{t_j\}_q x_1^{t_1} \cdots x_j^{t_j-1} \cdots x_n^{t_n} & \text{if } t_j \neq 0 \\ 0 & \text{if } t_j = 0 \end{cases}$$

It should be observed that the usual algebraic differentiation operator ∂_{x_j}, which is defined by

$$[\partial_{x_j} f](x_1, \ldots, x_n) = \sum_{k=0}^{\deg f} \sum_{\substack{\mathbf{t}\in\mathbb{N}^n \\ |\mathbf{t}|=k \\ t_j>0}} c_{\mathbf{t}} t_j \prod_{i=1}^{n} x_i^{t_i - \delta_{ij}},$$

appears as the special case $\partial_{x_j,1}$ of the operator $\partial_{x_j,q}$.

If $P[\mathfrak{X}]$ is considered as a set of functions on $\mathcal{K}^n$, there is an alternative pointwise definition of $\partial_{x_j,q}$ for $q \neq 1$. It turns out that

$$[\partial_{x_j,q} f](x_1, \ldots, x_n) = \frac{f(x_1, \ldots, x_n) - f(x_1, \ldots, qx_j, \ldots, x_n)}{(1-q)x_j}$$

for $x_j \neq 0$. For fields with a limit concept, such as $\mathbb{R}$ or $\mathbb{C}$, the $q \to 1$ limit of the above fraction is, if it exists, equal to the textbook expression for the partial derivative $\frac{\partial f}{\partial x_j}$.

It is a matter of straightforward calculations and induction to prove that the linear operators defined by (8.2), (8.3), and (8.4) have the following properties.

Property 8.1 For any $j \in \{1, \ldots, n\}$ and $r_j \in \mathbb{N}$:

$$[M_{x_j}^{r_j} f](x_1, \ldots, x_n) = x_j^{r_j} f(x_1, \ldots, x_n) = \sum_{k=0}^{\deg f} \sum_{\substack{\mathbf{t} \in \mathbb{N}^n \\ |\mathbf{t}|=k}} c_{\mathbf{t}} x_1^{t_1} \cdots x_j^{t_j + r_j} \cdots x_n^{t_n}. \tag{8.5}$$

Property 8.2 For any $(r_1, \ldots, r_n) \in \mathbb{N}^n$ and any permutation $\sigma \in S_n$:

$$\begin{aligned}
[M_{x_1}^{r_1} \cdots M_{x_n}^{r_n} f](x_1, \ldots, x_n) &= x_1^{r_1} \cdots x_n^{r_n} f(x_1, \ldots, x_n) = \\
&= \sum_{k=0}^{\deg f} \sum_{\substack{\mathbf{t} \in \mathbb{N}^n \\ |\mathbf{t}|=k}} c_{\mathbf{t}} x_1^{t_1 + r_1} \cdots x_n^{t_n + r_n} = x_{\sigma(1)}^{r_{\sigma(1)}} \cdots x_{\sigma(n)}^{r_{\sigma(n)}} f(x_1, \ldots, x_n) = \\
&= \left[M_{x_{\sigma(1)}}^{r_{\sigma(1)}} \cdots M_{x_{\sigma(n)}}^{r_{\sigma(n)}} f \right] (x_1, \ldots, x_n).
\end{aligned} \tag{8.6}$$

Property 8.3 For any $j \in \{1, \ldots, n\}$ and $r_j \in \mathbb{N}$:

$$\begin{aligned}
\left[T_{x_j, q}^{r_j} f \right] (x_1, \ldots, x_n) &= f(x_1, \ldots, q^{r_j} x_j, \ldots, x_n) = \\
&= \sum_{k=0}^{\deg f} \sum_{\substack{\mathbf{t} \in \mathbb{N}^n \\ |\mathbf{t}|=k}} c_{\mathbf{t}} q^{r_j t_j} x_1^{t_1} \ldots x_n^{t_n}.
\end{aligned} \tag{8.7}$$

Property 8.4 For any $(r_1, \ldots, r_n) \in \mathbb{N}^n$ and any permutation $\sigma \in S_n$:

$$\begin{aligned}
\left[T_{x_1, q_1}^{r_1} \cdots T_{x_n, q_n}^{r_n} f \right] (x_1, \ldots, x_n) &= f(q^{r_1} x_1, \ldots, q^{r_n} x_n) = \\
&= \sum_{k=0}^{\deg f} \sum_{\substack{\mathbf{t} \in \mathbb{N}^n \\ |\mathbf{t}|=k}} c_{\mathbf{t}} q_1^{r_1 t_1} \cdots q_n^{r_n t_n} x_1^{t_1} \cdots x_n^{t_n} = \\
&= \left[T_{x_{\sigma(1)}, q_{\sigma(1)}}^{r_{\sigma(1)}} \cdots T_{x_{\sigma(n)}, q_{\sigma(n)}}^{r_{\sigma(n)}} f \right] (x_1, \ldots, x_n).
\end{aligned} \tag{8.8}$$

Property 8.5 For any $j \in \{1, \ldots, n\}$ and $r_j \in \mathbb{N}$:

$$\left[\partial_{x_j,q}^{r_j} f\right](x_1, \ldots, x_n) =$$
$$= \sum_{k=0}^{\deg f} \sum_{\substack{\mathbf{t}\in\mathbb{N}^n \\ |\mathbf{t}|=k \\ t_j \geqslant r_j}} c_{\mathbf{t}} \cdot \left(\prod_{l=0}^{r_j-1} \{t_j - l\}_q \right) \cdot x_1^{t_1} \cdots x_j^{t_j - r_j} \cdots x_n^{t_n} =$$
$$= \sum_{k=0}^{\deg f} \sum_{\substack{\mathbf{t}\in\mathbb{N}^n \\ |\mathbf{t}|=k \\ t_j \geqslant r_j}} c_{\mathbf{t}} \{r_j\}_q! \binom{t_j}{r_j}_q x_1^{t_1} \cdots x_j^{t_j - r_j} \cdots x_n^{t_n} \tag{8.9}$$

Property 8.6 For any $(r_1, \ldots, r_n) \in \mathbb{N}^n$ and any permutation $\sigma \in S_n$:

$$\left[\partial_{x_1,q_1}^{r_1} \cdots \partial_{x_n,q_n}^{r_n} f\right](x_1, \ldots, x_n) =$$
$$= \sum_{k=0}^{\deg f} \sum_{\substack{\mathbf{t}\in\mathbb{N}^n \\ |\mathbf{t}|=k \\ t_1 \geqslant r_1, \ldots, t_n \geqslant r_n}} c_{\mathbf{t}} \cdot \left(\prod_{j=1}^{n} \prod_{l=0}^{r_j-1} \{t_j - l\}_{q_j} \right) \cdot x_1^{t_1 - r_1} \cdots x_n^{t_n - r_n} =$$
$$= \left[\partial_{x_{\sigma(1)},q_{\sigma(1)}}^{r_{\sigma(1)}} \cdots \partial_{x_{\sigma(n)},q_{\sigma(n)}}^{r_{\sigma(n)}} f\right](x_1, \ldots, x_n). \tag{8.10}$$

Property 8.7 For any $i, j \in \{1, \ldots, n\}$ and $\mathbf{x} = (x_1, \ldots, x_n)$:

$$[\partial_{x_i,q_i} M_{x_j} f](\mathbf{x}) = \sum_{k=0}^{\deg f} \sum_{\substack{\mathbf{t}\in\mathbb{N}^n \\ |\mathbf{t}|=k \\ t_i \geqslant 1-\delta_{ij}}} c_{\mathbf{t}} \{t_i + \delta_{ij}\}_{q_i} \prod_{l=1}^{n} x_l^{t_l - \delta_{il} + \delta_{jl}} \tag{8.11}$$

If furthermore $P[\mathfrak{X}]$ is considered as a set of functions on $\mathcal{K}^n$, $x_i \neq 0$, and $q_i \neq 1$ then

$$[\partial_{x_i,q_i} M_{x_j} f](\mathbf{x}) = \frac{x_j f(\mathbf{x}) - \big(1 + (q_i - 1)\delta_{ij}\big) x_j [T_{x_i,q} f](\mathbf{x})}{(1 - q_i) x_i}.$$

Property 8.8 For any $i, j \in \{1, \ldots, n\}$:

$$\left[M_{x_j} \partial_{x_i,q_i} f\right](x_1, \ldots, x_n) = \sum_{k=0}^{\deg f} \sum_{\substack{\mathbf{t}\in\mathbb{N}^n \\ |\mathbf{t}|=k \\ t_i \geqslant 1}} c_{\mathbf{t}} \{t_i\}_{q_i} \prod_{l=1}^{n} x_l^{t_l - \delta_{il} + \delta_{jl}} \tag{8.12}$$

If furthermore $P[\mathfrak{X}]$ is considered as a set of functions on $\mathcal{K}^n$, $x_i \neq 0$, and $q_i \neq 1$ then

$$\left[M_{x_j}\partial_{x_i,q_i} f\right](x_1,\ldots,x_n) = x_j \frac{f(x_1,\ldots,x_n) - f(x_1,\ldots,q_i x_i,\ldots,x_n)}{(1-q_i)x_i}.$$

Property 8.9 For any $i, j \in \{1,\ldots,n\}$ and $p, r \in \mathbb{N}$:

$$\left[\partial^p_{x_i,q_i} M^r_{x_j} f\right](x_1,\ldots,x_n) =$$

$$= \sum_{k=0}^{\deg f} \sum_{\substack{\mathbf{t}\in\mathbb{N}^n \\ |\mathbf{t}|=k \\ t_i \geqslant p - r\delta_{ij}}} c_{\mathbf{t}} \cdot \left(\prod_{l=0}^{p-1} \{t_i + r\delta_{ij} - l\}_{q_i}\right) \cdot \prod_{l=0}^{n} x_l^{t_l + r\delta_{jl} - p\delta_{il}} =$$

$$= \sum_{k=0}^{\deg f} \sum_{\substack{\mathbf{t}\in\mathbb{N}^n \\ |\mathbf{t}|=k \\ t_i \geqslant p - r\delta_{ij}}} c_{\mathbf{t}} \{p\}_{q_i}! \binom{t_i + r\delta_{ij}}{p}_{q_i} \prod_{l=0}^{n} x_l^{t_l + r\delta_{jl} - p\delta_{il}} \tag{8.13}$$

Property 8.10 For any $i, j \in \{1,\ldots,n\}$ and $p, r \in \mathbb{N}$:

$$\left[M^r_{x_j} \partial^p_{x_i,q_i} f\right](x_1,\ldots,x_n) = x_j^r \left[\partial^p_{x_i,q_i} f\right](x_1,\ldots,x_n) =$$

$$= \sum_{k=0}^{\deg f} \sum_{\substack{\mathbf{t}\in\mathbb{N}^n \\ |\mathbf{t}|=k \\ t_i \geqslant p}} c_{\mathbf{t}} \cdot \left(\prod_{l=0}^{p-1} \{t_i - l\}_{q_i}\right) \cdot \prod_{l=0}^{n} x_l^{t_l - p\delta_{il} + r\delta_{jl}} =$$

$$= \sum_{k=0}^{\deg f} \sum_{\substack{\mathbf{t}\in\mathbb{N}^n \\ |\mathbf{t}|=k \\ t_i \geqslant p}} c_{\mathbf{t}} \{p\}_{q_i}! \binom{t_i}{p}_{q_i} \prod_{l=0}^{n} x_l^{t_l - p\delta_{il} + r\delta_{jl}} \tag{8.14}$$

Theorem 8.1 *Let I denote the identity operator on $P[\mathfrak{X}]$. For any nonempty set J, a nonempty subset $\mathfrak{X}_0 = \{x_j\}_{j\in J} \subseteq \mathfrak{X}$ of indeterminates and any $(q_j)_{j\in J} = \mathbf{q} \in \mathcal{K}^J$, the operators $\{\partial_{x_j,q_j}, M_{x_j}\}_{j\in J}$ satisfy the relations*

$$\partial_{x_i,q_i}\partial_{x_j,q_j} - \partial_{x_j,q_j}\partial_{x_i,q_i} = 0 \qquad \text{if } i \neq j, \tag{8.15}$$

$$M_{x_i}M_{x_j} - M_{x_j}M_{x_i} = 0 \qquad \text{if } i \neq j, \tag{8.16}$$

$$\partial_{x_i,q_i}M_{x_j} - M_{x_j}\partial_{x_i,q_i} = 0 \qquad \text{if } i \neq j, \tag{8.17}$$

$$\partial_{x_i,q_i}M_{x_i} - q_i M_{x_i}\partial_{x_i,q_i} = I. \tag{8.18}$$

for all $i, j \in J$. In other words, the algebra generated by the operators $\{\partial_{x_j,q_j}, M_{x_j}\}_{j\in J}$ is a generalised q-deformed Heisenberg algebra by J.

Proof. The equalities (8.15) and (8.16) are obviously special cases of the properties (8.6) and (8.10). The equalities (8.17) and (8.18) are proved by subtraction of (8.12) from (8.11). Indeed, when $i \neq j$ we get

$$\begin{aligned}
&\big[(\partial_{x_i,q_i} M_{x_j} - M_{x_j}\partial_{x_i,q_i})f\big](x_1,\dots,x_n) = \\
&\qquad = \sum_{k=0}^{\deg f} \sum_{\substack{\mathbf{t}\in\mathbb{N}^n \\ |\mathbf{t}|=k \\ t_i \geqslant 1}} c_{\mathbf{t}}\big(\{t_i\}_{q_i} - \{t_i\}_{q_i}\big) \prod_{l=1}^{n} x_l^{t_l - \delta_{il} + \delta_{jl}} = 0,
\end{aligned}$$

and when $i = j$ it follows that

$$\begin{aligned}
&\big[(\partial_{x_j,q_j} M_{x_j} - q_j M_{x_j}\partial_{x_j,q_j})f\big](x_1,\dots,x_n) = \\
&= \sum_{k=0}^{\deg f} \sum_{\substack{\mathbf{t}\in\mathbb{N}^n \\ |\mathbf{t}|=k \\ t_j \geqslant 0}} c_{\mathbf{t}}\{t_j+1\}_{q_j} \prod_{l=1}^{n} x_l^{t_l} - q_j \sum_{k=0}^{\deg f} \sum_{\substack{\mathbf{t}\in\mathbb{N}^n \\ |\mathbf{t}|=k \\ t_j \geqslant 1}} c_{\mathbf{t}}\{t_j\}_{q_j} \prod_{l=1}^{n} x_l^{t_l} = \\
&= \sum_{k=0}^{\deg f} \Bigg(\sum_{\substack{\mathbf{t}\in\mathbb{N}^n \\ |\mathbf{t}|=k \\ t_j = 0}} c_{\mathbf{t}}\{1\}_{q_j} \prod_{l=1}^{n} x_l^{t_l} + \sum_{\substack{\mathbf{t}\in\mathbb{N}^n \\ |\mathbf{t}|=k \\ t_j \geqslant 1}} c_{\mathbf{t}}\big(\{t_j+1\}_{q_j} - q_j\{t_j\}_{q_j}\big) \prod_{l=1}^{n} x_l^{t_l} \Bigg) = \\
&= \sum_{k=0}^{\deg f} \Bigg(\sum_{\substack{\mathbf{t}\in\mathbb{N}^n \\ |\mathbf{t}|=k \\ t_j = 0}} c_{\mathbf{t}} \prod_{l=1}^{n} x_l^{t_l} + \sum_{\substack{\mathbf{t}\in\mathbb{N}^n \\ |\mathbf{t}|=k \\ t_j \geqslant 1}} c_{\mathbf{t}} \prod_{l=1}^{n} x_l^{t_l} \Bigg) = f(x_1,\dots,x_n),
\end{aligned}$$

where we have used that $\{t_j+1\}_{q_j} - q_j\{t_j\}_{q_j} = \{1\}_{q_j} = 1$ (cf. (C.4)). □

A consequence of Theorem 8.1 is that the mapping sending A_j to ∂_{x_j,q_j} and B_j to M_{x_j} for all $j \in J$, can be extended in unique way to the representation $\psi_{\mathbf{q},\mathfrak{x}_0} \colon \mathcal{H}(\mathbf{q}, J) \longrightarrow \mathrm{End}\big(P[\mathfrak{X}]\big)$ of the algebra $\mathcal{H}(\mathbf{q}, J)$ on the linear space $P[\mathfrak{X}]$.

The following observation turns out to be useful in various contexts.

Theorem 8.2 *If q_j is of torsion type for some $j \in J$ and $n \in \mathbb{Z}^+$ is the order of q_j, then*

$$\partial^n_{x_j,q_j} = 0. \tag{8.19}$$

Proof. Any element in $P[\mathfrak{X}]$ is a linear combination of monomials, hence in order to prove (8.19) it is sufficient to show that

$$\partial^n_{x_j,q_j}\mu = 0. \tag{8.20}$$

for any monomial $\mu = x_1^{m_1}\cdots x_j^{m_j}\cdots x_r^{m_r} \in P[\mathfrak{X}]$. By (8.9) we have

$$\partial^n_{x_j,q_j}\mu = \begin{cases} \{n\}_{q_j}! \binom{m_j}{n}_{q_j} x_1^{m_1}\cdots x_j^{m_j-n}\cdots x_r^{m_r} & \text{for } m_j \geqslant n, \\ 0 & \text{for } m_j < n. \end{cases}$$

Since q_j is of torsion type and n is its order, $\{n\}_{q_j} = 0$. Therefore $\{n\}_{q_j}! = 0$, and hence $\partial^n_{x_j,q_j}\mu = 0$. □

A representation $\psi\colon \mathcal{A} \to \mathrm{End}(V)$ of an algebra $\mathcal{A}$ on a vector space V is said to be **faithful** if

$$\ker\psi = \{0\},$$

or equivalently

$$\psi(\alpha)\, f = 0 \quad \text{in } V$$

for all $f \in V$ and some $\alpha \in \mathcal{A}$ if and only if $\alpha = 0$ in $\mathcal{A}$.

In the next theorem we will give a necessary and sufficient condition for the faithfulness of the representations $\psi_{q,\mathfrak{X}_0}$ of q-deformed Heisenberg algebras $\mathcal{H}(\mathbf{q}, J)$.

Theorem 8.3 *The representation $\psi_{\mathbf{q},\mathfrak{X}_0}\colon \mathcal{H}(\mathbf{q}, J) \longrightarrow \mathrm{End}\big(P[\mathfrak{X}_0]\big)$ is faithful if and only if the set $\{q_j\}_{j\in J}$ contains no element of torsion type.*

Proof. We begin by considering the 'if' case. Let $\mathbf{q} \in \mathcal{K}^{\square\supseteq J}$ satisfy that no element of the set $\{q\}_{j\in J}$ is of torsion type. For brevity, denote $\psi_{\mathbf{q},\mathfrak{X}_0}$ by ψ. By Theorem 4.4,

$$\mathcal{H}(\mathbf{q}, J) = \bigoplus_{w\in\mathbb{Z}(J)} \mathrm{Span}(K_w).$$

So, any $\alpha \in \mathcal{H}(\mathbf{q}, J)$ can be represented in the form

$$\alpha = \sum_{w\in\Omega} \alpha_w, \tag{8.21}$$

where Ω is a finite subset of $\mathbb{Z}(J)$ and $\alpha_w = \alpha \sqcap K_w \sqsubseteq K_w$ for all $w \in \Omega$. Applying ψ to α we get

$$\psi(\alpha) = \sum_{w \in \Omega} \psi(\alpha_w) \tag{8.22}$$

which means that

$$\psi(\alpha)\, f = \sum_{w \in \Omega} \psi(\alpha_w)\, f \tag{8.23}$$

for all $f \in P[\mathfrak{X}]$.

Every $f \in P[\mathfrak{X}]$ is a linear combination of monomials in indeterminates from $\mathfrak{X}$. Therefore $\psi(\alpha) = 0$ if and only if

$$\psi(\alpha)\, \mu = 0 \tag{8.24}$$

for every monomial $\mu \in P[\mathfrak{X}]$. Let $\{x_{j_1}, \ldots, x_{j_m}\}$ be the set of indeterminates appearing either in μ or in $\psi(\alpha)$. This set is finite by the choice of μ and the construction of $\psi(\alpha)$. Write the monomial μ in the form

$$\mu = x_{j_1}^{r_1} \cdots x_{j_m}^{r_m},$$

where $\mathbf{r} = (r_1, \ldots, r_m) \in \mathbb{N}^m$, and $r_j = 0$ if and only if the indeterminate x_j is not appearing in μ. By the properties (8.6) and (8.10) of the operators $\psi(A_j) = \partial_{x_j, q_j}$ and $\psi(B_j) = M_{x_j, q_j}$,

$$\psi(\alpha_w)\, \mu = \begin{cases} a_w(\mathbf{r}) x_{j_1}^{r_1 + w(j_1)} \cdots x_{j_m}^{r_m + w(j_m)} & \text{if } r_i + w(j_i) \geqslant 0 \\ & \text{for } 1 \leqslant i \leqslant m, \\ 0 & \text{otherwise.} \end{cases}$$

for some $a_w(\mathbf{r}) \in \mathcal{K}$ and any $w \in \Omega$. Thus substitution of $f = \mu$ into (8.23) yields

$$\psi(\alpha_w)\, \mu = \sum_{w \in \Omega'} a_w(\mathbf{r}) x_{j_1}^{r_1 + w(j_1)} \cdots x_{j_m}^{r_m + w(j_m)}.$$

where $\Omega' = \{ w \in \Omega \,|\, r_i + w(j_i) \geqslant 0 \text{ for } 1 \leqslant i \leqslant m \}$.

Since the monomials of the form $x_{j_1}^{s_1} \cdots x_{j_m}^{s_m} \in P[\mathfrak{X}]$ form a linearly independent set, equation (8.24) is satisfied for a monomial μ if and only if $a_w(\mathbf{r}) = 0$ for all $w \in \Omega'$. Thus (8.24) is satisfied for all monomials μ if and only if

$$\psi(\alpha_w)\, \mu = 0$$

for all $w \in \Omega$ and all monomials $\mu \in P[\mathfrak{X}]$.

Now assume that $\alpha_0 \sqsubseteq K_0$ is nonzero. Then the question is whether it could be that

$$\psi(\alpha_0)\,\mu = 0 \tag{8.25}$$

for all monomials $\mu \in P[\mathfrak{X}]$. The theorem claims it cannot, and this will now be shown.

To this end, note that according to Theorem 3.1 any $\alpha_0 \sqsubseteq K_0$ can be represented in the form

$$\alpha_0 = \sum_{\mathbf{t}\in T} c_{\mathbf{t}} B_{j_1}^{t_1} A_{j_1}^{t_1} \cdots B_{j_m}^{t_m} A_{j_m}^{t_m}$$

where the set T is a finite subset of $\mathbb{N}^m$ and $c_{\mathbf{t}} \neq 0$ for all $\mathbf{t} \in T$. Let μ be the monomial

$$\mu = x_{j_1}^{s_1} \cdots x_{j_m}^{s_m}$$

where $\mathbf{s} = (s_1, \dots, s_m)$ is defined by the recursive rule

$$\begin{aligned} s_m &= \min\{\, t_m \mid \mathbf{t} \in T \,\}, \\ s_k &= \min\Big\{\, t_k \;\Big|\; \mathbf{t} \in T, t_{k+1} = s_{k+1}, \dots, t_m = s_m \,\Big\} \end{aligned}$$

for $k = 1, \dots, m-1$. Then

$$\begin{aligned} \psi(\alpha_0)\,\mu &= \sum_{\mathbf{t}\in T} c_{\mathbf{t}} \psi(B_{j_1})^{t_1} \psi(A_{j_1})^{t_1} \cdots \psi(B_{j_m})^{t_m} \psi(A_{j_m})^{t_m}\,\mu = \\ &= \sum_{\mathbf{t}\in T} c_{\mathbf{t}} M_{x_{j_1}}^{t_1} \partial_{x_{j_1},q_{j_1}}^{t_1} \cdots M_{x_{j_m}}^{t_m} \partial_{x_{j_m},q_{j_m}}^{t_m}\,\mu = \\ &= \sum_{\substack{\mathbf{t}\in T \\ t_m = s_m}} c_{\mathbf{t}} \{s_m\}_{q_{j_m}}!\, M_{x_{j_1}}^{t_1} \partial_{x_{j_1},q_{j_1}}^{t_1} \cdots M_{x_{j_m}}^{t_m} x_{j_1}^{s_1} \cdots x_{j_{m-1}}^{s_{m-1}} = \\ &= \sum_{\substack{\mathbf{t}\in T \\ t_m = s_m}} c_{\mathbf{t}} \{s_m\}_{q_{j_m}}!\, M_{x_{j_1}}^{t_1} \partial_{x_{j_1},q_{j_1}}^{t_1} \cdots M_{x_{j_{m-1}}}^{t_{m-1}} \partial_{x_{j_{m-1}},q_{j_{m-1}}}^{t_{m-1}}\,\mu = \\ &= \cdots = \\ &= c_{\mathbf{s}} \{s_1\}_{q_{j_1}}! \cdots \{s_m\}_{q_{j_m}}!\, x_{j_1}^{s_1} \cdots x_{j_m}^{s_m} \end{aligned}$$

and since $c_{\mathbf{s}} \neq 0$, equality (8.25) can only hold when $\{s_k\}_{q_{j_k}}! = 0$ for some $k \in \{1, \dots, m\}$—that is, when $\{l\}_{q_{j_k}} = 0$ for some $l \in \{1, \dots, s_k\}$. That, in turn, is equivalent to that some q_{j_k} is of torsion type, which would be in contradiction with the choice of $\mathbf{q}$.

Thus the claim about α_w for $w = 0$ has been verified, but what of all other w? Well, by Corollary 4.12, there exist a $\beta \sqsubseteq K_0$ such that

$$\alpha_w = B_{j_m}^{u_m} \cdots B_{j_1}^{u_1} \beta A_{j_1}^{v_1} \cdots A_{j_m}^{v_m}$$

for $u_i = \max\{0, w(j_i)\}$ and $v_i = \max\{0, -w(j_i)\}$. This β can be treated just as α_0 was above and doing so produces an $\mathbf{s}$ such that, for $\mu = x_{j_1}^{s_1} \cdots x_{j_m}^{s_m}$,

$$\psi(\beta)\,\mu = b\,\{s_1\}_{q_{j_1}}! \cdots \{s_m\}_{q_{j_m}}!\, x_{j_1}^{s_1} \cdots x_{j_m}^{s_m}$$

and $b \neq 0$. Hence for $\nu = x_{j_1}^{s_1+v_1} \cdots x_{j_m}^{s_m+v_m}$,

$$\begin{aligned}
\psi(\alpha_w)\,\nu &= \psi(B_{j_m}^{u_m} \cdots B_{j_1}^{u_1})\psi(\beta)\psi(A_{j_1}^{v_1} \cdots A_{j_m}^{v_m})\,\nu = \\
&= \psi(B_{j_m}^{u_m} \cdots B_{j_1}^{u_1})\psi(\beta)\left(\prod_{i=1}^{m}\prod_{l=1}^{v_i}\{s_i + l\}_{q_{j_i}}\right)\mu = \\
&= \psi(B_{j_m}^{u_m} \cdots B_{j_1}^{u_1})\, b\,\{s_1 + v_1\}_{q_{j_1}}! \cdots \{s_m + v_m\}_{q_{j_m}}!\, x_{j_1}^{s_1} \cdots x_{j_m}^{s_m} = \\
&= b\,\{s_1 + v_1\}_{q_{j_1}}! \cdots \{s_m + v_m\}_{q_{j_m}}!\, x_{j_1}^{s_1+u_1} \cdots x_{j_m}^{s_m+u_m} \neq 0
\end{aligned}$$

and it has thus been shown that no nonzero α satisfy (8.24). This is equivalent to that the representation $\psi = \psi_{\mathbf{q},\mathfrak{x}_0}$ is faithful, so the 'if' case of the theorem has been shown.

To show the 'only if' case, suppose that there is $j \in J$ such that q_j is of torsion type and let n be the order of that q_j. Then, according to Theorem 8.2,

$$\psi_{\mathbf{q},\mathfrak{x}_0}(A_j^n) = \partial_{x_j,q_j}^n = 0.$$

Therefore, since $A_j^n \neq 0$ in $\mathcal{H}(\mathbf{q}, J)$, the representation $\psi_{\mathbf{q},\mathfrak{x}_0}$ of such $\mathcal{H}(\mathbf{q}, J)$ is not faithful. □

We complete this chapter with a theorem that is a direct corollary to Theorem 7.4, obtained by replacing all elements of $\mathcal{H}(q)$ by their images under the representation $\pi_{q,x}$ and then using the criterion for faithfulness of $\pi_{q,x}$ obtained in Theorem 8.3.

Theorem 8.4 *Let $q \in \mathcal{K}^*$ be of free type, and*

$$\alpha = \sum_{0 \le j \le n} a_j(M_x)\partial^j_{x,q}$$

$$\beta = \sum_{0 \le k \le m} b_k(M_x)\partial^k_{x,q}$$

be two q-difference operators on $P\big[\{x\}\big]$ with $\{a_j\}_{j=0}^n, \{b_k\}_{k=0}^m \subset \mathcal{K}[x]$.

If $[\alpha, \beta] = 0$ then there exists a nonzero polynomial $Q \in \mathcal{K}[y_1, y_2]$ such that $Q(\alpha, \beta) = 0$.

Theorem 5.1 [illegible] of finite type, and

$$[illegible]$$

$$[illegible]$$

be [illegible] [illegible] $\{\gamma_i\}$ [illegible] $\in A[x]$.
[illegible] then exists a nonzero polynomial $Q \in A[x_i, y_j]$ such that $Q(\ldots) = 0$ [illegible]

Appendix A
The Diamond Lemma

The main results in this paper are trivial. But what is trivial when described in the abstract can be far from clear in the context of a complicated situation where it is needed. Hence it seems worthwhile to set down explicit formulations and proofs of these results. — from [32] by G. M. BERGMAN

When Švejk subsequently described life in the lunatic asylum, he did so in exceptionally eulogistic terms: 'I really don't know why those loonies get so angry when they're kept there. You can crawl naked on the floor, howl like a jackal, rage and bite. If anyone did this anywhere on the promenade people would be astonished, but there it's the most common or garden thing to do. ...'

— from *The Good Soldier Švejk* by J. HAŠEK,
translation by C. PARROTT

The Diamond Lemma for Ring Theory is a powerful tool with numerous applications throughout the theory of associative rings. The main advantage with using this theorem in proofs, rather than relying directly on elementary methods, is that it often takes care of most of the technicalities. Thus one's attention can be focused on the parts of the proofs that actually are more specific to the problem in question.

The foremost source of information about the Diamond Lemma is Bergman's paper [32], in which the Lemma is not only proved but also applied to a wide variety of problems. Our presentation of the Lemma is mainly aimed at making this book self-contained. Hence we do give a complete proof of the Lemma, but restrict examples of how it is used to our own applications of it in Theorems 3.1–3.3.

It should be observed that the notations used in [32] differ from ours—there are even a few insignificant differences in the definitions—but, to aid readers familiar with [32], we have explained all of these differences in

Section A.2. We recommend such readers to skip to that section right now, as that will probably help them to understand our notations faster than would reading through all the definitions and proofs in Section A.1.

Finally, a few words about the name "Diamond Lemma". 'Diamond' is as in "diamond-shaped", so it has more to do with the symbol $\diamondsuit$ (used for example to mark one of the suits of cards) than with the jewel (even though the Lemma itself can sure be a veritable gem at times). The name Diamond Lemma is actually inherited from an earlier, more general but for ring theory less practical, result by Newman [297].

A.1 Definitions and proofs

For this section, let $\mathcal{R}$ be an associative and commutative ring with unit, and let X be a nonempty set. What will be considered is $\mathcal{R}\langle X\rangle$—the free associative $\mathcal{R}$-algebra on X with unit 1. Let Y be the set of monomials of $\mathcal{R}\langle X\rangle$ and note that Y is a basis for $\mathcal{R}\langle X\rangle$ seen as an $\mathcal{R}$-module. Finally, let the function $M\colon \mathcal{R}\langle X\rangle \longrightarrow 2^Y = \{\, Z \mid Z \subseteq Y \,\}$ be defined by that

$$M(\alpha) = \bigcap_{\substack{Z\subseteq Y\\ \alpha\in\mathrm{Span}(Z)}} Z = \text{the least } Z \subseteq Y \text{ such that } \alpha \in \mathrm{Span}(Z)$$

for all $\alpha \in \mathcal{R}\langle X\rangle$. Note that $M(\alpha)$ is always finite and that $M(\alpha) = \varnothing$ if and only if $\alpha = 0$.

Definition A.1 Let $S \subseteq Y \times \mathcal{R}\langle X\rangle$. Such a set S is called a **reduction system** for $\mathcal{R}\langle X\rangle$ and the elements of S are called **rules**. Let s be a rule. Then the first component of s will be written μ_s and the second component will be written α_s. Thus $s = (\mu_s, \alpha_s)$.

For each reduction system S there exists a corresponding ideal $\mathcal{I}(S)$, which is the two-sided ideal that is generated by $\{\, \alpha_s - \mu_s \mid s \in S \,\}$.

Definition A.2 Let S be a reduction system. Let $s \in S$. Let $\lambda, \nu \in \mathcal{R}\langle X\rangle$ be monomials.

Denote by $t_{\lambda s\nu}$ the module homomorphism $\mathcal{R}\langle X\rangle \longrightarrow \mathcal{R}\langle X\rangle$ that satisfies

$$t_{\lambda s\nu}(\mu) = \begin{cases} \lambda\alpha_s\nu & \text{if } \mu = \lambda\mu_s\nu \\ \mu & \text{otherwise} \end{cases}$$

for all monomials $\mu \in \mathcal{R}\langle X\rangle$.

Let $T_0(S) = \{\mathrm{id}\}$, where $\mathrm{id}\colon \mathcal{R}\langle X\rangle \longrightarrow \mathcal{R}\langle X\rangle$ is the identity map. Let

$$T_1(S) = \left\{\, t_{\lambda s \nu} \,\middle|\, \lambda, \nu \in \mathcal{R}\langle X\rangle \text{ are monomials and } s \in S \,\right\}. \tag{A.1}$$

Recursively define

$$T_{n+1}(S) = \left\{\, t_1 \circ t_2 \,\middle|\, t_1 \in T_1(S) \text{ and } t_2 \in T_n(S) \,\right\} \tag{A.2}$$

for all $n \in \mathbb{Z}^+$. Set

$$T(S) = \bigcup_{n\in\mathbb{N}} T_n(S). \tag{A.3}$$

The elements of $T(S)$ are called **reductions**.

Definition A.3 Let S be a reduction system. Let $\alpha \in \mathcal{R}\langle X\rangle$.

If $t(\alpha) = \alpha$ for all $t \in T(S)$ then α is called **irreducible**. The set of all irreducible elements in $\mathcal{R}\langle X\rangle$ is denoted $\mathrm{Irr}(S)$, or $\mathrm{Irr}\bigl(S, \mathcal{R}\langle X\rangle\bigr)$ if one wants to stress the algebra.

If for all $t_1 \in T(S)$ there exists some $t_2 \in T(S)$ such that $t_2\bigl(t_1(\alpha)\bigr) \in \mathrm{Irr}(S)$, then α is called **persistently reducible**. If $t_1(\alpha) = t_2(\alpha)$ for all $t_1, t_2 \in T(S)$ such that both $t_1(\alpha)$ and $t_2(\alpha)$ belong to $\mathrm{Irr}(S)$, then α is called **uniquely reducible**. The set of all elements in $\mathcal{R}\langle X\rangle$ that is both uniquely and persistently reducible is denoted $\mathrm{Red}(S)$, or $\mathrm{Red}\bigl(S, \mathcal{R}\langle X\rangle\bigr)$ if one wants to stress the algebra.

The map $t^S\colon \mathrm{Red}(S) \longrightarrow \mathrm{Irr}(S)$ is defined by that, for any $\alpha \in \mathrm{Red}(S)$, the element $t^S(\alpha)$ is the unique element in $\mathrm{Irr}(S)$ that α can be mapped to by a function $t \in T(S)$. The element $t^S(\alpha)$ is called the **normal form** of $\alpha \in \mathcal{R}\langle X\rangle$.

Lemma A.1 *Let S be a reduction system. Then* $\mathrm{Irr}(S)$ *and* $\mathrm{Red}(S)$ *are both submodules of* $\mathcal{R}\langle X\rangle$. *Furthermore* $t^S\colon \mathrm{Red}(S) \longrightarrow \mathrm{Irr}(S)$ *is a module homomorphism.*

Proof. Let $\alpha, \beta \in \mathrm{Irr}(S)$, $r \in \mathcal{R}$, and $t \in T(S)$ be arbitrary. Then

$$\begin{aligned} t(\alpha + \beta) &= t(\alpha) + t(\beta) = \alpha + \beta, \\ t(r\alpha) &= rt(\alpha) = r\alpha. \end{aligned}$$

Hence $\mathrm{Irr}(S)$ is a submodule of $\mathcal{R}\langle X\rangle$.

Let $\alpha, \beta \in \mathrm{Red}(S)$ and $r \in \mathcal{R}$ be arbitrary. Let $t_1 \in T(S)$ be arbitrary. Let $t_2 \in T(S)$ be such that $(t_2 \circ t_1)(\alpha) \in \mathrm{Irr}(S)$. Let $t_3 \in T(S)$ be such that $(t_3 \circ t_2 \circ t_1)(\beta) \in \mathrm{Irr}(S)$. Then

$$\begin{aligned}(t_3 \circ t_2 \circ t_1)(\alpha + \beta) &= (t_3 \circ t_2 \circ t_1)(\alpha) + (t_3 \circ t_2 \circ t_1)(\beta) = \\ &= t_3\bigl((t_2 \circ t_1)(\alpha)\bigr) + (t_3 \circ t_2 \circ t_1)(\beta) = \\ &= (t_2 \circ t_1)(\alpha) + (t_3 \circ t_2 \circ t_1)(\beta) = \\ &= t^S(\alpha) + t^S(\beta) \in \mathrm{Irr}(S) \qquad \text{(A.4)}\end{aligned}$$

$$(t_2 \circ t_1)(r\alpha) = r(t_2 \circ t_1)(\alpha) = rt^S(\alpha) \in \mathrm{Irr}(S) \qquad \text{(A.5)}$$

and hence both the sum of two elements in $\mathrm{Red}(S)$ and any scalar multiple of an element in $\mathrm{Red}(S)$ are always persistently reducible.

Now let $t_1 \in T(S)$ be any reduction for which $t_1(\alpha + \beta) \in \mathrm{Irr}(S)$. For these reductions, (A.4) implies that

$$t_1(\alpha + \beta) = (t_3 \circ t_2 \circ t_1)(\alpha + \beta) = t^S(\alpha) + t^S(\beta)$$

and hence $\alpha + \beta$ is uniquely reducible. If instead the reductions t_1 such that $t_1(r\alpha) \in \mathrm{Irr}(S)$ are considered, equation (A.5) will imply that

$$t_1(r\alpha) = (t_2 \circ t_1)(r\alpha) = rt^S(\alpha),$$

which proves that $r\alpha$ is uniquely reducible. Thus $\mathrm{Red}(S)$ is indeed a submodule of $\mathcal{R}\langle X \rangle$.

Once it has been verified that reduction is unique, the above equations also imply that $t^S(\alpha + \beta) = t^S(\alpha) + t^S(\beta)$ and $t^S(r\alpha) = rt^S(\alpha)$. Hence t^S is a module homomorphism. □

Definition A.4 Let S be a reduction system and let $s_1, s_2 \in S$.

If $\nu_1, \nu_2, \nu_3 \in \mathcal{R}\langle X \rangle$ are monomials not equal to 1 such that $\mu_{s_1} = \nu_1\nu_2$ and $\mu_{s_2} = \nu_2\nu_3$ then $(s_1, s_2, \nu_1, \nu_2, \nu_3)$ is said to be an **overlap ambiguity**. An overlap ambiguity $(s_1, s_2, \nu_1, \nu_2, \nu_3)$ is said to be **resolvable** if there exists $t_1, t_2 \in T(S)$ such that $t_1(\alpha_{s_1}\nu_3) = t_2(\nu_1\alpha_{s_2})$.

If $\nu_1, \nu_2, \nu_3 \in \mathcal{R}\langle X \rangle$ are monomials such that $\mu_{s_1} = \nu_1\nu_2\nu_3$ and $\mu_{s_2} = \nu_2$ then $(s_1, s_2, \nu_1, \nu_2, \nu_3)$ is said to be an **inclusion ambiguity**. An inclusion ambiguity $(s_1, s_2, \nu_1, \nu_2, \nu_3)$ is said to be **resolvable** if there exists $t_1, t_2 \in T(S)$ such that $t_1(\alpha_{s_1}) = t_2(\nu_1\alpha_{s_2}\nu_3)$.

The concepts defined so far are the ones which the Diamond Lemma makes claims about. The concepts that follow ensure that there is enough structure on, in particular, the reduction system for the proof of the Lemma to go through.

Definition A.5 Let Z be an arbitrary set and let $\preceq$ be a partial order on Z. Then $\preceq$ is said to satisfy the **descending chain condition (DCC)** if, for every sequence $\{z_i\}_{i=0}^{\infty} \subseteq Z$ such that $z_i \succeq z_{i+1}$ for all $i \in \mathbb{N}$, there is an $n \in \mathbb{N}$ such that $z_{i+1} = z_i$ for all $i \geqslant n$. Considered as a partially ordered set, Z is said to be **Artinian** if $\preceq$ satisfies the descending chain condition.

A partial order $\preceq$ that satisfies the descending chain condition has some properties in common with a well-order, most notably that every nonempty subset of the domain of $\preceq$ has a $\preceq$-minimal element. To see this, assume that Z is the domain of $\preceq$ and that $Z' \subseteq Z$ is nonempty and has no $\preceq$-minimal element. Since Z' is nonempty, there is a $z_0 \in Z'$. Now if a $z_i \in Z'$ is not minimal in Z' then there exists a $z_{i+1} \in Z'$ such that $z_{i+1} \prec z_i$. Repeating the last step for $i = 0, 1, 2, \ldots$ generates an infinite strictly descending chain $z_0 \succ z_1 \succ z_2 \succ z_3 \succ \cdots$, quite in contradiction to $\preceq$ satisfying the descending chain condition. Hence the descending chain condition on $\preceq$ is enough to guarantee the existence of $\preceq$-minimal elements.

The main use of this minimal element property is that it allows one to perform induction. The argument for this is the normal deduction of the principle of mathematical induction from the principle of the least counter-example, but it gets a slightly different form due to that $\preceq$ need not be a total order: Let Z be an arbitrary set and let $\preceq$ be a partial order on Z that satisfies the DCC. Assume that W is a subset of Z with the property that

$$\{\, z \in Z \mid z \prec z_0 \,\} \subseteq W \implies z_0 \in W$$

for all $z_0 \in Z$. Then $W = Z$, because otherwise $Z \setminus W$ is nonempty and hence there is a $\preceq$-minimal element $z_0 \in Z \setminus W$. Since z_0 is minimal in $Z \setminus W$, it follows that all $z \in Z$ satisfying $z \prec z_0$ are in W. Hence by the property of W assumed above, it follows that $z_0 \in W$, in contradiction with the choice of z_0. Indeed, the only possibility is the above claimed $W = Z$.

Definition A.6 Let $\mathcal{Z}$ be a semigroup. Then a partial order $\leqslant$ on $\mathcal{Z}$ is said to be a **semigroup partial order** if for all $\lambda, \mu, \nu \in \mathcal{Z}$ such that $\mu \leqslant \nu$, it holds that $\lambda\mu \leqslant \lambda\nu$ and $\mu\lambda \leqslant \nu\lambda$.

Definition A.7 Let S be a reduction system. Let $\leqslant$ be a semigroup partial order on Y seen as the semigroup of all monomials in the free algebra $\mathcal{R}\langle X\rangle$. Then $\leqslant$ is said to be **compatible** with S if $\nu < \mu_s$ for all $\nu \in M(\alpha_s)$ for every $s \in S$.

Lemma A.2 *Let S be a reduction system for $\mathcal{R}\langle X\rangle$. Let $\leqslant$ be a semigroup partial order that is compatible with S and satisfies the descending chain condition. Then every element in $\mathcal{R}\langle X\rangle$ is persistently reducible.*

Proof. The proof is by induction over the monomials of $\mathcal{R}\langle X\rangle$. Once it has been shown that every monomial in $\mathcal{R}\langle X\rangle$ is persistently reducible then it will follow that all elements of $\mathcal{R}\langle X\rangle$ are persistently reducible since the set of persistently reducible elements is closed under addition and multiplication by a scalar.

Let $\mu \in Y$ be arbitrary and assume that every $\nu \in Y$ such that $\nu < \mu$ is persistently reducible; it will be shown that this implies the persistent reducibility of μ. To this end, let $t_1 \in T(S)$ be given; it will now be shown that there exists some $t_2 \in T(S)$ such that $t_2\bigl(t_1(\mu)\bigr) \in \operatorname{Irr}(S)$. If $\mu \in \operatorname{Irr}(S)$ then $t_1(\mu) \in \operatorname{Irr}(S)$ as well and one may choose $t_2 = \mathrm{id}$. Otherwise there exists some $t_{2,0} \in T_1(S)$ which acts nontrivially on μ.

Let $\{\nu_1, \dots, \nu_n\} = M\bigl((t_{2,0} \circ t_1)(\mu)\bigr)$. By the compatibility of S with $\leqslant$, it must be the case that $\nu_i < \mu$ for $i = 1, \dots, n$, since either $t_{2,0}$ or some simple reduction in t_1 will have acted nontrivially on μ. Hence by the induction hypothesis, all the ν_i are persistently reducible. This implies that there is some $t_{2,1} \in T(S)$ such that $t_{2,1}(\nu_1) \in \operatorname{Irr}(S)$ and similarly that there is some $t_{2,i} \in T(S)$ such that $t_{2,i}\bigl((t_{2,i-1} \circ \cdots \circ t_{2,1})(\nu_i)\bigr) \in \operatorname{Irr}(S)$ for $i = 2, \dots, n$.

Let $t_2 = t_{2,n} \circ \cdots \circ t_{2,1} \circ t_{2,0}$ and let $\{r_i\}_{i=1}^n \subseteq \mathcal{R}$ be such that

$$(t_{2,0} \circ t_1)(\mu) = \sum_{i=1}^{n} r_i \nu_i.$$

Then

$$\begin{aligned} t_2\bigl(t_1(\mu)\bigr) &= (t_{2,n} \circ \cdots \circ t_{2,1})\left(\sum_{i=1}^{n} r_i \nu_i\right) = \\ &= \sum_{i=1}^{n} r_i (t_{2,n} \circ \cdots \circ t_{2,1})(\nu_i) = \end{aligned}$$

$$= \sum_{i=1}^{n} r_i(t_{2,n} \circ \cdots \circ t_{2,i+1})\big(\underbrace{(t_{2,i} \circ \cdots \circ t_{2,1})(\nu_i)}_{\in \mathrm{Irr}(S)}\big) =$$

$$= \sum_{i=1}^{n} r_i(t_{2,i} \circ \cdots \circ t_{2,1})(\nu_i) \in \mathrm{Irr}(S)$$

and hence μ is indeed persistently reducible. This completes the inductive step, and thus, by induction, the lemma follows. □

Theorem A.3 (the Diamond Lemma) *Let S be a reduction system for $\mathcal{R}\langle X\rangle$. Let $\leqslant$ be a semigroup partial order that is compatible with S and satisfies the descending chain condition.*

Then the following three conditions are equivalent:

(i) *All ambiguities of S are resolvable.*
(ii) $\mathrm{Red}(S) = \mathcal{R}\langle X\rangle$.
(iii) $\mathcal{R}\langle X\rangle = \mathrm{Irr}(S) \oplus \mathcal{I}(S)$.

Note: The implications (ii) $\Longrightarrow$ (iii) and (ii) $\Longrightarrow$ (i) hold even if there is no partial order with which S is compatible.

Proof. Assume condition (ii), that $\mathrm{Red}(S) = \mathcal{R}\langle X\rangle$. Then t^S is a projection of $\mathcal{R}\langle X\rangle$ onto $\mathrm{Irr}(S)$. It is quite clear that the kernel of t^S is a subset of $\mathcal{I}(S)$, since every $t_{\lambda s \nu} \in T_1(S)$ can change its argument only by a scalar multiple of $\lambda(\alpha_s - \mu_s)\nu$ and all such elements belong to $\mathcal{I}(S)$. On the other hand, the kernel of t^S is a submodule of $\mathcal{R}\langle X\rangle$, and in particular the kernel is a submodule that contains all elements of the form $\lambda(\alpha_s - \mu_s)\nu$, where $s \in S$ and $\lambda, \nu \in \mathcal{R}\langle X\rangle$ are monomials. Since it can be shown that these elements span the ideal $\mathcal{I}(S)$, it follows that the kernel of t^S is $\mathcal{I}(S)$. Hence $\mathrm{Irr}(S) \oplus \mathcal{I}(S) = \mathcal{R}\langle X\rangle$. Thus condition (ii) implies condition (iii).

Again assume condition (ii). Let $(s_1, s_2, \nu_1, \nu_2, \nu_3)$ be any overlap ambiguity. Then persistent reducibility implies that there exists $t_1, t_2 \in T(S)$ such that both $t_1(\alpha_{s_1}\nu_3)$ and $t_2(\nu_1\alpha_{s_2})$ belong to $\mathrm{Irr}(S)$. Furthermore, unique reducibility of $\nu_1\nu_2\nu_3$ implies that

$$t_1(\alpha_{s_1}\nu_3) = (t_1 \circ t_{1 s_1 \nu_3})(\nu_1\nu_2\nu_3) = (t_2 \circ t_{\nu_1 s_2 1})(\nu_1\nu_2\nu_3) = t_2(\nu_1\alpha_{s_2}),$$

hence $(s_1, s_2, \nu_1, \nu_2, \nu_3)$ is resolvable.

Now let $(s_1, s_2, \nu_1, \nu_2, \nu_3)$ be any inclusion ambiguity. Then the same arguments imply the existence of $t_1, t_2 \in T(S)$ such that both $t_1(\alpha_{s_1})$ and $t_2(\nu_1\alpha_{s_2}\nu_3)$ belong to $\mathrm{Irr}(S)$. Again unique reducibility implies that

$$t_1(\alpha_{s_1}) = (t_1 \circ t_{1s_11})(\nu_1\nu_2\nu_3) = (t_2 \circ t_{\nu_1 s_2 \nu_3})(\nu_1\nu_2\nu_3) = t_2(\nu_1\alpha_{s_2}\nu_2),$$

hence $(s_1, s_2, \nu_1, \nu_2, \nu_3)$ is resolvable. Thus condition (ii) implies condition (i).

To prove the reversed implications though, one need to use that every element of $\mathcal{R}\langle X\rangle$ is persistently reducible by Lemma A.2.

Next assume condition (iii). Let $\alpha \in \mathcal{R}\langle X\rangle$ be arbitrary. Let $t_1, t_2 \in T(S)$ be such that both $t_1(\alpha)$ and $t_2(\alpha)$ belong to $\mathrm{Irr}(S)$. Then

$$t_1(\alpha) - t_2(\alpha) = \big(t_1(\alpha) - \alpha\big) - \big(t_2(\alpha) - \alpha\big) \in \mathrm{Irr}(S) \cap \mathcal{I}(S) = \{0\}.$$

Hence α is uniquely reducible and it follows that $\mathrm{Red}(S) = \mathcal{R}\langle X\rangle$. Thus condition (iii) implies condition (ii).

Finally assume condition (i). Since $\mathrm{Red}(S)$ by Lemma A.1 is a submodule of $\mathcal{R}\langle X\rangle$, condition (ii) would follow if it could be proved that all monomials of $\mathcal{R}\langle X\rangle$ belong to $\mathrm{Red}(S)$. This in turn can be proved using induction, thanks to the fact that $\leqslant$ satisfies DCC.

Consider a monomial $\mu \in \mathcal{R}\langle X\rangle$, and assume that all monomials μ' that satisfy $\mu' < \mu$ also satisfy $\mu' \in \mathrm{Red}(S)$. If there is no more than one reduction $t \in T_1(S)$ that acts nontrivially on μ, then μ is uniquely reducible. Hence assume that there are two different $s_1, s_2 \in S$ and monomials $\lambda, \lambda', \nu, \nu' \in \mathcal{R}\langle X\rangle$ such that $\lambda\mu_{s_1}\nu' = \mu = \lambda'\mu_{s_2}\nu$. Then both $t_{\lambda s_1 \nu'}$ and $t_{\lambda' s_2 \nu}$ will act nontrivially on μ.

Without loss of generality, s_1 and s_2 can be chosen such that $\lambda' = \lambda\kappa$ for some monomial κ (the alternative would be $\lambda = \lambda'\kappa$). The inductive hypothesis implies that both $t_{\lambda s_1 \nu'}(\mu)$ and $t_{\lambda' s_2 \nu}(\mu)$ belong to $\mathrm{Red}(S)$. What shall be proved is that

$$t^S\big(t_{\lambda s_1 \nu'}(\mu)\big) = t^S\big(t_{\lambda' s_2 \nu}(\mu)\big).$$

It turns out there are three cases.

Case 1. $\mu = \lambda\kappa_1\kappa_2\kappa_3\nu$, where κ_1, κ_2, and κ_3 are monomials such that $\mu_{s_1} = \kappa_1\kappa_2$ and $\mu_{s_2} = \kappa_2\kappa_3$; in other words there is an overlap ambiguity. Condition (i) states that this ambiguity is resolvable, hence $t_{\lambda s_1 \nu'}(\mu) = \lambda\alpha_{s_1}\kappa_3\nu$ and $t_{\lambda' s_2 \nu}(\mu) = \lambda\kappa_1\alpha_{s_2}\nu$ can be reduced to some common element

$\beta \in \mathrm{Red}(S)$. But then

$$t^S(\lambda\alpha_{s_1}\kappa_3\nu) = t^S(\beta) = t^S(\lambda\kappa_1\alpha_{s_2}\nu)$$

and both $t^S \circ t_{\lambda s_1 \nu'}$ and $t^S \circ t_{\lambda' s_2 \nu}$ reduce μ to the same irreducible element.

Case 2. $\mu = \lambda\kappa_1\kappa_2\kappa_3\nu'$, where κ_1, κ_2, and κ_3 are monomials such that $\mu_{s_1} = \kappa_1\kappa_2\kappa_3$ and $\mu_{s_2} = \kappa_2$; in other words there is an inclusion ambiguity. Condition (i) states that this ambiguity is resolvable, hence $t_{\lambda s_1 \nu'}(\mu) = \lambda\alpha_{s_1}\nu'$ and $t_{\lambda' s_2 \nu}(\mu) = \lambda\kappa_1\alpha_{s_2}\kappa_3\nu'$ can be reduced to some common element $\beta \in \mathrm{Red}(S)$. But then

$$t^S(\lambda\alpha_{s_1}\nu') = t^S(\beta) = t^S(\lambda\kappa_1\alpha_{s_2}\kappa_3\nu')$$

and both $t^S \circ t_{\lambda s_1 \nu'}$ and $t^S \circ t_{\lambda' s_2 \nu}$ reduce μ to the same irreducible element.

Case 3. $\mu = \lambda\kappa_1\kappa_2\kappa_3\nu$, where κ_1, κ_2, and κ_3 are monomials such that $\mu_{s_1} = \kappa_1$ and $\mu_{s_2} = \kappa_3$. In this case it is clear that there are $t_1, t_2 \in T(S)$ such that

$$t_1(\lambda\alpha_{s_1}\kappa_2\kappa_3\nu) = \lambda\alpha_{s_1}\kappa_2\alpha_{s_2}\nu = t_2(\lambda\kappa_1\kappa_2\alpha_{s_2}\nu).$$

Therefore both $t^S \circ t_{\lambda s_1 \nu'}$ and $t^S \circ t_{\lambda' s_2 \nu}$ reduce μ to the same irreducible element, just as in the previous cases.

To finish off the line of reasoning, it follows that all reductions $t \in T(S)$ such that $t(\mu) \in \mathrm{Irr}(S)$ reduce μ to the same thing, i.e., μ is uniquely reducible. Hence $\mu \in \mathrm{Red}(S)$. Since this can be applied to any monomial μ, the principle of induction implies that all monomials $\mu \in \mathcal{R}\langle X\rangle$ satisfy $\mu \in \mathrm{Red}(S)$. Hence $\mathrm{Red}(S) = \mathcal{R}\langle X\rangle$.

Thus condition (i) implies condition (ii), and it follows that conditions (i)–(iii) are equivalent. □

A.2 A short key to the notations

The notations in this book are directly based on those in the earlier paper [182] by one of us. Therefore readers familiar with Bergman's notation will probably find Table A.1 useful, as it compares the two notation systems.

There are also some terms which are almost equivalent, but where the very real differences might appear somewhat obscure. This explains these differences more precisely:

Our symbol	Description	Symbol in [32]
$\mathcal{R}$	a commutative associative ring with unit	k
Y	the set of monomials in $\mathcal{R}\langle X\rangle$	$\langle X\rangle$
$s = (\mu_s, \alpha_s)$	element in a reduction system	$\sigma = (W_\sigma, f_\sigma)$
$\mathcal{I}(S)$	ideal generated by $\{\, a_s - \mu_s \mid s \in S \,\}$	I
κ, λ, μ, ν	monomials	A, B, C, K, L
α, β	elements in $\mathcal{R}\langle X\rangle$	a, b, c
r	element in $\mathcal{R}$	α
$t_{\lambda s \nu}$	reduction mapping $\lambda\mu_s\nu$ to $\lambda\alpha_s\nu$ (or $AW_\sigma B$ to $Af_\sigma B$)	$r_{A\sigma B}$
$T(S)$	set of all reductions	—
$T_1(S)$	set of all simple reductions	—
$\mathrm{Irr}(S)$	the set of all irreducible elements in $\mathcal{R}\langle X\rangle$	$k\langle X\rangle_{\mathrm{irr}}$
$\mathrm{Red}(S)$	the set of all persistently and uniquely reducible elements in $\mathcal{R}\langle X\rangle$	—
t^S	the map that takes everything to its normal form	r_S
—	the submodule of $\mathcal{R}\langle X\rangle$ that is spanned by all elements $A(W_\sigma - f_\sigma)C$ such that A and C are monomials, $\sigma \in S$, and $AW_\sigma C < D$	I_D

The symbol — in the table means that there is no symbol for this concept in that particular paper.

Table A.1 A comparison of our notations and the notations in [32]

reduction In [32], a *reduction* is a map $r_{A\sigma B}$ (Bergman's notation); in our book, a *reduction* can be any finite composition of such maps. When we need to specify that a reduction t is such a simple reduction, then this is written as $t \in T_1(S)$.

persistently reducible There is no concept in [32] which is equivalent to this, instead *reduction-finite* is used. These two concepts fill the same function in the theory, namely to ensure that $\mathrm{Red}(S)$ and *the set of reduction-unique elements of* $k\langle X\rangle$ respectively are modules, but *persistently reducible* is weaker.

A reduction system S for which the differences are displayed is

$$S = \bigcup_{n=0}^{\infty} \{(\mathsf{a}_n, \mathsf{a}_{n+1}), (\mathsf{a}_n, \mathsf{b})\}.$$

With this reduction system, a_0 will be persistently reducible, but not reduction-finite. It should however be noticed that the conditions in Lemma A.2, as well as those in the Diamond Lemma itself, are sufficiently strong to imply that every element in $\mathcal{R}\langle X \rangle$ is reduction-finite. Thus Lemma A.2 could alternatively be proved by first showing that every element in $\mathcal{R}\langle X \rangle$ is reduction-finite and then using that this implies persistent reducibility.

uniquely reducible This is basically the same thing as *reduction-unique*, as found in [32], i.e., there is only one irreducible element to which a uniquely reducible element can be reduced. The difference is that *reduction-unique* is defined so that it implies reduction-finiteness, whilst *uniquely reducible* does not imply anything like that, not even persistent reducibility.
We define $\mathrm{Red}(S)$ to be the set of all persistently and uniquely reducible elements in $\mathcal{R}\langle X \rangle$.

Some yet other terminology has been used by Roitman [326], which presents a slightly different version of the Diamond Lemma. The word *terminal* is used instead of irreducible. A reduction system is called a *rewriting system* if every element of $\mathcal{R}\langle X \rangle$ is reduction-finite. A rewriting system is called *confluent* if $\mathrm{Red}(S) = \mathcal{R}\langle X \rangle$.

A.3 A few extra results

This section contains some results that are directly connected to the apparatus that has been developed around the Diamond Lemma. They are used to simplify some of the proofs and to strengthen some of the results that are proved using the Diamond Lemma.

Lemma A.4 *Let $\mathcal{R}$ be an associative and commutative ring with unit. Let X be a nonempty set, let $\mathcal{X}$ be a partition of X, and let $<$ be a total order on $\mathcal{X}$. Let S be a reduction system for $\mathcal{R}\langle X \rangle$ such that $(1, \alpha) \notin S$ for any α.*

If S can be partitioned as

$$S = C \cup \left(\bigcup_{Y \in \mathcal{X}} S_Y \right) \tag{A.6}$$

so that

- *every S_Y is a reduction system for $\mathcal{R}\langle Y \rangle$, and*
- $C = \{ (\mu_2\mu_1, \mu_1\mu_2) \mid \mu_1 \in Y_1 \in \mathcal{X},\ \mu_2 \in Y_2 \in \mathcal{X},$ *and* $Y_1 < Y_2 \}$,

then every ambiguity in S is resolvable if and only if every ambiguity in every S_Y is resolvable.

Proof. To begin with the 'if' part, it is easy to see that the possible cases of ambiguity fall into the following four categories: (i) between two elements of C, (ii) between an element of C and an element of S_Y, (iii) between two elements in different S_Y's, and (iv) between two elements of the same S_Y. The 'if' part of the lemma is clearly proved if all ambiguities in categories (i)–(iii) are proved resolvable, because resolvability of the fourth category is then an assumption of the lemma.

An ambiguity in category (i) has to be an overlap ambiguity, because all left hand sides are the same length and $s_1, s_2 \in C$ are equal if and only if $\mu_{s_1} = \mu_{s_2}$. Thus the only possible ambiguity is $(s_2, s_1, \nu_3, \nu_2, \nu_1)$, where

$$\begin{aligned} &\nu_1 \in Y_1 \in \mathcal{X} && \nu_2 \in Y_2 \in \mathcal{X} \\ &\nu_3 \in Y_3 \in \mathcal{X} && Y_1 < Y_2 < Y_3 \\ &s_1 = (\nu_2\nu_1, \nu_1\nu_2) && s_2 = (\nu_3\nu_2, \nu_2\nu_3). \end{aligned}$$

This ambiguity is resolved as follows

$$\begin{aligned} \nu_3\nu_2\nu_1 &\overset{s_1}{\longmapsto} \nu_3\nu_1\nu_2 \longmapsto \nu_1\nu_3\nu_2 \longmapsto \nu_1\nu_2\nu_3, \\ \nu_3\nu_2\nu_1 &\overset{s_2}{\longmapsto} \nu_2\nu_3\nu_1 \longmapsto \nu_2\nu_1\nu_3 \longmapsto \nu_1\nu_2\nu_3. \end{aligned}$$

An ambiguity in category (ii) must have one of the forms

$$(s_1, s_2, \nu_2, \nu_1, \lambda) \quad \text{and} \quad (s_2, s_1, \lambda, \nu_1, \nu_2),$$

where $s_1 \in C$, $s_2 \in S_{Y_1}$, $\nu_2 \in Y_2 \in \mathcal{X}$, $\nu_1 \in Y_1 \in \mathcal{X}$, and $\lambda \in \mathcal{R}\langle Y_1 \rangle$. Furthermore, in the left case $Y_1 < Y_2$ and in the right case $Y_1 > Y_2$. The ambiguity might be either an overlap ambiguity or an inclusion ambiguity, but in the latter case μ_{s_2} must be included in μ_{s_1} and λ must be 1.

The left kind of ambiguity is resolved as follows:

$$\nu_2\nu_1\lambda \overset{s_1}{\longmapsto} \nu_1\nu_2\lambda \overbrace{\longmapsto \cdots \longmapsto}^{\in C} \nu_1\lambda\nu_2 \overset{s_2}{\longmapsto} \alpha_{s_2}\nu_2,$$
$$\nu_2\nu_1\lambda \xmapsto{\qquad s_2 \qquad} \nu_2\alpha_{s_2} \overbrace{\longmapsto \cdots \longmapsto}^{\in C} \alpha_{s_2}\nu_2.$$

The right ambiguity is resolved similarly:

$$\lambda\nu_1\nu_2 \overset{s_1}{\longmapsto} \lambda\nu_2\nu_1 \overbrace{\longmapsto \cdots \longmapsto}^{\in C} \nu_2\lambda\nu_1 \overset{s_2}{\longmapsto} \nu_2\alpha_{s_2},$$
$$\lambda\nu_1\nu_2 \xmapsto{\qquad s_2 \qquad} \alpha_{s_2}\nu_2 \overbrace{\longmapsto \cdots \longmapsto}^{\in C} \nu_2\alpha_{s_2}.$$

There cannot be a category (iii) ambiguity, because the only monomial that is in both $\mathcal{R}\langle Y_1\rangle$ and $\mathcal{R}\langle Y_2\rangle$ for $Y_1 \neq Y_2$ is 1 which, by the assumptions, is not the μ_s for any $s \in S$. Thus all ambiguities in S are resolvable if all category (iv) ambiguities in S are resolvable, which is just what the lemma states.

To prove the 'only if' part, let $Y \in \mathcal{X}$ and let $(s_1, s_2, \nu_1, \nu_2, \nu_3)$ be an ambiguity such that $s_1, s_2 \in S_Y$. Furthermore assume that this ambiguity cannot be resolved in S_Y. It is then easy to see that this ambiguity is unresolvable in S as well, because the only reductions in $T_1(S)$ that act nontrivially on an element in $\mathcal{R}\langle Y\rangle$, as for example $\nu_1\nu_2\nu_3$, are the reductions in $T_1(S_Y)$ and these reductions map the element to another element in $\mathcal{R}\langle Y\rangle$. □

Lemma A.5 *Let $\mathcal{R}$ be an associative and commutative ring with unit. Let X and $X' \subseteq X$ be nonempty sets. Let S be a reduction system for $\mathcal{R}\langle X\rangle$ and let S' be defined by*

$$S' = \left\{ (\mu_s, \alpha_s) \in S \,\middle|\, \mu_s \in \mathcal{R}\langle X'\rangle \right\}. \tag{A.7}$$

Then $\operatorname{Red}\bigl(S', \mathcal{R}\langle X'\rangle\bigr) \subseteq \operatorname{Red}\bigl(S, \mathcal{R}\langle X\rangle\bigr)$ *if S' is a reduction system for $\mathcal{R}\langle X'\rangle$, i.e., if $\alpha_s \in \mathcal{R}\langle X'\rangle$ for all $s \in S'$.*

Proof. If $\lambda, \nu \in \mathcal{R}\langle X\rangle$ are monomials, $s \in S \setminus S'$, and $\mu \in \mathcal{R}\langle X'\rangle$ is a monomial then $t_{\lambda s\nu}(\mu) = \mu$, because $\lambda\mu_s\nu \notin \mathcal{R}\langle X'\rangle$ and hence most certainly $\lambda\mu_s\nu$ cannot be equal to μ. Hence $t(\alpha) = \alpha$ for all $\alpha \in \mathcal{R}\langle X'\rangle$ and $t \in T_1(S) \setminus T_1(S')$. It follows from this that for each $\alpha \in \mathcal{R}\langle X'\rangle$ and $t \in T(S)$ there is some $t' \in T(S')$ such that $t'(\alpha) = t(\alpha)$.

This has some consequences. An $\alpha \in \mathcal{R}\langle X'\rangle$ is irreducible by S if and only if it is irreducible by S', hence $\mathrm{Irr}(S', \mathcal{R}\langle X'\rangle) = \mathrm{Irr}(S, \mathcal{R}\langle X\rangle) \cap \mathcal{R}\langle X'\rangle$. Let $\alpha \in \mathrm{Red}(S', \mathcal{R}\langle X'\rangle)$. For each $t_1 \in T(S)$, there is some $t_1' \in T(S')$ such that $t_1'(\alpha) = t_1(\alpha)$, and hence there is also some $t_2 \in T(S')$ such that $t_2(t_1'(\alpha)) \in \mathrm{Irr}(S', \mathcal{R}\langle X'\rangle) \subseteq \mathrm{Irr}(S, \mathcal{R}\langle X\rangle)$. Thus every $\alpha \in \mathrm{Red}(S', \mathcal{R}\langle X'\rangle)$ is persistently reducible by S. Also, for every two reductions $t_1, t_2 \in T(S)$ such that both $t_1(\alpha)$ and $t_2(\alpha)$ belong to $\mathrm{Irr}(S, \mathcal{R}\langle X\rangle)$, there are reductions $t_1', t_2' \in T(S')$ such that $t_1'(\alpha) = t_1(\alpha)$ and $t_2'(\alpha) = t_2(\alpha)$. Hence $t_1'(\alpha)$ and $t_2'(\alpha)$ both belong to $\mathrm{Irr}(S', \mathcal{R}\langle X'\rangle)$, and it follows that

$$t_1(\alpha) = t_1'(\alpha) = t_2'(\alpha) = t_2(\alpha).$$

Thus $\alpha \in \mathrm{Red}(S, \mathcal{R}\langle X\rangle)$, and hence $\mathrm{Red}(S', \mathcal{R}\langle X'\rangle) \subseteq \mathrm{Red}(S, \mathcal{R}\langle X\rangle)$. □

Appendix B

Degree functions and gradations

> First the bad news: TEX is a large and complicated program that goes to extraordinary lengths to produce attractive typeset material. This very complication can cause unexpected things to happen at times. Now the good news: straightforward text is very easy to typeset using TEX. So it's possible to start with easier text and work up to more complicated situations.
>
> — from *A Gentle Introduction to TEX* by M. DOOB

In studying commutative polynomial rings, a frequently used concept is the *degree* of a polynomial. This appendix describes a construction by which a concept of degree can be defined for arbitrary associative algebras. This concept is defined without reference to a particular basis for the algebra, but it is also proved that given a suitable basis, the degree of an element in the algebra can be determined simply by inspection.

Several of the more advanced theorems are based on the apparatus around the Diamond Lemma, although not necessarily on the Lemma itself. Many of the results concerning degree in $\mathcal{H}(\mathbf{q}, J)$ in the main body of the book also make this combination. The notations used are the same as in Appendix A.

This appendix ends with some general results on gradations, a concept which prove to be just as important as degree throughout the book. That part is independent of the material on degree functions.

B.1 General theory of degree functions

This section develops a theory of degree functions that can be used in any associative algebra. In order to do this, recall the following definitions.

Definition B.1 Let $\mathcal{G}$ be a commutative semigroup (let $+$ denote its operation) such that if $g+h = g+k$ for some $g,h,k \in \mathcal{G}$, then $h = k$. Let $\leqslant$ be a semigroup partial order on $\mathcal{G}$. Let $\mathcal{R}$ be an associative and commutative ring with unit, and let $\mathcal{A}$ be an $\mathcal{R}$-algebra.

Let $\{\mathcal{A}_g\}_{g\in\mathcal{G}}$ be a family of submodules of $\mathcal{A}$. If

- $g \leqslant h$ implies $\mathcal{A}_g \subseteq \mathcal{A}_h$, and
- $\mathcal{A}_g \cdot \mathcal{A}_h \subseteq \mathcal{A}_{g+h}$

for all $g,h \in \mathcal{G}$ then $\{\mathcal{A}_g\}_{g\in\mathcal{G}}$ is called an **ascending $\mathcal{G}$-filtration in $\mathcal{A}$**. Furthermore, if it is also the case that

- $\displaystyle\bigcup_{g\in\mathcal{G}} \mathcal{A}_g = \mathcal{A}$

then $\{\mathcal{A}_g\}_{g\in\mathcal{G}}$ is called an **ascending $\mathcal{G}$-filtration on $\mathcal{A}$**, or alternatively an ascending $\mathcal{G}$-filtration *of* $\mathcal{A}$.

Definition B.2 Let $\mathcal{A}$ be an algebra and $X \subseteq \mathcal{A}$ a nonempty set. If $\mathcal{A}$ itself is the only subalgebra of $\mathcal{A}$ that contains X, then X is called a **generating set** for $\mathcal{A}$. Furthermore, an element in $\mathcal{A}$ is called a **monomial** (with respect to the generating set X) if it can be expressed as a (finite) product of elements in X.

Rather than repeating all prerequisites in each definition, lemma, and theorem below, we shall let some symbols keep their definitions throughout this section. Thus let $\mathcal{R}$ be an associative and commutative ring with unit, let $\mathcal{A}$ be an associative $\mathcal{R}$-algebra, and let X be a generating set for $\mathcal{A}$. Finally choose an arbitrary function $v\colon X \longrightarrow \mathbb{Z}$.

B.1.1 *A generalisation of the Bernstein filtration*

Here we describe the construction of the degree function and prove that it is well-defined. Construct the sets

$$P_i^1 = \{\, x \in X \mid v(x) \leqslant i \,\} \tag{B.1}$$

for all $i \in \mathbb{Z}$. Recursively construct the sets

$$P_i^{n+1} = P_i^n \cup \left(\bigcup_{x \in X} P_{i-v(x)}^n \cdot x \right) \tag{B.2}$$

for all $i \in \mathbb{Z}$ and $n \in \mathbb{Z}^+$. Finally let

$$P_i = \bigcup_{n \in \mathbb{Z}^+} P_i^n \tag{B.3}$$

for all $i \in \mathbb{Z}$.

Lemma B.1 *The following hold for all $i, j, k \in \mathbb{Z}$ and $n, m \in \mathbb{Z}^+$:*

(i) *If $i \leqslant j$ then $P_i^n \subseteq P_j^n$ and $P_i \subseteq P_j$.*
(ii) *If $i + j = k$ then $P_i^m \cdot P_j^n \subseteq P_k^{m+n}$ and $P_i \cdot P_j \subseteq P_k$.*

Proof. First, the statements about P_i^n will be proved by induction on n. The case $n = 1$ of (i) is trivial;

$$P_i^1 = \{\, x \in X \mid v(x) \leqslant i \,\} \subseteq \{\, x \in X \mid v(x) \leqslant j \,\} = P_j^1.$$

The case $n = l + 1 > 1$ follows from the case $n = l$ by

$$\begin{gathered} P_i^l \subseteq P_j^l, \\ P_{i-v(x)}^l \subseteq P_{j-v(x)}^l, \\ P_{i-v(x)}^l \cdot x \subseteq P_{j-v(x)}^l \cdot x, \\ \bigcup_{x \in X} P_{i-v(x)}^l \cdot x \subseteq \bigcup_{x \in X} P_{j-v(x)}^l \cdot x, \\ P_i^n = P_i^l \cup \left(\bigcup_{x \in X} P_{i-v(x)}^l \cdot x \right) \subseteq P_j^l \cup \left(\bigcup_{x \in X} P_{j-v(x)}^l \cdot x \right) = P_j^n. \end{gathered}$$

Thus (i) is holds for all P_i^n.

To see that (ii) holds for $n = 1$, one might observe that

$$P_{i+j}^{m+1} = P_{i+j}^m \cup \left(\bigcup_{x \in X} P_{i+j-v(x)}^m \cdot x \right) \supseteq \bigcup_{\substack{x \in X \\ v(x) \leqslant j}} P_{i+(j-v(x))}^m \cdot x \overset{(*)}{\supseteq}$$

$$\overset{(*)}{\supseteq} \bigcup_{\substack{x \in X \\ v(x) \leqslant j}} P_i^m \cdot x = P_i^m \cdot \{\, x \in X \mid v(x) \leqslant j \,\} = P_i^m \cdot P_j^1,$$

where what happens at the inclusion relation marked by $(*)$ simply is an application of (i).

The proof for the case $n = l+1 > 1$ is probably less confusing, it suffices to observe that

$$\begin{aligned} P_i^m \cdot P_j^n &= P_i^m \cdot \left(P_j^l \cup \left(\bigcup_{x\in X} P_{j-v(x)}^l \cdot x \right) \right) = \\ &= (P_i^m \cdot P_j^l) \cup \left(\bigcup_{x\in X} P_i^m \cdot P_{j-v(x)}^l \cdot x \right) \overset{(*)}{\subseteq} \\ &\overset{(*)}{\subseteq} P_{i+j}^{m+l} \cup \left(\bigcup_{x\in X} P_{i+j-v(x)}^{m+l} \cdot x \right) = P_{i+j}^{m+n}, \end{aligned}$$

where the induction hypothesis was applied at the inclusion relation marked by $(*)$. Thus (ii) holds for all P_i^n.

The proofs of (i) and (ii) for P_i should be no surprise. Starting with (i), one might note that for each $\mu \in P_i$ there is an $n \in \mathbb{Z}^+$ such that $\mu \in P_i^n$. Since $P_i^n \subseteq P_j^n \subseteq P_j$, it follows that $P_i \subseteq P_j$ if $i \leqslant j$. The proof of (ii) is similar; for each $\mu \in P_i$ there is an $m \in \mathbb{Z}^+$ such that $\mu \in P_i^m$ and for each $\nu \in P_j$ there is an $n \in \mathbb{Z}^+$ such that $\nu \in P_j^n$. Thus for each $\mu\nu \in P_i \cdot P_j$ there are n and m such that $\mu\nu \in P_i^m \cdot P_j^n \subseteq P_k^{n+m} \subseteq P_k$, and hence $P_i \cdot P_j \subseteq P_k$ if $i + j = k$. □

The next step is to construct a filtration from these sets. Let

$$M_i = \operatorname{Span}(P_i) \tag{B.4}$$

for all $i \in \mathbb{Z}$. Also let

$$M_{-\infty} = \bigcap_{i\in\mathbb{Z}} M_i. \tag{B.5}$$

Lemma B.2 *The family of submodules $\{M_i\}_{i\in\mathbb{Z}}$ is an ascending $\mathbb{Z}$-filtration on $\mathcal{A}$. In particular, for each $\alpha \in \mathcal{A}$ there is an $i \in \mathbb{Z}$ such that $M_i \ni \alpha$. Also, for each $\alpha \in \mathcal{A} \setminus M_{-\infty}$ there is a least $i \in \mathbb{Z}$ such that $\alpha \in M_i$.*

Proof. If $i \leqslant j$ then

$$M_i = \operatorname{Span}(P_i) \subseteq \operatorname{Span}(P_j) = M_j,$$

hence $M_i \subseteq M_j$. If $i + j = k$ then

$$M_i \cdot M_j = \mathrm{Span}(P_i) \cdot \mathrm{Span}(P_j) \subseteq \mathrm{Span}(P_i \cdot P_j) \subseteq \mathrm{Span}(P_k) = M_k,$$

hence $M_i \cdot M_j \subseteq M_k$. Thus $\{M_i\}_{i\in\mathbb{Z}}$ is an ascending $\mathbb{Z}$-filtration in $\mathcal{A}$.

Let

$$\mathcal{B} = \bigcup_{i\in\mathbb{Z}} M_i.$$

Then $\mathcal{B}$ is an $\mathcal{R}$-algebra containing X. Since X is a generating set for $\mathcal{A}$, it follows that $\mathcal{A} = \mathcal{B} = \bigcup_{i\in\mathbb{Z}} M_i$. Thus $\{M_i\}_{i\in\mathbb{Z}}$ is an ascending $\mathbb{Z}$-filtration *on* $\mathcal{A}$, and for each $\alpha \in \mathcal{A}$ there is an $i \in \mathbb{Z}$ such that $\alpha \in M_i$.

To prove the final claim, note that if for some $\alpha \in \mathcal{A}$ there is no least $i \in \mathbb{Z}$ such that $\alpha \in M_i$ then it must be the case that $\alpha \in M_j$ for all $j \in \mathbb{Z}$. Therefore $\alpha \in \bigcap_{i\in\mathbb{Z}} M_i = M_{-\infty}$; the claim follows. □

With this filtration established, the degree function can be formally defined.

Definition B.3 Let $\hat{\mathbb{Z}} = \mathbb{Z} \cup \{-\infty\}$. As usual, $-\infty$ has the following properties

$$-\infty + n = -\infty \qquad -\infty \leqslant n \qquad -\infty = \inf \mathbb{Z}$$

for all $n \in \hat{\mathbb{Z}}$. Note that every subset of $\mathbb{Z}$ has an infimum in $\hat{\mathbb{Z}}$, as $-\infty$ is a greatest lower bound of those subsets that do not have a minimal element.

Now define $d_v \colon \mathcal{A} \longrightarrow \hat{\mathbb{Z}}$ by letting

$$d_v(\alpha) = \begin{cases} -\infty & \text{if } \alpha \in M_{-\infty} \\ n & \text{otherwise, where } n \in \mathbb{Z} \text{ is such} \\ & \text{that } \alpha \in M_n \setminus M_{n-1} \end{cases} \tag{B.6}$$

for all $\alpha \in \mathcal{A}$. The function d_v is called the **degree function** for v.

In the common case when there is a set $V \subseteq X$ such that

$$v(x) = \begin{cases} 1 & \text{if } x \in V \\ 0 & \text{if } x \notin V, \end{cases} \tag{B.7}$$

d_v can alternatively be denoted d_V. In the important special case where V is the whole of X—or rather X minus the unit of $\mathcal{A}$, if $\mathcal{A}$ has a unit—the degree function d_V can be written as the more familiar deg.

The filtration $\{M_i\}_{i\in\mathbb{N}}$ (the "non-negative half" of the filtration used to define deg) is sometimes, for example in [38], referred to as the **Bernstein filtration** on $\mathcal{A}$. This is the reason for the name of this section.

B.1.2 *The basic properties*

Some properties of the degree function follow directly from the definition. These are stated below.

Lemma B.3 *For all $\alpha \in \mathcal{A}$,*

$$d_v(\alpha) = \inf\{\, n \in \mathbb{Z} \mid \alpha \in M_n \,\}. \tag{B.8}$$

Proof. If $\alpha \in M_{-\infty}$ then $\alpha \in M_n$ for all $n \in \mathbb{Z}$, and thus the infimum is $-\infty$ as claimed. Otherwise $\alpha \in M_n \setminus M_{n-1}$ for some $n \in \mathbb{Z}$, and hence $\alpha \notin M_k$ for any $k < n$. Thus the infimum is $n = d_v(\alpha)$ as claimed. □

Lemma B.4 *Let $B \subseteq \mathcal{A}$ be finite. If $\alpha \in \operatorname{Span}(B)$ then*

$$d_v(\alpha) \leqslant \max_{\beta\in B} d_v(\beta). \tag{B.9}$$

Now let $\alpha \in \mathcal{A}$ be arbitrary. If $C \subseteq \mathcal{A}$ is such that $d_v(\gamma) < d_v(\alpha)$ for all $\gamma \in C$, then

$$d_v(\alpha + \gamma) = d_v(\alpha) \tag{B.10}$$

for all $\gamma \in \operatorname{Span}(C)$.

Proof. Let $n = \max_{\beta\in B} d_v(\beta)$. Now $B \subseteq M_n$ and hence $\operatorname{Span}(B) \subseteq M_n$. Thus $\alpha \in M_n$, which is equivalent to $d_v(\alpha) \leqslant n$. This proves (B.9).

Now consider (B.10). Note that $d_v(\alpha)$ cannot be $-\infty$, since $-\infty \leqslant d_v(\gamma) < d_v(\alpha)$. Let $n = d_v(\alpha)$. By definition of d_v and since n is finite, $\alpha \in M_n \setminus M_{n-1}$. It is clear that $\gamma \in M_{n-1}$, hence $\alpha + \gamma \in M_n$. It can furthermore not be the case that $\alpha + \gamma \in M_{n-1}$, since that would imply $\alpha = (\alpha+\gamma) - \gamma \in M_{n-1}$. Thus $\alpha+\gamma \in M_n \setminus M_{n-1}$, which proves (B.10). □

Theorem B.5 *Let $\alpha, \beta \in \mathcal{A}$. Then*

$$d_v(\alpha\beta) \leqslant d_v(\alpha) + d_v(\beta). \tag{B.11}$$

Furthermore $M_{-\infty}$ is a two-sided ideal of $\mathcal{A}$.

Proof. To begin with, assume that $\alpha, \beta \notin M_{-\infty}$. Let $n = d_v(\alpha)$ and $m = d_v(\beta)$. Then $\alpha\beta \in M_n \cdot M_m \subseteq M_{n+m}$, and hence the degree of $\alpha\beta$ is at most $n + m$.

To prove (B.11) in the remaining case, when α or β happens to belong to $M_{-\infty}$, is equivalent to proving that $M_{-\infty}$ is an ideal. As $M_{-\infty}$ is known to be a submodule, all that is necessary to prove is that $\alpha\beta$ and $\beta\alpha$ belong to $M_{-\infty}$ when $\alpha \in M_{-\infty}$. The $\alpha\beta$ case follows from

$$\begin{aligned} M_{-\infty} \cdot \mathcal{A} &= \Bigl(\bigcap_{m\in\mathbb{Z}} M_m\Bigr) \cdot \Bigl(\bigcup_{n\in\mathbb{Z}} M_n\Bigr) = \bigcup_{n\in\mathbb{Z}} \Bigl(\bigcap_{m\in\mathbb{Z}} M_m\Bigr) \cdot M_n \subseteq \\ &\subseteq \bigcup_{n\in\mathbb{Z}} \bigcap_{m\in\mathbb{Z}} (M_m \cdot M_n) \subseteq \bigcup_{n\in\mathbb{Z}} \bigcap_{m\in\mathbb{Z}} M_{m+n} = \bigcup_{n\in\mathbb{Z}} \bigcap_{k\in\mathbb{Z}+n} M_k = \\ &= \bigcup_{n\in\mathbb{Z}} M_{-\infty} = M_{-\infty} \end{aligned}$$

and the $\beta\alpha$ case is proved analogously. □

Given a "good" choice of v, the ideal $M_{-\infty}$ will be $\{0\}$, but unless some care is taken, unexpected things can happen. In an algebra generated by A, B, A^{-1}, and B^{-1}, and in which A^{-1} and B^{-1} are the multiplicative inverses of A and B respectively, one would expect that $v(A) = v(B) = 1$ and $v(A^{-1}) = v(B^{-1}) = -1$ would make a perfectly good degree function. In many cases, this is indeed what it is.

But if $A^{15} = B^{14}$ also holds in the algebra, then drastic things occur. As the unit $I = A^{15}(A^{-1})^{15} = B^{14}(A^{-1})^{15}$, which clearly belongs to P_{-1}^{29}, a little induction leads to the conclusion that $I \in P_{-n}^{29n}$ for every $n \in \mathbb{Z}^+$. Hence I belongs to every P_j, every M_j, and therefore also to $M_{-\infty}$. Since the latter is an ideal, it follows that $M_{-\infty} = \mathcal{A}$. Obviously, this particular d_v is not very informative.

A better choice of v for this particular algebra would be $v(A) = 14$, $v(B) = 15$, $v(A^{-1}) = -14$, and $v(B^{-1}) = -15$, as this assigns the same degree to both sides of all defining equations.

Another conclusion of this example is that one should not expect this to be a useful method of constructing non-trivial ideals. In most cases, the structure of $M_{-\infty}$ turns out to be pretty simple. What one usually wants is to have $M_{-\infty} = \{0\}$, and the next lemma helps in proving this.

Lemma B.6 *If v is non-negative then $M_{-\infty} = \{0\}$.*

Proof. Obviously $P_j^1 = \varnothing$ for $j < 0$. If any P_j^n for $j < 0$ is to contain anything, there has to be an $x \in X$ such that $j - v(x) \geqslant 0$, but since v is non-negative, there are no such x. Thus $P_j^n = \varnothing$ and $P_j = \varnothing$ for all $j < 0$ and $n \in \mathbb{Z}^+$. As $\operatorname{Span}(\varnothing) = \{0\}$, it follows that $M_{-\infty} = \{0\}$. □

Lemma B.7 *Let $\alpha \in \mathcal{A} \setminus M_{-\infty}$ and let $n = d_v(\alpha)$. Then there exists a finite subset P of P_n and coefficients $\{a_\mu\}_{\mu \in P} \subseteq \mathcal{R} \setminus \{0\}$ such that*

$$\alpha = \sum_{\mu \in P} a_\mu \, \mu. \tag{B.12}$$

Furthermore any finite subset P of P_n for which there exist coefficients $\{a_\mu\}_{\mu \in P} \subseteq \mathcal{R} \setminus \{0\}$ such that (B.12) *is fulfilled also has the property that at least one $\mu \in P$ satisfies $d_v(\mu) = n$.*

Proof. First note that since α is nonzero and belongs to $M_n = \operatorname{Span}(P_n)$, there does indeed exist at least one finite subset of P_n that satisfies (B.12). Now let P be such a set. As $\alpha \notin M_{-\infty}$, it follows that $\alpha \in M_n \setminus M_{n-1}$. Had $P \subseteq M_{n-1}$ then $\alpha \in M_{n-1}$, since M_{n-1} is a module. Therefore $P \nsubseteq M_{n-1}$, and thus there is some $\mu \in P \cap (M_n \setminus M_{n-1})$. Such μ satisfy $d_v(\mu) = n$. □

Note that it may well happen that $P_n \subseteq M_{n-1}$ for some n, especially in finite-dimensional algebras, but in these cases $M_n = M_{n-1}$.

The concept from this section that will be used in other parts of this book is the degree function, but the modules $\{M_i\}_{i \in \mathbb{Z}}$ and sets of monomials $\{P_i\}_{i \in \mathbb{Z}}$ and $\{P_i^n\}_{i \in \mathbb{Z};\, n \in \mathbb{Z}^+}$ are occasionally used, mainly in proofs of properties of degree functions, and in these cases the constructions will not be repeated. It is sometimes even necessary to keep track of two different families of M_i, P_i, and P_i^n at the same time. This is then done by adding the function v or set V, put in parenthesis, to the name of the set, as in

$$M_i(v), \quad P_i(v), \quad P_i^n(v) \qquad \text{or} \qquad M_i(V), \quad P_i(V), \quad P_i^n(V).$$

B.2 Degree and free algebras

As almost all algebras in this book are either free algebras or constructed as the quotient of a free algebra by an ideal, it is only natural that the properties of degree functions in free algebras are studied more closely. The first part of this section studies the behaviour of the degree functions within free algebras, concentrating on various addition properties. The second part shows a few technical lemmas that will be useful when studying degree in quotients. The third part studies the relation between degree in a quotient and the corresponding degree in the free algebra.

Given that the free algebra is $\mathcal{R}\langle X\rangle$ (the free associative $\mathcal{R}$-algebra with set of generators X and unit 1), there are a few things that can be said. Clearly, the set $X \cup \{1\}$ is a generating set (1 has to be in the generating set because the definition of generating set does not require the subalgebras to contain a unit) and no proper subset of this set is a generating set. It is not the only possible generating set, or even a necessary subset of every generating set, but it is clearly the generating set that has the closest relation to the definition of the algebra. *It will therefore be assumed throughout this book that $X \cup \{1\}$ is the generating set used for a free algebra $\mathcal{R}\langle X\rangle$. It will also be assumed that $\bigl(X \cup \{1\}\bigr) + \mathcal{I}$ is the generating set used for any algebra constructed as $\mathcal{R}\langle X\rangle/\mathcal{I}$.*

B.2.1 *Additivity of degree in free algebras*

Lemma B.8 *Let $\mathcal{R}$ be an associative and commutative ring with unit. Let X be a nonempty set. Let P be a nonempty finite set of monomials in $\mathcal{R}\langle X\rangle$ and let $\{a_\mu\}_{\mu\in P} \subseteq \mathcal{R}\setminus\{0\}$. Finally let $v\colon \bigl(X \cup \{1\}\bigr) \longrightarrow \mathbb{Z}$. Then*

$$d_v\left(\sum_{\mu\in P} a_\mu \mu\right) = \max_{\mu\in P} d_v(\mu) \tag{B.13}$$

Proof. Let $\alpha = \sum_{\mu\in P} a_\mu\mu$. Since the set of monomials in $\mathcal{R}\langle X\rangle$ is a basis for $\mathcal{R}\langle X\rangle$, one cannot express α as a linear combination of monomials in more than one way. This means in particular that

$$\begin{aligned} d_v(\alpha) = \inf\{\, n\in\mathbb{Z} \mid \alpha\in M_n \,\} &= \inf\{\, n\in\mathbb{Z} \mid P\subseteq P_n \,\} = \\ &= \max_{\mu\in P} \inf\{\, n\in\mathbb{Z} \mid \mu\in P_n \,\} \geqslant \\ &\geqslant \max_{\mu\in P} \inf\{\, n\in\mathbb{Z} \mid \mu\in M_n \,\} = \\ &= \max_{\mu\in P} d_v(\mu). \end{aligned}$$

Conversely, $d_v(\alpha) \leqslant \max_{\mu\in P} d_v(\mu)$ by Lemma B.4. □

Theorem B.9 *Let $\mathcal{R}$ be an associative and commutative ring with unit. Let X be a nonempty set. Let $v\colon \bigl(X\cup\{1\}\bigr) \longrightarrow \mathbb{Z}$ be a function such that*

$v(1) \geqslant 0$. *Then*

$$\prod_{i=1}^{n} x_i \in P_j^m(v) \iff j \geqslant \sum_{i=1}^{n} v(x_i) \text{ and } n \leqslant m \tag{B.14a}$$

$$1 \in P_j^m(v) \iff j \geqslant v(1) \tag{B.14b}$$

for all $n, m \in \mathbb{Z}^+$, $j \in \mathbb{Z}$, *and* $\{x_i\}_{i=1}^n \subseteq X$;

$$\prod_{i=1}^{n} x_i \in P_j(v) \iff j \geqslant \sum_{i=1}^{n} v(x_i) \tag{B.15a}$$

$$1 \in P_j(v) \iff j \geqslant v(1) \tag{B.15b}$$

for all $n \in \mathbb{Z}^+$, $j \in \mathbb{Z}$, *and* $\{x_i\}_{i=1}^n \subseteq X$; *and*

$$d_v\left(\prod_{i=1}^{n} x_i\right) = \sum_{i=1}^{n} v(x_i) \tag{B.16a}$$

$$d_v(1) = v(1) \tag{B.16b}$$

for all $n \in \mathbb{Z}^+$ *and* $\{x_i\}_{i=1}^n \subseteq X$.

Proof. Observe that for every monomial $\mu \in \mathcal{R}\langle X\rangle$ except 1 there exists a unique $n \in \mathbb{Z}^+$ and $\{x_i\}_{i=1}^n \subseteq X$ such that

$$\mu = \prod_{i=1}^{n} x_i.$$

This fact will be used frequently in what follows.

The main problem is to prove (B.14), and this problem will be tackled by induction on m. Uniqueness of factorisation of monomials in free algebras implies that the equivalences (B.14) for $m = 1$ are just rephrased versions of the definition of P_j^1.

Next assume that (B.14) holds for $m = k \in \mathbb{Z}^+$ and consider it for $m = k + 1$. If the left hand side of (B.14a) is true then

$$\prod_{i=1}^{n} x_i \in P_j^m = P_j^k \cup \left(\bigcup_{x\in X} P_{j-v(x)}^k \cdot x\right) \cup (P_{j-v(1)}^k \cdot 1).$$

If $\prod_{i=1}^n x_i \in P_j^k$ then $j \geqslant \sum_{i=1}^n v(x_i)$ and $n \leqslant k < m$ by assumption, so in this case the right hand side is true. Since $P_{j-v(1)}^k \cdot 1 = P_{j-v(1)}^k \subseteq P_j^k$, the case $\prod_{i=1}^{n} x_i \in P_{j-v(1)}^{k} \cdot 1$ is a subcase of the case already treated. What remains is the case $\prod_{i=1}^{n} x_i \in P_{j-v(x)} \cdot x$ for some $x \in X$.

By uniqueness of factorisation $x = x_n$. If $n > 1$ then $\prod_{i=1}^{n-1} x_i \in P_{j-v(x_n)}^{k}$. By assumption, this implies $j - v(x_n) \geqslant \sum_{i=1}^{n-1} v(x_i)$ and $n-1 \leqslant k$, which is equivalent to $j \geqslant \sum_{i=1}^{n} v(x_i)$ and $n \leqslant m$ as required. If $n = 1$ then $1 \in P_{j-v(x_n)}^{k}$. Again by assumption, this implies $j - v(x_n) \geqslant v(1) \geqslant 0$, and hence $j \geqslant v(x_1)$ and $n = 1 \leqslant m$ as required. The $\Longrightarrow$ part of (B.14a) has thus been shown.

To prove the reverse implication, note that if $n < m$ and $\sum_{i=1}^{n} v(x_i) \leqslant j$ then trivially $\prod_{i=1}^{n} x_i \in P_j^k \subseteq P_j^m$. If instead $n = m$ and $\sum_{i=1}^{n} v(x_i) \leqslant j$ then clearly $\prod_{i=1}^{k} x_i \in P_{j-v(x_m)}^{k}$ and hence $\prod_{i=1}^{m} x_i \in P_{j-v(x_m)}^{k} \cdot x_m \subseteq P_j^m$, just as required.

Part (B.14b) is easier. As $1 = \mu\nu$ for μ and ν monomials implies $\mu = \nu = 1$, the left hand side of the equivalence is equivalent to $1 \in P_j^k \cup (P_{j-v(1)}^{k} \cdot 1) = P_j^k$, which by assumption is equivalent to $j \geqslant v(1)$. As both equivalences of (B.14) has now been shown to hold for $m = k+1$, it follows by induction that (B.14) holds for all $m \in \mathbb{Z}^+$.

The two equivalences (B.15) are easy consequences of (B.14). For any $m \geqslant n$ the right hand side of (B.15a) is known to be equivalent with $\prod_{i=1}^{n} x_i \in P_j^m$, and that trivially implies the left hand side. If one instead starts with the left hand side then since $\prod_{i=1}^{n} x_i \in P_j = \bigcup_{m \in \mathbb{Z}^+} P_j^m$, then for some m large enough $\prod_{i=1}^{n} x_i \in P_j^m$, which implies the right hand side. The same argument gives (B.15b).

Now let $j = \sum_{i=1}^{n} v(x_i)$. As $\prod_{i=1}^{n} x_i \in P_j \subseteq \operatorname{Span}(P_j) = M_j$, it follows that the left hand side of (B.16a) is bounded above by the right hand side. To prove equality, one needs to show that $\prod_{i=1}^{n} x_i \notin M_{j-1}$. If that would not be the case then there would exist some $P \subseteq P_{j-1}$ and $\{a_\mu\}_{\mu \in P} \subseteq \mathcal{R}$ such that

$$\prod_{i=1}^{n} x_i = \sum_{\mu \in P} a_\mu \mu.$$

As $\prod_{i=1}^{n} x_i \notin P$, this would be a linear dependence between the linearly independent monomials of $\mathcal{R}\langle X \rangle$, therefore $\prod_{i=1}^{n} x_i \notin M_{j-1}$ and equality in (B.16a) follows. A very similar argument, with $j = v(1)$, is used to prove (B.16b). $\square$

By combining Theorem B.9 and Lemma B.8, it is possible to determine

$d_v(\alpha)$ for any $\alpha \in \mathcal{R}\langle X\rangle$ and $v\colon \bigl(X \cup \{1\}\bigr) \longrightarrow \mathbb{Z}$ such that $v(1) \geqslant 0$ in $O(l \log l)$ time, where l is the number of characters it would take to write down α in fully expanded form.

Corollary B.10 *Let $\mathcal{R}$ be an associative and commutative ring with unit. Let X be a nonempty set and consider the algebra $\mathcal{R}\langle X\rangle$. Let $v\colon \bigl(X \cup \{1\}\bigr) \longrightarrow \mathbb{Z}$ be arbitrary. Then $d_v(1) = v(1)$ and $d_v|_X = v|_X$ if and only if $v(1) \geqslant 0$.*

Proof. The 'if' part is a special case of Theorem B.9. Now consider the case $v(1) < 0$. Clearly then, $1 \in P^m_{mv(1)}$ for all $m \in \mathbb{Z}^+$. Thus $1 \in M_{-\infty}(v)$, and as this makes $d_v(1) = -\infty \notin \mathbb{Z}$; the 'only if' part is proved. □

The problem of determining $d_v(\alpha)$ when $v(1) < 0$ is trivial: $d_v(\alpha)$ is always $-\infty$. This is partly a consequence of Theorem B.5, since by the proof of Corollary B.10, it holds, for v such that $v(1) < 0$, that $M_{-\infty}(v)$ contains the ideal generated by 1, which is the whole of $\mathcal{R}\langle X\rangle$. As remarked in the previous section, such degree functions are not very useful.

Corollary B.11 *Let $\mathcal{R}$ be an associative and commutative ring with unit. Let X be a nonempty set and let μ be a monomial in $\mathcal{R}\langle X\rangle$. If*

$$v, w\colon \bigl(X \cup \{1\}\bigr) \longrightarrow \mathbb{Z}$$

are functions such that $v(1) \geqslant 0$ and $w(1) \geqslant 0$ then

$$d_{v+w}(\mu) = d_v(\mu) + d_w(\mu). \tag{B.17}$$

Proof. If $\mu = 1$ then $d_{v+w}(\mu) = (v+w)(\mu) = v(\mu)+w(\mu) = d_v(\mu)+d_w(\mu)$ by (B.16b). If $\mu \neq 1$ then there exists a unique $n \in \mathbb{Z}^+$ and $\{x_i\}_{i=1}^n \subseteq X$ such that

$$\mu = \prod_{i=1}^{n} x_i.$$

Then by (B.16a),

$$d_{v+w}(\mu) = \sum_{i=1}^{n}(v + w)(x_i) = \sum_{i=1}^{n} v(x_i) + \sum_{i=1}^{n} w(x_i) = d_v(\mu) + d_w(\mu).$$

□

Corollary B.12 *Let $\mathcal{R}$ be an associative and commutative ring with unit. Let X be a nonempty set. Let $v\colon \big(X \cup \{1\}\big) \longrightarrow \mathbb{Z}$ be such that $v(1) = 0$. Then*

$$d_v(\mu\nu) = d_v(\mu) + d_v(\nu) \tag{B.18}$$

for all monomials $\mu, \nu \in \mathcal{R}\langle X\rangle$.

If instead $v\colon \big(X \cup \{1\}\big) \longrightarrow \mathbb{Z}$ satisfies $v(1) > 0$ then

$$d_v(\mu\nu) = d_v(\mu) + d_v(\nu) \tag{B.19}$$

for all non-unit monomials $\mu, \nu \in \mathcal{R}\langle X\rangle$.

Proof. As $d_v(1) = 0$ by Theorem B.9, (B.18) is trivial if $\mu = 1$ or $\nu = 1$. Therefore assume that $\mu \neq 1$ and $\nu \neq 1$. Then there exist unique $m, n \in \mathbb{Z}^+$ and $\{x_i\}_{i=1}^m, \{y_j\}_{j=1}^n \subseteq X$ such that

$$\mu = \prod_{i=1}^m x_i \quad \text{and} \quad \nu = \prod_{j=1}^n y_i.$$

Now by Theorem B.9,

$$\begin{aligned} d_v(\mu\nu) &= d_v\left(\prod_{i=1}^m x_i \cdot \prod_{j=1}^n y_j\right) = \sum_{i=1}^m v(x_i) + \sum_{j=1}^n v(y_j) = \\ &= d_v\left(\prod_{i=1}^m x_i\right) + d_v\left(\prod_{j=1}^n y_j\right) = d_v(\mu) + d_v(\nu). \end{aligned}$$

□

Theorem B.13 *Let $\mathcal{R}$ be an integral domain and let X be a nonempty set. If*

$$v\colon \big(X \cup \{1\}\big) \longrightarrow \mathbb{Z}$$

satisfies $v(1) = 0$ then

$$d_v(\alpha\beta) = d_v(\alpha) + d_v(\beta) \tag{B.20}$$

for all $\alpha, \beta \in \mathcal{R}\langle X\rangle$.

Proof. Let $\alpha, \beta \in \mathcal{R}\langle X\rangle$ be arbitrary. If $\alpha = 0$ or $\beta = 0$ then (B.20) is trivial; hence assume that both α and β are nonzero.

Let $n = d_v(\alpha)$ and $m = d_v(\beta)$. Then Lemma B.7 implies that there exists $P \subseteq P_n$ and $Q \subseteq P_m$ such that

$$\alpha = \sum_{\mu \in P} a_\mu \mu \qquad \text{and} \qquad \beta = \sum_{\nu \in Q} b_\nu \nu$$

for some $\{a_\mu\}, \{b_\nu\} \subseteq \mathcal{R} \setminus \{0\}$. Set

$$P' = \{\, \mu \in P \mid d_v(\mu) = n \,\}, \qquad \alpha' = \sum_{\mu \in P'} a_\mu \mu,$$
$$Q' = \{\, \nu \in Q \mid d_v(\nu) = m \,\}, \quad \text{and} \quad \beta' = \sum_{\nu \in Q'} b_\nu \nu.$$

Let $n' = d_X(\alpha')$ and $m' = d_X(\beta')$. Choose some $\mu_0 \in P'$ that satisfies $d_X(\mu_0) = n'$ and some $\nu_0 \in Q'$ that satisfies $d_X(\nu_0) = m'$. By Corollary B.12, $d_v(\mu_0\nu_0) = n + m$.

It is clear that if $\mu_1\nu_1$ is another factorisation of $\mu_0\nu_0$, then either $d_X(\mu_1) < n'$ or $d_X(\nu_1) < m'$, which would imply $d_X(\nu_1) > m'$ or $d_X(\mu_1) > n'$ respectively. Thus $\mu_0\nu_0$ itself is the only factorisation of $\mu_0\nu_0$ as $\mu\nu$ where $\mu \in P'$ and $\nu \in Q'$. It is, in fact, the only factorisation of $\mu_0\nu_0$ as $\mu\nu$ where $\mu \in P$ and $\nu \in Q$ too, because if $\mu \in P \setminus P'$ and $\nu \in Q$ then by Theorem B.5, $d_v(\mu\nu) \leqslant d_v(\mu) + d_v(\nu) < n + m$, and hence $\mu\nu \neq \mu_0\nu_0$. A similar argument applies when $\mu \in P$ and $\nu \in Q \setminus Q'$.

Let $l = d_v(\alpha\beta)$. Then Lemma B.7 implies that there exists $R \subseteq P_l$ such that

$$\sum_{\substack{\mu \in P \\ \nu \in Q}} a_\mu b_\nu \mu\nu = \alpha\beta = \sum_{\lambda \in R} c_\lambda \lambda$$

for some $\{c_\lambda\}_{\lambda \in R} \subseteq \mathcal{R} \setminus \{0\}$. It follows from the above and the observation that $a_{\mu_0} b_{\nu_0} \neq 0$ that $\mu_0\nu_0 \in R$, and hence $l \geqslant n + m$.

Since α and β were both arbitrary, $d_v(\alpha) + d_v(\beta) \leqslant d_v(\alpha\beta)$ for all $\alpha, \beta \in \mathcal{R}\langle X \rangle$. Finally, by Theorem B.5, the reversed inequality is also true. Thus (B.20) is verified. □

B.2.2 *Some technical lemmas*

Although the material in this subsection is very technical, the idea behind it is pretty simple. Everything is really about the concept of most significant term, and how it should be defined in a polynomial algebra with more than one generator/variable in the polynomials.

Definition B.4 Let S be a set and let $T \subseteq S$ be finite and nonempty. Let

$$f_1, f_2, \ldots, f_n : S \longrightarrow \mathbb{Z}$$

be arbitrary functions. Let

$$T_1 = \left\{ t \in T \,\middle|\, f_1(t) = \max_{u \in T} f_1(u) \right\} \qquad \text{and}$$

$$T_m = \left\{ t \in T_{m-1} \,\middle|\, f_m(t) = \max_{u \in T_{m-1}} f_m(u) \right\}$$

for all m satisfying $1 < m \leqslant n$. Then the set T_n will be denoted

$$\operatorname{Maj}(T, f_1, \ldots, f_n).$$

If $\mathbf{f} = (f_1, \ldots, f_n)$ then T_n might also be denoted

$$\operatorname{Maj}(T, \mathbf{f}).$$

In this definition, the functions $\{f_i\}_{i=1}^n$ should be thought of as *measures of significance.* The operation of taking $\operatorname{Maj}(T, f_1)$ returns T stripped of all elements but the most significant (according to the view that f_1 has on what is significant and what is not). As it is often desirable to end up with one single element of T that, in some sense, is more significant than every other element, it is often necessary to apply this procedure repeatedly, which leads to the notation $\operatorname{Maj}(T, f_1, \ldots, f_n)$. Note that the order in which the f_i are applied is significant, since different orders in general give different results.

Definition B.5 Let $\mathcal{R}$ be an associative and commutative ring with unit. Let X be a nonempty set. Let Y be the set of all monomials (with respect to X) in $\mathcal{R}\langle X\rangle$. Choose some $\alpha \in \mathcal{R}\langle X\rangle$. Then the fact that Y is a basis for $\mathcal{R}\langle X\rangle$ implies that there exists a unique subset T of $\big(\mathcal{R} \setminus \{0\}\big) \times Y$ such that

(i) $\alpha = \sum_{(a,\mu) \in T} a\mu$,

(ii) if $(a, \mu) \in T$, $(b, \nu) \in T$, and $\mu = \nu$ then $(a, \mu) = (b, \nu)$.

This set T is called the **monomial decomposition** of α. The monomial decomposition of α is written symbolically as $\mathrm{Mon}(\alpha)$.

A function which comes in handy when dealing with monomial decompositions is $\mathrm{pr}_2 \colon \big(\mathcal{R} \setminus \{0\}\big) \times Y \longrightarrow Y$, $(a, \mu) \mapsto \mu$.

As this subsection deals with the concept of most significant *term*, but Maj as defined above acts on sets, there is a need to convert sums to sets in such a way that there is a 1–1 correspondence between the terms in the sum and the elements of the set. This is what Mon does with the elements of $\mathcal{R}\langle X\rangle$, considering the sums as being the unique linear combinations of monomials that yield each element of the algebra.

Note: The monomial decomposition of a monomial μ is the set $\big\{(1, \mu)\big\}$. The monomial decomposition of 0 is $\varnothing$.

The first lemma examines how these concepts relate to addition in $\mathcal{R}\langle X\rangle$. What the lemma does is that it answers the question "*Suppose I've got an element in* $\mathrm{Mon}(\beta)$ *that is more significant than any element of* $\mathrm{Mon}(\alpha)$. *Can I then conclude that the set of the most significant terms in* $\alpha + \beta$ *is exactly the set of the most significant terms in* β*?*" by "*Yes, you may.*"

Lemma B.14 *Let $\mathcal{R}$ be an associative and commutative ring with unit. Let X be a nonempty set. Let $\alpha, \beta \in \mathcal{R}\langle X\rangle$. Let Y be the set of monomials of $\mathcal{R}\langle X\rangle$ and $d_1, \dots, d_r \colon Y \longrightarrow \mathbb{Z}$ be functions. Let $f_i = d_i \circ \mathrm{pr}_2$ for $i = 1, \dots, r$.*

If there exists some $(b, \nu) \in \mathrm{Mon}(\beta)$ such that:

(i) *there is an $i \leqslant r$ for each $(a, \mu) \in \mathrm{Mon}(\alpha)$ such that $d_i(\mu) < d_i(\nu)$,*
(ii) *if $(a, \mu) \in \mathrm{Mon}(\alpha)$ satisfies $d_i(\mu) > d_i(\nu)$ for some $i \leqslant r$ then there exists some $j < i$ such that $d_j(\mu) < d_j(\nu)$,*

then

$$\mathrm{Maj}\big(\mathrm{Mon}(\alpha + \beta), f_1, \dots, f_r\big) = \mathrm{Maj}\big(\mathrm{Mon}(\beta), f_1, \dots, f_r\big). \tag{B.21}$$

Proof. First note that $(b, \nu) \in \mathrm{Mon}(\alpha+\beta)$, since $\mathrm{Mon}(\alpha)$ would otherwise have to contain an element (a, ν), which it obviously cannot since there is no i for which $d_i(\nu) < d_i(\nu)$.

Now let s be the largest integer not greater than r such that

$$(b, \nu) \in \mathrm{Maj}\big(\mathrm{Mon}(\beta), f_1, \dots, f_s\big).$$

The pair (b, ν) can always be chosen so that s is at least 1.

Then it is in fact the case that $(b, \nu) \in \operatorname{Maj}\bigl(\operatorname{Mon}(\alpha + \beta), f_1, \dots, f_s\bigr)$. This is because otherwise there would be a least $i \leqslant s$ for which

$$(b, \nu) \notin \operatorname{Maj}\bigl(\operatorname{Mon}(\alpha + \beta), f_1, \dots, f_i\bigr),$$

and that would have to be caused by an

$$(a, \mu) \in \operatorname{Maj}\bigl(\operatorname{Mon}(\alpha + \beta), f_1, \dots, f_i\bigr) \tag{B.22}$$

such that $d_i(\mu) > d_i(\nu)$. It cannot be the case that there is such an $(a, \mu) \in \operatorname{Mon}(\beta)$, as that would make $s < i$, contrary to the above. Thus the only way (B.22) can hold is that $(a, \mu) \in \operatorname{Mon}(\alpha)$, but then there would also be some $j < i$ such that $d_j(\mu) < d_j(\nu)$. That would imply

$$(a, \mu) \notin \operatorname{Maj}\bigl(\operatorname{Mon}(\alpha + \beta), f_1, \dots, f_j\bigr) \supseteq \operatorname{Maj}\bigl(\operatorname{Mon}(\alpha + \beta), f_1, \dots, f_i\bigr),$$

which would contradict the defining property of (a, μ). Therefore

$$(b, \nu) \in \operatorname{Maj}\bigl(\operatorname{Mon}(\alpha + \beta), f_1, \dots, f_s\bigr).$$

It now follows that

$$\begin{aligned} &\operatorname{Maj}\bigl(\operatorname{Mon}(\beta), f_1, \dots, f_s\bigr) = \\ &\qquad = \bigl\{\, (a, \mu) \in \operatorname{Mon}(\beta) \bigm| d_i(\mu) = d_i(\nu) \text{ for all } i = 1, \dots, s \,\bigr\}, \\ &\operatorname{Maj}\bigl(\operatorname{Mon}(\alpha + \beta), f_1, \dots, f_s\bigr) = \\ &\qquad = \bigl\{\, (a, \mu) \in \operatorname{Mon}(\alpha + \beta) \bigm| d_i(\mu) = d_i(\nu) \text{ for all } i = 1, \dots, s \,\bigr\}. \end{aligned}$$

Let $P = \operatorname{pr}_2\bigl(\operatorname{Mon}(\alpha) \cup \operatorname{Mon}(\beta)\bigr)$. Then there are unique elements $\{c_\mu\}_{\mu \in P}$, $\{d_\mu\}_{\mu \in P} \subseteq \mathcal{R}$ such that

$$\alpha = \sum_{\mu \in P} c_\mu \mu, \qquad \beta = \sum_{\mu \in P} d_\mu \mu.$$

These elements of $\mathcal{R}$ have some important properties:

- $(d_\mu, \mu) \in \operatorname{Mon}(\beta)$ if and only if $d_\mu \neq 0$.
- $(c_\mu + d_\mu, \mu) \in \operatorname{Mon}(\alpha + \beta)$ if and only if $c_\mu + d_\mu \neq 0$.
- $c_\mu = 0$ for all $\mu \in P$ which satisfy $d_i(\mu) = d_i(\nu)$ for $i = 1, \dots, s$ (a consequence of condition (i)).

Hence

$$\begin{aligned}
&\operatorname{Maj}\bigl(\operatorname{Mon}(\beta), f_1, \dots, f_s\bigr) = \\
&\quad = \bigl\{ (a,\mu) \in \operatorname{Mon}(\beta) \bigm| d_i(\mu) = d_i(\nu) \text{ for all } i = 1, \dots, s \bigr\} = \\
&\quad = \bigl\{ (d_\mu,\mu) \bigm| \mu \in P \text{ and } d_i(\mu) = d_i(\nu) \text{ for all } i = 1, \dots, s \bigr\} = \\
&\quad = \bigl\{ (c_\mu + d_\mu,\mu) \bigm| \mu \in P \text{ and } d_i(\mu) = d_i(\nu) \text{ for all } i = 1, \dots, s \bigr\} = \\
&\quad = \bigl\{ (a,\mu) \in \operatorname{Mon}(\alpha+\beta) \bigm| d_i(\mu) = d_i(\nu) \text{ for all } i = 1, \dots, s \bigr\} = \\
&\quad = \operatorname{Maj}\bigl(\operatorname{Mon}(\alpha+\beta), f_1, \dots, f_s\bigr).
\end{aligned}$$

Therefore

$$\begin{aligned}
&\operatorname{Maj}\bigl(\operatorname{Mon}(\alpha+\beta), f_1, \dots, f_r\bigr) = \\
&\qquad\qquad = \operatorname{Maj}\Bigl(\operatorname{Maj}\bigl(\operatorname{Mon}(\alpha+\beta), f_1, \dots, f_s\bigr), f_{s+1}, \dots, f_r\Bigr) = \\
&\qquad\qquad = \operatorname{Maj}\Bigl(\operatorname{Maj}\bigl(\operatorname{Mon}(\beta), f_1, \dots, f_s\bigr), f_{s+1}, \dots, f_r\Bigr) = \\
&\qquad\qquad = \operatorname{Maj}\bigl(\operatorname{Mon}(\beta), f_1, \dots, f_r\bigr)
\end{aligned}$$

as (B.21) claims. □

The next lemma gives sufficient conditions for that the most significant term of a product can be computed as the most significant term of the product of the respective most significant terms in the two factors. As it is aimed at proving things about degree in quotient algebras however, there are a few extra twists to it. There is for example a linear mapping π which is applied to each product, and when one is dealing with quotient algebras of the form $\mathcal{R}\langle X\rangle/\mathcal{I}(S)$, this mapping is usually the conversion to normal form t^S. (See Theorem 4.2 for an example of this.)

Lemma B.15 *Let $\mathcal{R}$ be an integral domain. Let X be a nonempty set. Let $\{w_i\}_{i=1}^r$ be a family of functions $X \cup \{1\} \longrightarrow \mathbb{Z}$ such that $w_i(1) = 0$ for all $i = 1, \dots, r$. Define the family of functions $\{f_i\}_{i=1}^r$ through*

$$f_i = d_{w_i} \circ \mathrm{pr}_2,$$

where pr_2 is defined by $\mathrm{pr}_2(a,\mu) = \mu$. Set $\mathbf{f} = (f_1, \dots, f_r)$.

Let $\pi\colon \mathcal{R}\langle X\rangle \longrightarrow \mathcal{R}\langle X\rangle$ be a linear mapping which satisfies

$$d_{w_i}\bigl(\pi(\alpha)\bigr) \leqslant d_{w_i}(\alpha)$$

for all $\alpha \in \mathcal{R}\langle X\rangle$ and $i = 1, \dots, r$.

Finally let $\alpha, \beta \in \mathcal{R}\langle X\rangle \setminus \{0\}$ *be such that*

$$\begin{aligned}\{(a_0,\mu_0)\} &= \operatorname{Maj}\big(\operatorname{Mon}(\alpha),\mathbf{f}\big),\\ \{(b_0,\nu_0)\} &= \operatorname{Maj}\big(\operatorname{Mon}(\beta),\mathbf{f}\big),\ \textit{and}\\ \{(c_0,\lambda_0)\} &= \operatorname{Maj}\Big(\operatorname{Mon}\big(\pi(a_0b_0\mu_0\nu_0)\big),\mathbf{f}\Big)\end{aligned}$$

for some $a_0, b_0, c_0 \in \mathcal{R}$ *and monomials* $\mu_0, \nu_0, \lambda_0 \in \mathcal{R}\langle X\rangle$.

Then

$$\operatorname{Maj}\Big(\operatorname{Mon}\big(\pi(\alpha\beta)\big),\mathbf{f}\Big) = \{(c_0,\lambda_0)\}$$

if $f_i(c_0,\lambda_0) = f_i(a_0,\mu_0) + f_i(b_0,\nu_0)$ *for all* $i = 1, 2, \ldots, r$.

Proof. Let

$$\begin{aligned}U_0 &= \operatorname{Mon}(\alpha) & U_i &= \operatorname{Maj}\big(\operatorname{Mon}(\alpha), f_1, \ldots, f_i\big)\\ V_0 &= \operatorname{Mon}(\beta) & V_i &= \operatorname{Maj}\big(\operatorname{Mon}(\beta), f_1, \ldots, f_i\big)\end{aligned}$$

for all $i = 1, 2, \ldots, r$ and

$$\alpha_i = \sum_{(a,\mu)\in U_i} a\mu, \qquad \beta_i = \sum_{(b,\nu)\in V_i} b\nu$$

for all $i = 0, 1, \ldots, r$. This makes $\alpha_0 = \alpha$, $\beta_0 = \beta$, $\alpha_r = a_0\mu_0$, and $\beta_r = b_0\nu_0$. Also let $\alpha_i' = \alpha_{i-1} - \alpha_i$, and $\beta_i' = \beta_{i-1} - \beta_i$. Now $\alpha_{i-1}\beta_{i-1} - \alpha_i\beta_i = \alpha_i\beta_i' + \alpha_i'\beta_{i-1}$.

If i and j are such that $1 \leqslant i \leqslant j \leqslant r$ then $f_i(t_1) = f_i(t_2)$ both for all $t_1, t_2 \in U_j$ and for all $t_1, t_2 \in V_j$. In the case $i = j$, this is merely the defining property of U_j (and V_j) as a subset of U_{j-1} (and V_{j-1} respectively). In the case $i < j$, this follows from $U_j \subseteq U_i$ and $V_j \subseteq V_i$. A consequence of this is that

$$\begin{aligned}d_{w_i}(\alpha_j\beta_j' + \alpha_j'\beta_{j-1}) &\leqslant \max\{d_{w_i}(\alpha_j\beta_j'), d_{w_i}(\alpha_j'\beta_{j-1})\} =\\ &= \max\{d_{w_i}(\alpha_j) + d_{w_i}(\beta_j'), d_{w_i}(\alpha_j') + d_{w_i}(\beta_{j-1})\} \leqslant\\ &\leqslant \max\{d_{w_i}(\alpha_j), d_{w_i}(\alpha_j')\} + \max\{d_{w_i}(\beta_j'), d_{w_i}(\beta_{j-1})\} \leqslant\\ &\leqslant \max_{u\in U_{j-1}} f_i(u) + \max_{v\in V_{j-1}} f_i(v) = f_i(a_0,\mu_0) + f_i(b_0,\nu_0) = f_i(c_0,\lambda_0),\end{aligned}$$

for all i and j satisfying $1 \leqslant i < j \leqslant r$.

This inequality is actually strict if $i = j$. As $f_i(t_1) < f_i(t_2)$ for all $t_1 \in U_{i-1} \setminus U_i$ and $t_2 \in U_i$, it follows that

$$d_{w_i}(\alpha'_i) < d_{w_i}(\alpha_i) = d_{w_i}(\alpha_{i-1})$$

and, as $f_i(t_1) < f_i(t_2)$ for all $t_1 \in V_{i-1} \setminus V_i$ and $t_2 \in V_i$, it follows that

$$d_{w_i}(\beta'_i) < d_{w_i}(\beta_i) = d_{w_i}(\beta_{i-1}).$$

This means that $d_{w_i}(\alpha'_i\beta_{i-1}) < d_{w_i}(\alpha_i\beta_i)$ and $d_{w_i}(\alpha_i\beta'_i) < d_{w_i}(\alpha_i\beta_i)$. Therefore

$$d_{w_i}(\alpha_i\beta'_i + \alpha'_i\beta_{i-1}) \leqslant \max\{d_{w_i}(\alpha_i\beta'_i), d_{w_i}(\alpha'_i\beta_{i-1})\} < \\ < d_{w_i}(\alpha_i\beta_i) = d_{w_i}(\alpha_i) + d_{w_i}(\beta_i) = d_{w_i}(\mu_0) + d_{w_i}(\nu_0) = d_{w_i}(\lambda_0).$$

Now let

$$T_0 = \mathrm{Mon}\big(\pi(\alpha_0\beta_0)\big) \qquad \text{and}$$
$$T_i = \mathrm{Maj}\Big(\mathrm{Mon}\big(\pi(\alpha_i\beta_i)\big), f_1, \ldots, f_i\Big)$$

for all $i = 1, 2, \ldots, r$. It will now be shown that

$$\mathrm{Maj}(T_{n-1}, f_n, \ldots, f_r) = T_r \tag{B.23}$$

for all $n = 1, \ldots, r$. The proof of this will be by induction on n, starting at the case $n = r$ and *descending* towards the case $n = 1$.

Let k satisfy $1 \leqslant k \leqslant r$. The induction hypothesis is that (B.23) holds for all n satisfying $r \geqslant n > k$ and the induction step will prove that this claim holds for $n = k$ as well.

One way to do this is to apply Lemma B.14 at the equality marked $(*)$ in

$$\mathrm{Maj}(T_{k-1}, f_k) = \mathrm{Maj}\Big(\mathrm{Mon}\big(\pi(\alpha_{k-1}\beta_{k-1})\big), f_1, \ldots, f_k\Big) = \\ = \mathrm{Maj}\Big(\mathrm{Mon}\big(\pi(\alpha_k\beta_k) + \pi(\alpha_k\beta'_k + \alpha'_k\beta_{k-1})\big), f_1, \ldots, f_k\Big) \overset{(*)}{=} \\ \overset{(*)}{=} \mathrm{Maj}\Big(\mathrm{Mon}\big(\pi(\alpha_k\beta_k)\big), f_1, \ldots, f_k\Big) = T_k \tag{B.24}$$

with the α in that lemma being $\pi(\alpha_k\beta'_k + \alpha'_k\beta_{k-1})$ and the β being $\pi(\alpha_k\beta_k)$.

The element (b, ν) in $\mathrm{Mon}\big(\pi(\alpha_k\beta_k)\big)$, which appears in the conditions of Lemma B.14, can be taken to be (c_0, λ_0). To prove its presence in

$\mathrm{Mon}(\pi(\alpha_k\beta_k))$ for $k = r$ is easy, as

$$(c_0, \lambda_0) \in \{(c_0, \lambda_0)\} = T_r = \mathrm{Maj}\Big(\mathrm{Mon}\big(\pi(\alpha_r\beta_r)\big), \mathbf{f}\Big) \subseteq \\ \subseteq \mathrm{Mon}\big(\pi(\alpha_r\beta_r)\big);$$

and for $k < r$ it is an easy consequence of the induction hypothesis, as

$$(c_0, \lambda_0) \in T_r = \mathrm{Maj}(T_k, f_{k+1}, \ldots, f_r) \subseteq T_k = \\ = \mathrm{Maj}\Big(\mathrm{Mon}\big(\pi(\alpha_k\beta_k)\big), f_1, \ldots, f_k\Big) \subseteq \mathrm{Mon}\big(\pi(\alpha_k\beta_k)\big).$$

It was shown above that if $1 \leqslant i \leqslant k$ then

$$d_{w_i}(\lambda_0) \geqslant d_{w_i}(\alpha_k\beta'_k + \alpha'_k\beta_{k-1}) \geqslant d_{w_i}\big(\pi(\alpha_k\beta'_k + \alpha'_k\beta_{k-1})\big)$$

and this fulfills condition (ii) on (c_0, λ_0). In particular it was shown that

$$d_{w_k}(\lambda_0) > d_{w_k}(\alpha_k\beta'_k + \alpha'_k\beta_{k-1}) \geqslant d_{w_k}\big(\pi(\alpha_k\beta'_k + \alpha'_k\beta_{k-1})\big)$$

and this fulfills condition (i). Thus (B.24) holds.

If $k = r$ then (B.24) is simply (B.23), and if $k \neq r$ then the latter can be proved by noticing that

$$T_r = \mathrm{Maj}(T_k, f_{k+1}, \ldots, f_r) = \mathrm{Maj}(T_{k-1}, f_k, \ldots, f_r),$$

where the first equality holds by the induction hypothesis.

Now, by the principle of induction, it follows that (B.23) holds for all $n = 1, \ldots, r$. In particular this is true for $n = 1$, and hence

$$\mathrm{Maj}\Big(\mathrm{Mon}\big(\pi(\alpha\beta)\big), \mathbf{f}\Big) = \mathrm{Maj}(T_0, \mathbf{f}) = T_r = \{(c_0, \lambda_0)\}.$$

Thus the claim of the lemma is proved. □

B.2.3 *Degree in a free algebra versus degree in a quotient*

Lemma B.16 *Let $\mathcal{R}$ be an associative and commutative ring with unit. Let X be a nonempty set. Let $\mathcal{I}$ be a two-sided ideal of the free associative algebra $\mathcal{R}\langle X\rangle$ and let $f\colon \mathcal{R}\langle X\rangle \longrightarrow \mathcal{R}\langle X\rangle/\mathcal{I}$ be the natural homomorphism.*

Let $X' = X \cup \{1\}$. Let $w\colon f(X') \longrightarrow \mathbb{Z}$ be arbitrary and let $v = w \circ f|_{X'}$. Then

$$P_j^n(w) = f\big(P_j^n(v)\big), \tag{B.25}$$
$$P_j(w) = f\big(P_j(v)\big), \tag{B.26}$$
$$M_j(w) = f\big(M_j(v)\big), \tag{B.27}$$
$$M_{-\infty}(w) \supseteq f\big(M_{-\infty}(v)\big) \tag{B.28}$$

for all $j \in \mathbb{Z}$ and $n \in \mathbb{Z}^+$.

Proof. Equation (B.25) is proved by induction on n. It holds for $n = 1$ since

$$\begin{aligned} f\big(P_j^1(v)\big) &= \{\, f(x) \,|\, x \in X' \text{ and } v(x) \leqslant j \,\} = \\ &= \Big\{\, f(x) \,\Big|\, x \in X' \text{ and } w\big(f(x)\big) \leqslant j \,\Big\} = \\ &= \{\, y \in f(X') \,|\, w(y) \leqslant j \,\} = P_j^1(w). \end{aligned}$$

Assuming that (B.25) holds for $n = k$, it follows that

$$\begin{aligned} f\big(P_j^{k+1}(v)\big) &= f\left(P_j^k(v) \cup \left(\bigcup_{x \in X'} P_{j-v(x)}^k(v) \cdot x\right)\right) = \\ &= f\big(P_j^k(v)\big) \cup \left(\bigcup_{x \in X'} f\big(P_{j-v(x)}^k(v) \cdot x\big)\right) = \\ &= P_j^k(w) \cup \left(\bigcup_{x \in X'} P_{j-w(f(x))}^k(w) \cdot f(x)\right) = \\ &= P_j^k(w) \cup \left(\bigcup_{y \in f(X')} P_{j-w(y)}^k(w) \cdot y\right) = P_j^{k+1}(w). \end{aligned}$$

Therefore, by the principle of induction, (B.25) holds for all $n \in \mathbb{Z}^+$.

Equation (B.26) follows from

$$P_j(w) = \bigcup_{n \in \mathbb{Z}^+} P_j^n(w) = \bigcup_{n \in \mathbb{Z}^+} f\big(P_j^n(v)\big) = f\big(P_j(v)\big),$$

and (B.27) follows from

$$\begin{aligned} M_j(w) &= \operatorname{Span}\big(P_j(w)\big) = \operatorname{Span}\Big(f\big(P_j(v)\big)\Big) = \\ &= f\Big(\operatorname{Span}\big(P_j(v)\big)\Big) = f\big(M_j(v)\big). \end{aligned}$$

Finally, (B.28) follows from

$$f\big(M_{-\infty}(v)\big) = f\Big(\bigcap_{j\in\mathbb{Z}} M_j(v)\Big) \subseteq \bigcap_{j\in\mathbb{Z}} f\big(M_j(v)\big) = \\ = \bigcap_{j\in\mathbb{Z}} M_j(w) = M_{-\infty}(w).$$

□

In general, equality does not hold in (B.28). One example of this would be $\mathcal{R} = \mathbb{Q}$, $X = \{\mathsf{a},\mathsf{b}\}$, $\mathcal{I} = \langle \mathsf{ab} - 1, \mathsf{ba} - 1\rangle$, and having w defined by $w(\mathsf{a} + \mathcal{I}) = 1$, $w(1 + \mathcal{I}) = 0$, and $w(\mathsf{b} + \mathcal{I}) = -2$. In that case, $M_{-\infty}(v) = \{0\}$ by Theorem B.9 as all monomials in $\mathcal{R}\langle X\rangle$ have finite degree. On the other hand, $1 + \mathcal{I} = (\mathsf{ab})^n + \mathcal{I} \in P_{-n}(w)$ for all $n \in \mathbb{N}$ and hence $M_{-\infty}(w) = \mathcal{R}\langle X\rangle/\mathcal{I}$. In the cases of interest to us here however, equality does hold.

Lemma B.17 *Let $\mathcal{R}$ be an associative and commutative ring with unit. Let X be a nonempty set. Let $\mathcal{I}$ be a two-sided ideal of the free associative algebra $\mathcal{R}\langle X\rangle$ and let $f\colon \mathcal{R}\langle X\rangle \longrightarrow \mathcal{R}\langle X\rangle/\mathcal{I}$ be the natural homomorphism.*

Let $X' = X \cup \{1\}$. Let $w\colon f(X') \longrightarrow \mathbb{Z}$ be arbitrary and let $v = w \circ f|_{X'}$. Then

$$d_w(\beta) = \inf\big\{\, d_v(\alpha) \,\big|\, \alpha \in \mathcal{R}\langle X\rangle \text{ and } \beta = f(\alpha) \,\big\} \tag{B.29}$$

for all $\beta \in \mathcal{R}\langle X\rangle/\mathcal{I}$.

Proof. Let $\alpha \in \mathcal{R}\langle X\rangle$ be arbitrary. Let $n = d_v(\alpha)$. Then $\alpha \in M_n(v)$ and $f(\alpha) \in f\big(M_n(v)\big) \subseteq M_n(w)$ (note that n might be $-\infty$ here). Hence

$$d_w\big(f(\alpha)\big) \leqslant d_v(\alpha)$$

for all $\alpha \in \mathcal{R}\langle X\rangle$.

Now let $\beta \in \big(\mathcal{R}\langle X\rangle/\mathcal{I}\big) \setminus M_{-\infty}(w)$ be arbitrary. Set $n = d_w(\beta)$. Since $\beta \in M_n(w) = f\big(M_n(v)\big)$, there exists an $\alpha \in M_n(v)$ such that $\beta = f(\alpha)$. But then

$$n = d_w(\beta) = d_w\big(f(\alpha)\big) \leqslant d_v(\alpha) \leqslant n,$$

and hence $d_w(\beta) = d_v(\alpha)$. This means α has the property $d_v(\alpha) \leqslant d_v(\gamma)$ for all $\gamma \in \mathcal{R}\langle X\rangle$ such that $f(\gamma) = \beta$. Thus

$$d_w(\beta) = d_v(\alpha) = \min\big\{\, d_v(\gamma) \,\big|\, \gamma \in \mathcal{R}\langle X\rangle \text{ and } \beta = f(\gamma) \,\big\}$$

for all $\beta \in \big(\mathcal{R}\langle X\rangle/\mathcal{I}\big) \setminus M_{-\infty}(w)$.

Finally consider an arbitrary $\beta \in M_{-\infty}(w)$ and $n \in \mathbb{Z}$. Then $\beta \in M_n(w)$ and hence there must exist an $\alpha \in M_n(v)$ such that $\beta = f(\alpha)$. This means that

$$\mathbb{Z} \cap \{ d_v(\gamma) \mid \gamma \in \mathcal{R}\langle X \rangle \text{ and } \beta = f(\gamma) \}$$

has no minimal element and therefore

$$\inf \{ d_v(\gamma) \mid \gamma \in \mathcal{R}\langle X \rangle \text{ and } \beta = f(\gamma) \} = -\infty = d_w(\beta)$$

as the lemma claims. □

Theorem B.18 *Let $\mathcal{R}$ be an associative and commutative ring with unit, and let X be a nonempty set. Let the reduction system S for $\mathcal{R}\langle X \rangle$ be such that* $\operatorname{Red}(S) = \mathcal{R}\langle X \rangle$. *Let $f \colon \mathcal{R}\langle X \rangle \longrightarrow \mathcal{R}\langle X \rangle / \mathcal{I}(S)$ be the natural homomorphism.*

Let $X' = X \cup \{1\}$. Let $w \colon f(X') \longrightarrow \mathbb{Z}$ be arbitrary and let $v = w \circ f|_{X'}$. If $d_v(\alpha_s) \leqslant d_v(\mu_s)$ for all $s \in S$ then

$$d_w\bigl(f(\alpha)\bigr) = d_v(\alpha) \tag{B.30}$$

for all $\alpha \in \operatorname{Irr}(S)$.

Proof. If $v(1) < 0$ then (B.30) is trivial, as both sides are $-\infty$. Hence assume $v(1) \geqslant 0$. In this case the infimum in (B.29) will be attained by the normal form.

First consider $d_v(\lambda\alpha_s\nu)$ and $d_v(\lambda\mu_s\nu)$ for non-unit monomials $\lambda, \nu \in \mathcal{R}\langle X \rangle$ and arbitrary $s \in S$. As there, for each $s \in S$, exists a set of monomials $P \subset \mathcal{R}\langle X \rangle$ and $\{a_\mu\}_{\mu \in P} \subseteq \mathcal{R} \setminus \{0\}$ such that $\alpha_s = \sum_{\mu \in P} a_\mu \mu$, it follows that

$$\begin{aligned} d_v(\lambda\alpha_s\nu) &\leqslant \max_{\mu \in P} d_v(\lambda\mu\nu) \leqslant \max_{\mu \in P}\bigl(d_v(\lambda) + d_v(\mu) + d_v(\nu)\bigr) = \\ &= d_v(\lambda) + \max_{\mu \in P} d_v(\mu) + d_v(\nu) = d_v(\lambda) + d_v(\alpha_s) + d_v(\nu) \leqslant \\ &\leqslant d_v(\lambda) + d_v(\mu_s) + d_v(\nu) = d_v(\lambda\mu_s\nu), \end{aligned}$$

where the last equality is by Corollary B.12. (There is a somewhat silly case in which this need not work, namely if $\mu_s = 1$, but then (B.30) holds anyway since $\operatorname{Irr}(S)$ must be $\{0\}$, because $\mu_s = 1$ makes every monomial reducible.) If λ or ν is 1 then the inequality $d_v(\lambda\alpha_s\nu) \leqslant d_v(\lambda\mu_s\nu)$ can be shown in a similar way just by considering the above calculations with this factor (term) removed.

Next note that $d_v\bigl(t_{\lambda s\nu}(\mu)\bigr) \leqslant d_v(\mu)$ for all monomials $\lambda, \mu, \nu \in \mathcal{R}\langle X\rangle$ and $s \in S$, as the only nontrivial case is $\mu = \lambda\mu_s\nu$ when

$$d_v\bigl(t_{\lambda s\nu}(\mu)\bigr) = d_v(\lambda\alpha_s\nu) \leqslant d_v(\lambda\mu_s\nu) = d_v(\mu).$$

Now fix an $\alpha \in \mathcal{R}\langle X\rangle$. Let the set P be as in Lemma B.7. From that lemma and Lemma B.4, it follows that

$$d_v\bigl(t_{\lambda s\nu}(\alpha)\bigr) \leqslant \max_{\mu\in P} d_v\bigl(t_{\lambda s\nu}(\mu)\bigr) \leqslant \max_{\mu\in P} d_v(\mu) = d_v(\alpha)$$

for all monomials $\lambda, \nu \in \mathcal{R}\langle X\rangle$ and $s \in S$. Hence $d_v\bigl(t(\alpha)\bigr) \leqslant d_v(\alpha)$ for all $\alpha \in \mathcal{R}\langle X\rangle$ and $t \in T(S)$.

Every $\alpha \in \mathcal{R}\langle X\rangle$ is, according to the assumptions, persistently and uniquely reducible, hence it is also the case that $d_v\bigl(t^S(\alpha)\bigr) \leqslant d_v(\alpha)$ for all $\alpha \in \mathcal{R}\langle X\rangle$. Thus if $\alpha \in \mathrm{Irr}(S)$ and $\beta \in \mathcal{R}\langle X\rangle$ are such that $f(\alpha) = f(\beta)$, then $d_v(\alpha) \leqslant d_v(\beta)$. But then by Lemma B.17,

$$d_w\bigl(f(\alpha)\bigr) = \inf\bigl\{\, d_v(\beta) \bigm| \beta \in \mathcal{R}\langle X\rangle \text{ and } f(\beta) = f(\alpha) \,\bigr\} = d_v(\alpha),$$

which proves the theorem. □

B.3 Gradations

Definition B.6 Let $\mathcal{G}$ be a commutative semigroup and use $+$ to denote the operation of $\mathcal{G}$. Let $\mathcal{R}$ be an associative and commutative ring with unit. Let $\mathcal{A}$ be an $\mathcal{R}$-algebra such that there exists a direct decomposition

$$\mathcal{A} = \bigoplus_{g\in\mathcal{G}} \mathcal{A}_g \tag{B.31}$$

of $\mathcal{A}$, where each $\mathcal{A}_g$ is an $\mathcal{R}$-submodule of $\mathcal{A}$. Then $\{\mathcal{A}_g\}_{g\in\mathcal{G}}$ is called a **$\mathcal{G}$-gradation** of $\mathcal{A}$ if

$$\mathcal{A}_g \cdot \mathcal{A}_h \subseteq \mathcal{A}_{g+h} \tag{B.32}$$

for all $g, h \in \mathcal{G}$.

If an $\mathcal{R}$-algebra $\mathcal{B}$ has a $\mathcal{G}$-gradation, then $\mathcal{B}$ is called a **$\mathcal{G}$-graded** algebra.

Theorem B.19 *Let $\mathcal{R}$ be an associative and commutative ring with unit. Let $\mathcal{A}$ be an associative $\mathcal{R}$-algebra. Let $\mathcal{G}$ be a commutative semigroup, and let $\{\mathcal{A}_g\}_{g\in\mathcal{G}}$ be a $\mathcal{G}$-gradation of $\mathcal{A}$. Let $J \subseteq \mathcal{A}$ be arbitrary, and let $\mathcal{I}$ be* *the two-sided ideal of* $\mathcal{A}$ *that is generated by* J. *Let* $f\colon \mathcal{A} \longrightarrow \mathcal{A}/\mathcal{I}$ *be the natural homomorphism.*

If $J \subseteq \bigcup_{g\in\mathcal{G}} \mathcal{A}_g$ *then*

$$\{f(\mathcal{A}_g)\}_{g\in\mathcal{G}} \tag{B.33}$$

is a $\mathcal{G}$*-gradation of* $\mathcal{A}/\mathcal{I}$.

Proof. It is easy to see that $f(\mathcal{A}_g)\cdot f(\mathcal{A}_h) = f(\mathcal{A}_g\cdot\mathcal{A}_h) \subseteq f(\mathcal{A}_{g+h})$ for all $g,h\in\mathcal{G}$. It is also easy to see that

$$\mathcal{A}/\mathcal{I} = f(\mathcal{A}) = f\Big(\sum_{g\in\mathcal{G}}\mathcal{A}_g\Big) = \sum_{g\in\mathcal{G}} f(\mathcal{A}_g).$$

What remains to show is that $f(\mathcal{A}_g)\cap f\big(\sum_{h\in\mathcal{G}\setminus\{g\}}\mathcal{A}_h\big) = \{0\}$ for all $g\in\mathcal{G}$.

To prove this, let $g_0\in\mathcal{G}$ be arbitrary. Let $\alpha\in\mathcal{A}_{g_0}$ and $\beta\in\sum_{h\neq g_0}\mathcal{A}_h$ be such that $f(\alpha)=f(\beta)$. Then $\alpha-\beta\in\mathcal{I}$, and thus

$$\alpha-\beta = \sum_{i=1}^{n}\gamma_i\delta_i\varepsilon_i$$

for some $n\in\mathbb{N}$, $\{\gamma_i\}_{i=1}^n, \{\varepsilon_i\}_{i=1}^n \subseteq \mathcal{A}$, and $\{\delta_i\}_{i=1}^n\subseteq J$. By assumption, there exist $\{g_i\}_{i=1}^n\subseteq\mathcal{G}$ such that $\delta_i\in\mathcal{A}_{g_i}$ for $i=1,\ldots,n$. The gradation $\{\mathcal{A}_g\}_{g\in\mathcal{G}}$ gives decompositions

$$\gamma_i = \sum_{h\in\mathcal{G}}\gamma_{ih}, \qquad \gamma_{ih}\in\mathcal{A}_h \text{ for all } h\in\mathcal{G},$$

$$\varepsilon_i = \sum_{h\in\mathcal{G}}\varepsilon_{ih}, \qquad \varepsilon_{ih}\in\mathcal{A}_h \text{ for all } h\in\mathcal{G},$$

for $i=1,\ldots,n$. Hence

$$\alpha-\beta = \sum_{i=1}^{n}\sum_{h,k\in\mathcal{G}}\gamma_{ih}\delta_i\varepsilon_{ik} = \sum_{g\in\mathcal{G}}\sum_{i=1}^{n}\sum_{\substack{h,k\in\mathcal{G}\\ h+g_i+k=g}}\gamma_{ih}\delta_i\varepsilon_{ik}.$$

After projecting this equality onto $\mathcal{A}_{g_0}$ using the projection associated with the gradation $\{\mathcal{A}_g\}_{g\in\mathcal{G}}$, one gets

$$\alpha = \sum_{i=1}^{n}\sum_{\substack{h,k\in\mathcal{G}\\ h+g_i+k=g_0}}\gamma_{ih}\delta_i\varepsilon_{ik}\in\mathcal{I}.$$

Therefore $0 = f(\alpha) = f(\beta)$, which proves that $f(\mathcal{A}_{g_0}) \cap f\big(\sum_{h \neq g_0} \mathcal{A}_h\big) = \{0\}$ for all $g_0 \in \mathcal{G}$. □

Lemma B.20 *Let $\mathcal{A}$ be a $\mathcal{G}$-graded algebra, and let $\{\mathcal{A}_g\}_{g \in \mathcal{G}}$ be the gradation. For each $g \in \mathcal{G}$, let $\pi_g \colon \mathcal{A} \longrightarrow \mathcal{A}_g$ be the projection of $\mathcal{A}$ onto $\mathcal{A}_g$ that is associated with the decomposition $\mathcal{A} = \bigoplus_{g \in \mathcal{G}} \mathcal{A}_g$.*

Let $\alpha, \gamma \in \mathcal{A}$, $f, g \in \mathcal{G}$, and $\beta \in \mathcal{A}_f$. Then

$$\pi_{f+g}(\alpha\beta) = \pi_g(\alpha)\beta, \tag{B.34}$$

$$\pi_{f+g}(\beta\alpha) = \beta\pi_g(\alpha), \tag{B.35}$$

$$\pi_f(\alpha\gamma) = \sum_{\substack{g,h \in \mathcal{G} \\ g+h=f}} \pi_g(\alpha)\,\pi_h(\gamma), \tag{B.36}$$

$$\pi_f\big([\alpha, \gamma]\big) = \sum_{\substack{g,h \in \mathcal{G} \\ g+h=f}} \big[\pi_g(\alpha), \pi_h(\gamma)\big]. \tag{B.37}$$

Proof. Let $\alpha_g = \pi_g(\alpha)$ for all $g \in \mathcal{G}$. Then $\alpha = \sum_{g \in \mathcal{G}} \alpha_g$. Thus

$$\pi_{f+g}(\alpha\beta) = \pi_{f+g}\left(\sum_{h \in \mathcal{G}} \alpha_h\beta\right) = \sum_{h \in \mathcal{G}} \pi_{f+g}(\alpha_h\beta) = \alpha_g\beta = \pi_g(\alpha)\beta$$

which proves (B.34). Equation (B.35) is proved similarly.

Let $\gamma_g = \pi_g(\gamma)$ for all $g \in \mathcal{G}$. Then

$$\begin{aligned}\pi_f(\alpha\gamma) &= \pi_f\left(\left(\sum_{g \in \mathcal{G}} \alpha_g\right)\left(\sum_{h \in \mathcal{G}} \gamma_h\right)\right) = \pi_f\left(\sum_{g,h \in \mathcal{G}} \alpha_g\gamma_h\right) = \\ &= \sum_{g,h \in \mathcal{G}} \pi_f(\alpha_g\gamma_h) = \sum_{\substack{g,h \in \mathcal{G} \\ g+h=f}} \alpha_g\gamma_h = \sum_{\substack{g,h \in \mathcal{G} \\ g+h=f}} \pi_g(\alpha)\,\pi_h(\gamma),\end{aligned}$$

which proves (B.36). Equation (B.37) is then proved by

$$\begin{aligned}\pi_f\big([\alpha, \gamma]\big) &= \pi_f(\alpha\gamma - \gamma\alpha) = \pi_f(\alpha\gamma) - \pi_f(\gamma\alpha) = \\ &= \sum_{\substack{g,h \in \mathcal{G} \\ g+h=f}} \alpha_g\gamma_h - \sum_{\substack{g,h \in \mathcal{G} \\ g+h=f}} \gamma_h\alpha_g = \sum_{\substack{g,h \in \mathcal{G} \\ g+h=f}} [\alpha_g, \gamma_h] = \\ &= \sum_{\substack{g,h \in \mathcal{G} \\ g+h=f}} \big[\pi_g(\alpha), \pi_h(\gamma)\big].\end{aligned}$$

□

Appendix C
q-special combinatorics

> "Q is sincere this time, Number One," Picard had told him, preparing transporter coordinates, the validity of which Riker did not know. "This thing is bigger than both of us ... but it's mine, Number One, and you are not to interfere."
>
> — from *Q-squared* by P. DAVID

In order to state some of the reordering formulae in Section 2.2 in a reasonably compact form, some rather unusual functions have had to be employed. They are referred to as "q-combinatorial functions", and for a good reason. Not only are their notations similar to those of the ordinary elementary combinatorial functions, but their definitions and properties are similar as well. The q-combinatorial functions do in fact include their ordinary counterparts as special cases!

In general, if f is some function, f_q is a one-parameter (q) family of functions that nicely generalise f, and $f = f_1$, then one can say that f_q is a **q-analogue** of f. Most literature on the q-combinatorial functions treat them within the theory of q-analogues of hypergeometric functions, or more generally as q-analogues of special functions, i.e., as *q-special functions.* This approach certainly has its merits—for example it gives a method of finding closed formulae for sums with a varying number of terms, where the terms are expressions involving q-combinatorial functions—but it requires more structure in the field $\mathcal{K}$ than we have assumed in this book. The theory as such is also too large for a reasonable introduction to it to fit in this appendix. Instead we will present the q-combinatorial functions simply as q-analogues of the elementary combinatorial functions. All facts about q-combinatorial functions that are used in this book appear with proofs in this appendix.

The reader who wishes to follow the alternate route to familiarity with the q-combinatorial functions has a rich choice of literature on the subject. Some of the more complete references are [109; 151; 386]. Those who are new to the subject might also want to have an introduction to hypergeometric functions. For that purpose, [161, Ch. 5] and [234] could be good starting points.

C.1 Definitions and existence

Definition C.1 Let $q \in \mathcal{K}$, $k \in \mathbb{Z}$ and $n \in \mathbb{N}$. Then the nth **q-natural number** is

$$\{n\}_q = \sum_{l=0}^{n-1} q^l, \tag{C.1}$$

the nth **q-factorial** is defined by

$$\{n\}_q! = \prod_{l=1}^{n} \{l\}_q \tag{C.2}$$

and the **q-binomial coefficients** are defined recursively by

$$\binom{n}{k}_q = \begin{cases} 0 & \text{if } k < 0 \text{ or } k > n, \\ 1 & \text{if } k = 0 \text{ or } k = n, \\ \binom{n-1}{k-1}_q + q^k \binom{n-1}{k}_q & \text{otherwise.} \end{cases} \tag{C.3}$$

Treating first the matter of whether these things are well-defined, we call the reader's attention to the following conventions about sums and products

$$\sum_{i=n}^{n-1} a_i = \sum_{i \in \varnothing} a_i = 0, \qquad \prod_{i=n}^{n-1} a_i = \prod_{i \in \varnothing} a_i = 1$$

which have been used throughout this book. As special cases of these, $\{0\}_q = 0$ and $\{0\}_q! = 1$; apart from that, existence of $\{n\}_q$ and $\{n\}_q!$ for all $q \in \mathcal{K}$ and $n \in \mathbb{N}$ is obvious. The definition of the q-binomial coefficients is an ordinary recursive definition, and thus it presents no problem.

It should be observed that for fixed values of n and k, all these q-combinatorial functions can be viewed as polynomials in q. We shall use this fact in some of the proofs in this appendix, but it is not used in the main body of this book.

When the q-natural numbers, q-factorials, and q-binomial coefficients are defined within the framework of q-hypergeometric series, it is usually not through the formulae given in Definition C.1. A more common approach is to start with defining the so called *q-Pochhammer symbol* $(a;q)_n$ through

$$(a;q)_n = \prod_{i=0}^{n-1} (1 - aq^i),$$

or even

$$(a;q)_\infty = \prod_{i=0}^{\infty} (1 - aq^i) \qquad \text{and} \qquad (a;q)_n = \frac{(a;q)_\infty}{(aq^n;q)_\infty}.$$

Using this, the q-combinatorial functions can be defined by

$$\begin{aligned} \{n\}_q &= \frac{(q^n;q)_1}{1-q} \\ \{n\}_q! &= \frac{(q;q)_n}{(1-q)^n} \\ \binom{n}{k}_q &= \frac{(q;q)_n}{(q;q)_k (q;q)_{n-k}} \end{aligned}$$

The main advantage with using these formulae as definitions, apart from the fact that it makes them fit in better with the general framework of q-hypergeometric series, is that they can be taken as definitions of the q-combinatorial functions for non-integer values of n and k. As we have no need for such generalisations, and as the definitions using $(a;q)_n$ complicates matters in the important case $q = 1$, we found that the formulae in Definition C.1 are more suitable for our needs.

As a historical note, we remark that the q-binomial coefficients are known to have been studied by Gauss, and that they for that reason often are called the *Gaussian polynomials.** The history of q-hypergeometric series is even older and goes back to Euler, which is why these things are sometimes called *Eulerian series.* Euler, Gauss, and Jacobi discovered many of the fundamental results in the theory, but the first systematic study was undertaken by Heine around the middle of the 19th century.

C.2 Other properties

Most of the properties in this section are similar to properties of the normal factorial and binomial functions. This is not a coincidence. The map $n \mapsto \{n\}_1$ is a homomorphism $\mathbb{N} \longrightarrow \mathcal{K}$, so for this particular value of q some interesting equalities hold, as for example

$$\{n\}_1! = \{n!\}_1, \qquad \binom{n}{k}_1 = \left\{\binom{n}{k}\right\}_1.$$

An alternative point of view on this in the case when $\mathcal{K}$ has characteristic 0 would be to identify the subfield of $\mathcal{K}$ spanned by 1 with $\mathbb{Q}$. In that case these equalities would simply be

$$\{n\}_1 = n, \qquad \{n\}_1! = n!, \qquad \binom{n}{k}_1 = \binom{n}{k},$$

and the q-combinatorial functions would simplify to their usual counterparts.

If instead $\mathcal{K}$ has characteristic p then the subfield spanned by 1 can be identified with $\mathbb{Z}_p$ and the equalities can be written as

$$\{n\}_1 = n + p\mathbb{Z}, \qquad \{n\}_1! = n! + p\mathbb{Z}, \qquad \binom{n}{k}_1 = \binom{n}{k} + p\mathbb{Z}$$

(at least if $\mathbb{Z}_p$ is constructed as $\mathbb{Z}/p\mathbb{Z}$).

*Not to be confused with the Gauss polynomials of Definition C.3.

Yet another view would be to consider $\mathcal{K}$ as an $\mathbb{N}$-module, then one would have $\{n\}_1 = n1$ (the scalar n is multiplied by 1, which is an element of the module $\mathcal{K}$), but we think this should better be avoided as it probably would be mistaken for a misprint.

Leaving the special case $q = 1$ and going on to arbitrary $q \in \mathcal{K}$, a good relation to start with is that

$$q^k\{n\}_q + \{k\}_q = \{n+k\}_q \tag{C.4}$$

for all $n, k \in \mathbb{N}$ and $q \in \mathcal{K}$. Thus in general the mapping $n \mapsto \{n\}_q$ is not a homomorphism.

The equality (C.4) can easily be generalised to arbitrarily many terms, as is done in the following lemma.

Lemma C.1 *For all $r \in \mathbb{N}$, all $\{n_j\}_{j=1}^r \subset \mathbb{N}$, and $q \in \mathcal{K}$,*

$$\left\{\sum_{k=1}^r n_k\right\}_q = \sum_{k=1}^r \left(\prod_{j=k+1}^r q^{n_j}\right)\{n_k\}_q = \sum_{k=1}^r q^{N_k}\{n_k\}_q, \tag{C.5}$$

where $N_k = \sum_{j=k+1}^r n_j$ for $k = 1, 2, \ldots, r$.

Proof. When $r = 0$, the above equality simplifies to $\{0\}_q = 0$ (remember that an empty sum sums to zero). For all other values of r, it can be proved through a trivial induction on r. Assume that (C.5) holds for some $r \in \mathbb{N}$. Then by (C.4) and the assumption,

$$\left\{\sum_{k=1}^r n_k + n_{r+1}\right\}_q = q^{n_{r+1}}\left\{\sum_{k=1}^r n_k\right\}_q + \{n_{r+1}\}_q =$$

$$= q^{n_{r+1}}\sum_{k=1}^r \left(\prod_{j=k+1}^r q^{n_j}\right)\{n_k\}_q + \{n_{r+1}\}_q = \sum_{k=1}^{r+1}\left(\prod_{j=k+1}^{r+1} q^{n_j}\right)\{n_k\}_q.$$

Hence (C.5) remains true if r is replaced by $r+1$. Thus, by the principle of mathematical induction, (C.5) holds for all $r \in \mathbb{N}$. □

When $n_k = n$ for $k = 1, 2, \ldots, r$ in the above lemma, the formula (C.5) yields

$$\{rn\}_q = \sum_{k=1}^r q^{(r-k)n}\{n\}_q = \{n\}_q \sum_{k=1}^r (q^n)^{r-k} = \{n\}_q\{r\}_{q^n} \tag{C.6}$$

for all $n, r \in \mathbb{N}$ and $q \in \mathcal{K}$. This formula has a generalisation analogous to that of (C.4).

Lemma C.2 *For all $r \in \mathbb{N}$, all $\{n_j\}_{j=1}^r \subset \mathbb{N}$, and $q \in \mathcal{K}$,*

$$\left\{\prod_{k=1}^{r} n_k\right\}_q = \prod_{k=1}^{r} \{n_k\}_{Q_k(q,r)},$$

$$\text{where} \quad Q_k(q,r) = q^{N_k(r)} \quad \text{and} \quad N_k(r) = \prod_{j=k+1}^{r} n_j \quad \text{for } k = 1, \ldots, r. \tag{C.7}$$

Proof. When $r = 0$, the above equality simplifies to $\{1\}_q = 1$ (remember that a product in which the number of factors is zero takes as value the unit 1). For all other values of r, it can be proved through a trivial induction on r.

Assume that (C.7) holds for some $r \in \mathbb{N}$. Note that $Q(q^{n_{r+1}}, r) = (q^{n_{r+1}})^{N_k(r)} = q^{n_{r+1}N_k(r)} = q^{N_k(r+1)} = Q(q, r+1)$. Then by (C.6) and the assumption,

$$\begin{aligned}
\left\{\prod_{k=1}^{r} n_k \cdot n_{r+1}\right\}_q &= \{n_{r+1}\}_q \left\{\prod_{k=1}^{r} n_k\right\}_{q^{n_{r+1}}} = \\
&= \{n_{r+1}\}_q \prod_{k=1}^{r} \{n_k\}_{Q_k(q^{n_{r+1}},r)} = \\
&= \{n_{r+1}\}_{Q_{r+1}(q,r+1)} \prod_{k=1}^{r} \{n_k\}_{Q_k(q,r+1)} = \\
&= \prod_{k=1}^{r+1} \{n_k\}_{Q_k(q,r+1)}.
\end{aligned}$$

Hence (C.7) remains true if r is replaced by $r+1$. Thus, by the principle of mathematical induction, (C.7) holds for all $r \in \mathbb{N}$. □

The main attention from now on will be focused on the q-binomial coefficients and their relation to the two other q-combinatorial functions.

Lemma C.3 *For all $k, n \in \mathbb{N}$ satisfying $k \leqslant n$,*

$$\{k\}_q! \binom{n}{k}_q = \prod_{j=n-k+1}^{n} \{j\}_q. \tag{C.8}$$

as elements of $\mathcal{K}[q]$. Thus in particular the equality holds for all $q \in \mathcal{K}$ as well.

Proof. The proof is by induction on n. First observe that if $k = 0$ then

$$\text{LHS} = \{0\}_q! \binom{n}{0}_q = 1 \cdot 1 = 1 = \prod_{j=n+1}^{n} \{j\}_q = \text{RHS}$$

and if $k = n$ then

$$\text{LHS} = \{n\}_q! \binom{n}{n}_q = \{n\}_q! \cdot 1 = \{n\}_q! = \prod_{j=1}^{n} \{j\}_q = \text{RHS}.$$

This has laid the base for the induction.

Now assume (C.8) holds for $n = m \in \mathbb{N}$. It will be shown that this assumption implies that (C.8) holds for $n = m + 1$ as well. If $k = 0$ or $k = m + 1$ then (C.8) was shown above, hence let k satisfy $0 < k < n$. Then

$$\begin{aligned}\text{LHS} &= \{k\}_q! \left(\binom{m}{k-1}_q + q^k \binom{m}{k}_q \right) = \\ &= \{k\}_q \{k-1\}_q! \binom{m}{k-1}_q + q^k \{k\}_q! \binom{m}{k}_q = \\ &= \{k\}_q \prod_{j=m-k+2}^{m} \{j\}_q + q^k \prod_{j=m-k+1}^{m} \{j\}_q = \\ &= \left(\{k\}_q + q^k \{m-k+1\}_q \right) \prod_{j=m-k+2}^{m} \{j\}_q = \\ &= \{m+1\}_q \prod_{j=m+1-k+1}^{m} \{j\}_q = \text{RHS}.\end{aligned}$$

Thus, by the induction principle, the lemma follows. □

Corollary C.4 *Let $n, k \in \mathbb{N}$ and $q \in \mathcal{K}$. If $\{j\}_q \neq 0$ for all $j \in \mathbb{Z}^+$ and $k \leqslant n$ then*

$$\binom{n}{k}_q \neq 0 \quad \textit{and} \quad \{k\}_q! \neq 0. \tag{C.9}$$

Proof. The conditions imply that (C.8) holds and that its right hand side is nonzero. Thus all factors in the left hand side must be nonzero too. □

Corollary C.5 *Let $n \in \mathbb{N}$ and $q \in \mathcal{K}$. Then*

$$\binom{n}{1}_q = \{n\}_q. \tag{C.10}$$

Proof. Note that $\binom{0}{1}_q = 0 = \{0\}_q$ by definition. Apart from that, (C.10) is just the $k = 1$ case of (C.8). □

Theorem C.6 *Let $k, n \in \mathbb{N}$ satisfy $k \leqslant n$. Then*

$$\binom{n}{k}_q = \frac{\{n\}_q!}{\{k\}_q!\,\{n-k\}_q!} \tag{C.11}$$

as elements of $\mathcal{K}[q]$. Furthermore (C.11) *holds as an equality of elements in $\mathcal{K}$ for all $q \in \mathcal{K}$ such that $\{j\}_q \neq 0$ for all j satisfying $1 \leqslant j \leqslant n$.*

Proof. By Lemma C.3,

$$\{k\}_q!\binom{n}{k}_q = \prod_{j=n-k+1}^{n} \{j\}_q, \tag{C.12}$$

and hence $\{k\}_q! \;\Big|\; \prod_{j=n-k+1}^{n}\{j\}_q$ as elements of $\mathcal{K}[q]$. This means that if both sides of (C.12) are divided by $\{k\}_q!$, then both sides will still be elements of $\mathcal{K}[q]$. Hence

$$\binom{n}{k}_q = \frac{\prod_{j=n-k+1}^{n}\{j\}_q}{\{k\}_q!} = \frac{\{n\}_q!}{\{k\}_q!\,\{n-k\}_q!}$$

as elements of $\mathcal{K}[q]$, exactly as the lemma claims.

If $\{j\}_q \neq 0$ for all j satisfying $1 \leqslant j \leqslant n$ then $\{k\}_q! = \prod_{j=1}^{k}\{j\}_q \neq 0$ for all k satisfying $0 \leqslant k \leqslant n$. Thus in this case the right hand side of (C.11) is equally well defined when it is interpreted as a quotient of two elements of $\mathcal{K}$. Hence in that case the equality holds for such $q \in \mathcal{K}$. □

Note that (C.11) does not hold in general if q is interpreted as an element of $\mathcal{K}$. The reason for this is that when a specific value is given for q, the denominator of the right hand side may be zero, making the entire right hand side undefined. It is never the case that the left and right hand sides are both defined but not equal. Whenever it is defined, the right hand side has the same value as the left hand side.

The standard combinatorial interpretation of the q-binomial coefficients is that

$$\binom{n}{k}_q = \text{the number of } k\text{-dimensional subspaces of } \mathbb{F}_q^n,$$

where $\mathbb{F}_q$ is the field with q elements (the formula only makes sense if q is a prime power) and $\mathbb{F}_q^n$ accordingly is the n-dimensional vector space over $\mathbb{F}_q$. The formula is rather easy to prove by a straightforward counting argument and then applying Theorem C.6. Every i-dimensional subspace W of $\mathbb{F}_q^n$ contains q^i points (vectors), and hence there are $q^n - q^i = q^i(q-1)\{n-i\}_q$ points in $\mathbb{F}_q^n \setminus W$. Therefore there are $\prod_{i=0}^{k-1} q^i(q-1)\{n-i\}_q$ ordered sets of k vectors in $\mathbb{F}_q^n$ that span a k-dimensional subspace of $\mathbb{F}_q^n$, and in every k-dimensional subspace of $\mathbb{F}_q^n$ there are $\prod_{i=0}^{k-1} q^i(q-1)\{k-i\}_q$ different ordered sets of k vectors that span it.

A similar argument can be used to derive a combinatorial interpretation of the q-factorials—$\{n\}_q!$ is the number of maximal chains in the subspace lattice of the vector space $\mathbb{F}_q^n$. (Here we don't use the meaning of 'chain' that was given in Definition 4.3, but in the normal sense of a totally ordered subset of a partially ordered set.) To see why this is the case, imagine picking n linearly independent vectors $\mathbf{x}_1, \dots, \mathbf{x}_n \in \mathbb{F}_q^n$ one by one. Let $W_i = \operatorname{Span}(\{\mathbf{x}_1, \dots, \mathbf{x}_i\})$ for $i = 1, \dots, n$. Then $\{0\} \subset W_1 \subset W_2 \subset \cdots \subset W_n = \mathbb{F}_q^n$ will be a maximal chain in the subspace lattice, and all maximal chains can be constructed in this way.

When picking $\mathbf{x}_1$ there are $q^n - 1$ different vectors to choose from, but any nonzero vector in the resulting W_1 will produce the same space, and therefore there are $(q^n-1)/(q-1) = \{n\}_q$ choices for W_1. When picking $\mathbf{x}_{i+1}$, the vectors you may choose between are the $q^n - q^i$ different vectors in $\mathbb{F}_q^n \setminus W_i$. Any vector in the resulting $W_{i+1} \setminus W_i$ will produce the same W_{i+1}, and therefore there are $(q^n - q^i)/(q^{i+1} - q^i) = \{n-i\}_q$ choices for W_{i+1} when W_i has been fixed. Therefore there are in total $\prod_{i=0}^{n-1} \{n-i\}_q = \{n\}_q!$ maximal chains in $\mathbb{F}_q^n$.

Theorem C.7 *For all $n \in \mathbb{N}$, $k \in \mathbb{Z}$, and $q \in \mathcal{K}$:*

$$\binom{n}{k}_q = \binom{n}{n-k}_q. \tag{C.13}$$

Proof. If $k < 0$ or $k > n$ then both sides of the above equality are 0.

Otherwise Theorem C.6 applies and

$$\binom{n}{k}_q = \frac{\{n\}_q!}{\{k\}_q!\,\{n-k\}_q!} = \frac{\{n\}_q!}{\{n-k\}_q!\,\{n-(n-k)\}_q!} = \binom{n}{n-k}_q$$

as elements of $\mathcal{K}[q]$. Hence (C.13) holds as an equality of elements of $\mathcal{K}[q]$. Since both sides of (C.13) are defined for all $q \in \mathcal{K}$, the equality holds as an equality of elements of $\mathcal{K}$ as well. □

Corollary C.8 *Let $n \in \mathbb{Z}^+$, $k \in \mathbb{Z}$ and $q \in \mathcal{K}$. Then*

$$\binom{n-1}{k-1}_q + q^k \binom{n-1}{k}_q = \binom{n}{k}_q = q^{n-k}\binom{n-1}{k-1}_q + \binom{n-1}{k}_q. \tag{C.14}$$

both as an equality of elements of $\mathcal{K}[q]$ and as an equality of elements of $\mathcal{K}$.

Proof. If $0 < k < n$ then the first of these equalities is exactly the definition of the q-binomial coefficients. The proof of the second equality is easiest done using the first equality and equation (C.13):

$$q^{n-k}\binom{n-1}{k-1}_q + \binom{n-1}{k}_q = q^{n-k}\binom{n-1}{n-k}_q + \binom{n-1}{n-k-1}_q =$$
$$= \binom{n}{n-k}_q = \binom{n}{k}_q.$$

If $k = 0$ then (C.14) simplifies to

$$0 + q^0 \cdot 1 = 1 = q^n \cdot 0 + 1,$$

if $k = n$ then it simplifies to

$$1 + q^n \cdot 0 = 1 = q^0 \cdot 1 + 0,$$

and if $k < 0$ or $k > n$ then all three terms are 0 and equality is even more trivial. □

The next theorem gives a closed form formula for the q-binomial coefficients which can be used with any $q \in \mathcal{K}$. It's not very efficient, though.

Theorem C.9 *Let $n \in \mathbb{N}$ and $k \in \mathbb{Z}$. Then regardless of whether q is seen as an element of $\mathcal{K}$ or as a variable,*

$$\binom{n}{k}_q = \sum_{\substack{S \subseteq [n] \\ |S|=k}} q^{\sum S - k(k+1)/2} \tag{C.15}$$

where $\sum S$ is short for $\sum_{s \in S} s$ (cf. similar notation for union and intersection), $[n]$ denotes the set $\{1, \dots, n\}$, and in particular $[0] = \varnothing$.

Proof. The proof is by induction on n. If $n = 0$ then the only subset of $[n]$ is $\varnothing$ and $|\varnothing| = 0$. As furthermore $\sum \varnothing = \sum_{s \in \varnothing} s = 0$, it follows that

$$\sum_{\substack{S \subseteq [0] \\ |S|=k}} q^{\sum S - k(k+1)/2} = \begin{cases} 1 & \text{if } k = 0 \\ 0 & \text{if } k \neq 0 \end{cases} = \binom{0}{k}_q .$$

This takes care of the inductive base.

Now assume that (C.15) holds for $n = m-1$, where $m \in \mathbb{Z}^+$ is arbitrary; it will be proved that this implies that (C.15) holds for $n = m$ as well.

$$\begin{aligned} \sum_{\substack{S \subseteq [m] \\ |S|=k}} q^{\sum S - k(k+1)/2} &= \sum_{\substack{S \subseteq [m-1] \\ |S|=k-1}} q^{\sum S + m - k(k+1)/2} + \sum_{\substack{S \subseteq [m-1] \\ |S|=k}} q^{\sum S - k(k+1)/2} = \\ &= q^{m-k} \sum_{\substack{S \subseteq [m-1] \\ |S|=k-1}} q^{\sum S - (k(k+1)-2k)/2} + \binom{m-1}{k}_q = \\ &= q^{m-k} \binom{m-1}{k-1}_q + \binom{m-1}{k}_q = \binom{m}{k}_q , \end{aligned}$$

where the last step is by Corollary C.8. This completes the inductive step and thus, by the induction principle, the theorem follows. □

Theorem C.9 can be used to derive another combinatorial interpretation of the q-binomial coefficients, namely as the generating functions for certain kinds of partitions of natural numbers. Recall that an $\mathbf{r} \in (\mathbb{Z}^+)^k$ is said to be a *partition* of the number $n \in \mathbb{N}$ if $r_1 \geqslant r_2 \geqslant \cdots \geqslant r_k$ and $\sum_{i=1}^{k} r_i = n$. The numbers r_i are called the *parts* of the partition, and a partition with k parts is called a *k-partition.*

Let $\mathcal{K} \supseteq \mathbb{Z}$. Then by (C.15), the coefficient of q^m in $\binom{n}{k}_q$ is simply the number of sets $S \subseteq \{1, \dots, n\}$ such that $|S| = k$ and $\sum_{s \in S} s = m + \frac{1}{2}k(k+1)$. Each such set S corresponds bijectively to a k-partition of $m + \frac{1}{2}k(k+1)$ in which the parts are distinct and all parts are $\leqslant n$, and hence

$$f_1(q) = q^{k(k+1)/2} \binom{n}{k}_q$$

is the generating function for k-partitions in which all parts are distinct and $\leqslant n$, i.e., the coefficient of q^m in $f_1(q)$ gives the number of such partitions of m.

That all parts of a k-partition $\mathbf{r}$ are distinct and $\leqslant n$ can be expressed as

$$n \geqslant r_1 > r_2 > \cdots > r_k > 0; \tag{C.16}$$

any $\mathbf{r} \in \mathbb{Z}^k$ which satisfies (C.16) is such a partition. If one lets $s_i = r_i - (1 + k - i)$ for $i = 1, \ldots, k$, then $\mathbf{s} = (s_1, \ldots, s_k)$ satisfies $\sum_{i=1}^{k} r_i = \frac{1}{2}k(k+1) + \sum_{i=1}^{k} s_i$ and

$$n - k \geqslant s_1 \geqslant s_2 \geqslant \cdots \geqslant s_k \geqslant 0. \tag{C.17}$$

In fact, $s_i = r_i - (1 + k - i)$ for $i = 1, \ldots, k$ defines a bijection between the $\mathbf{r} \in \mathbb{Z}^k$ which satisfy (C.16) and the $\mathbf{s} \in \mathbb{Z}^k$ which satisfy (C.17). Furthermore each such $\mathbf{s}$ corresponds bijectively to a partition with *at most* k parts, all of which are $\leqslant n-k$; if $s_i > 0$ for $i \leqslant l$ and $s_i = 0$ for $i > l$, then $\mathbf{s}$ corresponds to the l-partition $(s_1, \ldots, s_l)$. Hence there is a bijection from the k-partitions in which all parts are distinct and $\leqslant n$, to the partitions with at most k parts, all of which are $\leqslant n - k$. This bijection maps a partition of m to a partition of $m - \frac{1}{2}k(k+1)$. Thus $f_2(q) = \binom{n}{k}_q$ is the generating function for the number of partitions with at most k parts, all of which are $\leqslant n - k$.

For each $\mathbf{s} \in \mathbb{Z}^k$ which satisfies (C.17), there is a $\mathbf{t} \in \mathbb{Z}^k$ which satisfies $\sum_{i=1}^{k} t_i = k(n-k) - \sum_{i=1}^{k} s_i$ and

$$n - k \geqslant t_1 \geqslant t_2 \geqslant \cdots \geqslant t_k \geqslant 0,$$

namely that which is defined by $t_i = (n-k) - s_{k+1-i}$ for $i = 1, \ldots, k$. This is a bijection from the partitions of m with at most k parts, all of which are $\leqslant n-k$, to the partitions of $k(n-k) - m$ with at most k parts, all of which are $\leqslant n - k$. As this implies that there are the same number of both kinds of partitions, the corresponding coefficients in the generating function must be equal, i.e., the coefficients of q^m and $q^{k(n-k)-m}$ in $\binom{n}{k}_q$ are equal.

This displays another symmetry of the q-binomial coefficients—seen as polynomials in q, you see the same coefficient sequence regardless of whether you start with the highest degree term or the lowest degree term. The following theorem proves this in a more general setting, by examining the relations between the q-combinatorial functions for q and for q^{-1}. It turns out that it is pretty simple to go from an expression including a q-combinatorial function in q^{-1} to the same q-combinatorial function in q.

Theorem C.10 *Let $q \in \mathcal{K}$ be nonzero, let $n \in \mathbb{N}$, and let $k \in \mathbb{Z}$. Then*

$$\{n\}_{q^{-1}} = q^{1-n}\{n\}_q, \tag{C.18}$$

$$\{n\}_{q^{-1}}! = q^{-\binom{n}{2}}\{n\}_q!, \tag{C.19}$$

$$\binom{n}{k}_{q^{-1}} = q^{-k(n-k)}\binom{n}{k}_q \tag{C.20}$$

Proof. Beginning with (C.18),

$$\{n\}_{q^{-1}} = \sum_{i=0}^{n-1} q^{-i} = q^{1-n}\sum_{i=0}^{n-1} q^{n-1-i} = q^{1-n}\sum_{j=0}^{n-1} q^j = q^{1-n}\{n\}_q.$$

With this in mind,

$$\{n\}_{q^{-1}}! = \prod_{i=1}^{n}\{i\}_{q^{-1}} = \prod_{i=1}^{n} q^{1-i}\{i\}_q = \prod_{j=0}^{n-1} q^{-j}\cdot\prod_{i=1}^{n}\{i\}_q = q^{-\binom{n}{2}}\{n\}_q!,$$

which proves (C.19).

The proof of (C.20) is a bit trickier, but not much. It is clearly true for $k < 0$ and $k > n$, since then both sides of the equality are 0. It is also true for $k = 0$ and $k = n$ since the exponent on q happens to be 0 in this case and the definitions of these q-binomial coefficients are independent of q. What is left is to prove it in the case $0 < k < n$.

This proof is easiest done by induction on n. Assume that (C.20) holds for $n = m - 1$, where $m > 0$, and all k satisfying $0 \leqslant k \leqslant m - 1$. It shall be proved that (C.20) holds for $n = m$ and all k satisfying $0 \leqslant k \leqslant m$ as well. The cases $k = 0$ and $k = m$ were taken care of above. By Theorem C.8,

$$\begin{aligned}\binom{m}{k}_{q^{-1}} &= \binom{m-1}{k-1}_{q^{-1}} + (q^{-1})^k\binom{m-1}{k}_{q^{-1}} = \\ &= q^{-(k-1)(m-k)}\binom{m-1}{k-1}_q + q^{-k}q^{-k(m-1-k)}\binom{m-1}{k}_q = \\ &= q^{-k(m-k)}q^{m-k}\binom{m-1}{k-1}_q + q^{-k(m-k)}\binom{m-1}{k}_q = \\ &= q^{-k(m-k)}\binom{m}{k}_q,\end{aligned}$$

which takes care of the remaining case. Thus by the principle of induction, it follows that (C.20) holds for all $k \in \mathbb{Z}$ and all $n \in \mathbb{N}$, as the theorem claims. □

Lemma C.11 *Let $q \in \mathcal{K}$, let $n, m, k \in \mathbb{Z}^+$, and assume $k < \min(n, m)$. Then*

$$\{m\}_q \{k-1\}_q! \binom{n-1}{k-1}_q \binom{m-1}{k-1}_q + q^k \{k\}_q! \binom{n-1}{k}_q \binom{m}{k}_q = \{k\}_q! \binom{n}{k}_q \binom{m}{k}_q \quad \text{(C.21)}$$

Proof. To begin with,

$$\{m\}_q \{k-1\}_q! \binom{m-1}{k-1}_q = \{m\}_q \prod_{j=m-k+1}^{m-1} \{j\}_q = \prod_{j=m-k+1}^{m} \{j\}_q = \{k\}_q! \binom{m}{k}_q$$

It thus follows that

$$\begin{aligned}\{m\}_q \{k-1\}_q! \binom{n-1}{k-1}_q \binom{m-1}{k-1}_q + q^k \{k\}_q! \binom{n-1}{k}_q \binom{m}{k}_q &= \\ = \{k\}_q! \binom{n-1}{k-1}_q \binom{m}{k}_q + q^k \{k\}_q! \binom{n-1}{k}_q \binom{m}{k}_q &= \\ = \{k\}_q! \binom{n}{k}_q \binom{m}{k}_q\end{aligned}$$

which proves the lemma. □

C.3 q-Stirling numbers

The *Stirling numbers* are two families of numbers called the Stirling numbers of the first and second kind respectively. These numbers occur as coefficients in several interesting formulae involving counting, polynomials, and generating functions. The Stirling numbers are named after the Scottish mathematician James Stirling† (1692–1770) who introduced these numbers in his *Methodus Differentialis* (1730).

†This is the same Stirling as in 'Stirling's formula' $n! \approx \sqrt{2\pi n}\,(n/e)^n$, but that formula was actually discovered by de Moivre.

The q-Stirling numbers relate to their ordinary counterparts pretty much like the three q-combinatorial functions defined above relate to their ordinary counterparts—they have an extra parameter named q, they can be seen as polynomials in this parameter, and for $q = 1$ one gets the ordinary counterparts back.

Definition C.2 By $\mathcal{K}(q)$ we mean the field of fractions of the polynomial ring $\mathcal{K}[q]$. The **q-deformed falling factorial polynomials** $\{F_n(x;q)\}_{n=0}^{\infty}$ are the elements of $\mathcal{K}(q)[x]$ which are defined by

$$F_n(x;q) = \prod_{i=0}^{n-1} \bigl(x - \{i\}_q\bigr) \qquad \text{for all } n \in \mathbb{N}. \tag{C.22}$$

Note that the q-deformed falling factorial polynomials constitute a basis of the vector space $\mathcal{K}(q)[x]$.

The **q-Stirling numbers of the first kind** $\left[{n \atop k}\right]_q$ are the elements of $\mathcal{K}(q)$ which satisfy

$$F_n(x;q) = \sum_{k=0}^{\infty} \begin{bmatrix} n \\ k \end{bmatrix}_q (-1)^{n-k} x^k \tag{C.23}$$

for all $n \in \mathbb{N}$. Similarly the **q-Stirling numbers of the second kind** $\left\{{n \atop k}\right\}_q$ are the elements of $\mathcal{K}(q)$ which satisfy

$$x^n = \sum_{k=0}^{\infty} \begin{Bmatrix} n \\ k \end{Bmatrix}_q F_k(x;q) \tag{C.24}$$

for all $n \in \mathbb{N}$.

Note: This is not the only q-analogue of the Stirling numbers that is known, but it happens to be sufficient for our needs. The reader may recognize it as the q-analogue studied by Carlitz [51] and Gould [160]. A comparison of this with a couple of other q-Stirling numbers can be found in Wagner [388], and yet other q-Stirling numbers are defined in Cigler [78], but these papers do not cover all q-analogues of Stirling numbers that are known in the literature.

The reader should also observe that we have defined the q-Stirling numbers so that they reduce to the *signless* Stirling numbers when $q = 1$. The Stirling numbers are otherwise often defined so that one of the kinds have alternating signs, as would be the case for those of the first kind if the factor $(-1)^{n-k}$ in (C.23) had been removed.

Our notation for q-Stirling numbers is based on the notation used by Knuth *et al.* in [161] and other books. The case for this notation is put forth in [232], which also contains several interesting observations about the history of the Stirling numbers, and of mathematical notation in general.

Since both $\{x^n\}_{n=0}^{\infty}$ and $\{F_n(x;q)\}_{n=0}^{\infty}$ are bases of the vector space $\mathcal{K}(q)[x]$, the equations (C.23) and (C.24) do really uniquely determine the q-Stirling numbers of the first and second kind for all $n, k \in \mathbb{N}$. Furthermore for each $n \in \mathbb{N}$, all but a finite number of terms in (C.23) and (C.24) are zero. These terms follow a very simple pattern, as the following lemma shows.

Lemma C.12 *If $n, k \in \mathbb{N}$ are such that $k > n$ then*

$$\begin{bmatrix} n \\ k \end{bmatrix}_q = 0 = \begin{Bmatrix} n \\ k \end{Bmatrix}_q. \tag{C.25}$$

Hence (C.23) *and* (C.24) *can be rephrased as the finite sums*

$$F_n(x;q) = \sum_{k=0}^{n} \begin{bmatrix} n \\ k \end{bmatrix}_q (-1)^{n-k} x^k, \tag{C.26}$$

$$x^n = \sum_{k=0}^{n} \begin{Bmatrix} n \\ k \end{Bmatrix}_q F_k(x;q). \tag{C.27}$$

Proof. Observe that $\deg F_n(x;q) = \deg x^n = n$. Since both sides of (C.23) and (C.24) have the same degree, all terms for $k > n$ in the right hand sides must be 0. □

The q-Stirling numbers can alternatively be defined recursively. The recursion formulae are derived in the following theorem.

Theorem C.13 *For all $n \in \mathbb{N}$ and $k \in \mathbb{Z}^+$,*

$$\begin{bmatrix} n+1 \\ k \end{bmatrix}_q = \begin{bmatrix} n \\ k-1 \end{bmatrix}_q + \{n\}_q \begin{bmatrix} n \\ k \end{bmatrix}_q, \tag{C.28}$$

$$\begin{Bmatrix} n+1 \\ k \end{Bmatrix}_q = \begin{Bmatrix} n \\ k-1 \end{Bmatrix}_q + \{k\}_q \begin{Bmatrix} n \\ k \end{Bmatrix}_q, \tag{C.29}$$

and furthermore

$$\begin{bmatrix}0\\0\end{bmatrix}_q = 1, \qquad \begin{bmatrix}k\\0\end{bmatrix}_q = 0, \qquad \begin{bmatrix}0\\k\end{bmatrix}_q = 0, \tag{C.30}$$

$$\begin{Bmatrix}0\\0\end{Bmatrix}_q = 1, \qquad \begin{Bmatrix}k\\0\end{Bmatrix}_q = 0, \qquad \begin{Bmatrix}0\\k\end{Bmatrix}_q = 0. \tag{C.31}$$

Proof. For $n = 0$, (C.23) becomes

$$\sum_{k=0}^{\infty} \begin{bmatrix}0\\k\end{bmatrix}_q (-1)^k x^k = F_0(x;q) = 1 = x^0.$$

Hence from identifying coefficients of x^k on both sides it follows that $\left[\begin{smallmatrix}0\\0\end{smallmatrix}\right]_q = 1$ and $\left[\begin{smallmatrix}0\\k\end{smallmatrix}\right]_q = 0$ for $k > 0$. By substituting $x = 0$ in (C.23) for $n > 0$, one gets

$$0 = F_n(0;q) = \sum_{k=0}^{\infty} \begin{bmatrix}n\\k\end{bmatrix}_q (-1)^{n-k} 0^k = \begin{bmatrix}n\\0\end{bmatrix}_q.$$

This has shown (C.30).

A consequence of the last equality is that for $n > 0$ the sum in (C.23) can be taken from 1 to ∞. Thus

$$\begin{aligned}
\sum_{k=1}^{\infty} \begin{bmatrix}n+1\\k\end{bmatrix}_q (-1)^{n+1-k} x^k &= \\
&= F_{n+1}(x;q) = \big(x - \{n\}_q\big) F_n(x;q) = \\
&= x \sum_{k=0}^{\infty} \begin{bmatrix}n\\k\end{bmatrix}_q (-1)^{n-k} x^k - \{n\}_q \sum_{k=0}^{\infty} \begin{bmatrix}n\\k\end{bmatrix}_q (-1)^{n-k} x^k = \\
&= \sum_{i=1}^{\infty} \begin{bmatrix}n\\i-1\end{bmatrix}_q (-1)^{n-(i-1)} x^i + \sum_{k=0}^{\infty} \{n\}_q \begin{bmatrix}n\\k\end{bmatrix}_q (-1)^{n+1-k} x^k = \\
&= \sum_{k=1}^{\infty} \left(\begin{bmatrix}n\\k-1\end{bmatrix}_q + \{n\}_q \begin{bmatrix}n\\k\end{bmatrix}_q \right) (-1)^{n+1-k} x^k,
\end{aligned}$$

where in the second last step $i := k+1$, but in the last step $k := i$. Equation (C.28) follows from the equality of coefficients of x^k in both sides.

For $n = 0$, (C.24) becomes

$$\sum_{k=0}^{\infty} \begin{Bmatrix} 0 \\ k \end{Bmatrix}_q F_k(x;q) = x^0 = 1 = F_0(x;q).$$

Hence from identifying coefficients of $F_k(x;q)$ on both sides it follows that $\left\{\begin{smallmatrix} 0 \\ 0 \end{smallmatrix}\right\}_q = 1$ and $\left\{\begin{smallmatrix} 0 \\ k \end{smallmatrix}\right\}_q = 0$ for $k > 0$. By substituting $x = 0$ in (C.24) for $n > 0$, one gets

$$0 = 0^n = \sum_{k=0}^{\infty} \begin{Bmatrix} n \\ k \end{Bmatrix}_q F_k(0;q) = \begin{Bmatrix} n \\ 0 \end{Bmatrix}_q.$$

This has shown (C.31).

A consequence of the last equality is that for $n > 0$ the sum in (C.24) can be taken from 1 to ∞. Thus for all $n \in \mathbb{N}$,

$$\begin{aligned}\sum_{k=1}^{\infty} \begin{Bmatrix} n+1 \\ k \end{Bmatrix}_q F_k(x;q) &= x^{n+1} = x \sum_{k=0}^{\infty} \begin{Bmatrix} n \\ k \end{Bmatrix}_q F_k(x;q) = \\ &= \sum_{k=0}^{\infty} \begin{Bmatrix} n \\ k \end{Bmatrix}_q \left(F_{k+1}(x;q) + \{k\}_q F_k(x;q)\right) = \\ &= \sum_{k=0}^{\infty} \begin{Bmatrix} n \\ k \end{Bmatrix}_q F_{k+1}(x;q) + \sum_{k=0}^{\infty} \begin{Bmatrix} n \\ k \end{Bmatrix}_q \{k\}_q F_k(x;q) = \\ &= \sum_{i=1}^{\infty} \begin{Bmatrix} n \\ i-1 \end{Bmatrix}_q F_i(x;q) + 0 + \sum_{k=1}^{\infty} \begin{Bmatrix} n \\ k \end{Bmatrix}_q \{k\}_q F_k(x;q) = \\ &= \sum_{k=0}^{\infty} \left(\begin{Bmatrix} n \\ k-1 \end{Bmatrix}_q + \{k\}_q \begin{Bmatrix} n \\ k \end{Bmatrix}_q \right) F_k(x;q),\end{aligned}$$

where as before $i := k + 1$ in the second step from the end, and $k := i$ in the last step. Equation (C.28) follows from the equality of coefficients of $F_k(x;q)$ on both sides. □

Corollary C.14 *All q-Stirling numbers are polynomials in q. Hence the q-Stirling number may also be viewed as functions $\mathcal{K} \longrightarrow \mathcal{K}$.*

That's really everything about the q-Stirling numbers that we use in this book, but since the Stirling numbers are often overlooked in elementary mathematical textbooks, there is little harm in us spending some more time with them here. Therefore we shall continue by mentioning some facts about the non-q-Stirling numbers $\left[\begin{smallmatrix} n \\ k \end{smallmatrix}\right] = \left[\begin{smallmatrix} n \\ k \end{smallmatrix}\right]_1$ and $\left\{\begin{smallmatrix} n \\ k \end{smallmatrix}\right\} = \left\{\begin{smallmatrix} n \\ k \end{smallmatrix}\right\}_1$.

$n \backslash k$	0	1	2	3	4	5	6	7	8
0	1	0	0	0	0	0	0	0	0
1	0	1	0	0	0	0	0	0	0
2	0	1	1	0	0	0	0	0	0
3	0	2	3	1	0	0	0	0	0
4	0	6	11	6	1	0	0	0	0
5	0	24	50	35	10	1	0	0	0
6	0	120	274	225	85	15	1	0	0
7	0	720	1764	1624	735	175	21	1	0
8	0	5040	13068	13132	6769	1960	322	28	1

Table C.1 The Stirling numbers of the first kind $\left[{n \atop k}\right]$.

$n \backslash k$	0	1	2	3	4	5	6	7	8
0	1	0	0	0	0	0	0	0	0
1	0	1	0	0	0	0	0	0	0
2	0	1	1	0	0	0	0	0	0
3	0	1	3	1	0	0	0	0	0
4	0	1	7	6	1	0	0	0	0
5	0	1	15	25	10	1	0	0	0
6	0	1	31	90	65	15	1	0	0
7	0	1	63	301	350	140	21	1	0
8	0	1	127	966	1701	1050	266	28	1

Table C.2 The Stirling numbers of the second kind $\left\{{n \atop k}\right\}$.

Tables C.1 and C.2 list the values of $\left[{n \atop k}\right]$ and $\left\{{n \atop k}\right\}$ respectively for $n, k \leqslant 8$. That the upper left half of the tables are all zeros is a direct consequence of Lemma C.12, and that the main $n = k$ diagonal is all ones is a direct consequence of the following corollary to Theorem C.13.

Corollary C.15 *For all $n \in \mathbb{N}$,*

$$\left[{n \atop n}\right]_q = \left\{{n \atop n}\right\}_q = 1. \tag{C.32}$$

Proof. Since $\left[{n \atop k}\right]_q = 0$ and $\left\{{n \atop k}\right\}_q = 0$ for $k > n$, it follows that $\left[{n \atop n}\right]_q = \left[{n-1 \atop n-1}\right]_q$ and $\left\{{n \atop n}\right\}_q = \left\{{n-1 \atop n-1}\right\}_q$ for all $n \in \mathbb{Z}^+$. That $\left[{0 \atop 0}\right]_q = \left\{{0 \atop 0}\right\}_q = 1$ was shown in the theorem. $\square$

Some other noteworthy features of these tables are that $\left[{n \atop 0}\right] = \left\{{n \atop 0}\right\} = 0$ for $n > 0$ and that $\left\{{n \atop 0}\right\} = 1$ and $\left[{n \atop 1}\right] = (n-1)!$ for $n > 0$. The first of these were shown in the theorem, and it is an easy exercise to show that $\left\{{n \atop 1}\right\}_q = 1$ and $\left[{n \atop 1}\right]_q = \{n-1\}_q!$ for $n > 0$.

Among combinatorists the Stirling numbers are best known for the fact that the Stirling numbers of the second kind $\left\{{n \atop k}\right\}$ gives the number of partitions with k blocks of a set with n elements—a consequence of this is that the number of surjections from a given set with n elements to a given set with k elements is exactly $k!\left\{{n \atop k}\right\}$. It is less well known that the Stirling numbers of the first kind $\left[{n \atop k}\right]$ also have a similar interpretation—they give the number of permutations with k cycles (cycles with length 1 included) of a set with n elements.

Both these interpretations generalise to properties of the q-Stirling numbers $\left[{n \atop k}\right]_q$ and $\left\{{n \atop k}\right\}_q$, but in quite different ways. The generalisation for $\left[{n \atop k}\right]_q$ will be treated next, whereas the generalisation for $\left\{{n \atop k}\right\}_q$ will have to wait until the end of the section.

In what follows, it will be convenient to view the set S_n of all permutations of $\{1,\ldots,n\}$ as as subset of S_{n+1}. Therefore let

$$S_n = \bigl\{\, \pi\colon \mathbb{Z}^+ \longrightarrow \mathbb{Z}^+ \bigm| \pi \text{ is a bijection and } \pi(i) = i \text{ for all } i > n \,\bigr\}$$

for all $n \in \mathbb{Z}^+$. The set of all permutations is $\bigcup_{n=1}^{\infty} S_n$, and two integers $i, j \in \mathbb{Z}^+$ are said to be *in the same cycle* of a permutation π if there is some $k \in \mathbb{N}$ such that $\pi^k(i) = j$. With this definition of S_n one cannot talk about the total number of cycles in a permutation and mean the same thing as with the standard definition, since every permutation has infinitely many cycles of length one, but one can talk about the number of cycles up to a bound. Therefore define *the number of cycles up to n* in a permutation π to be the maximal size of a subset of $\{1,\ldots,n\}$ in which no two elements belong to the same cycle of π. The cycle notation $(i\,j)$ will be used for denoting the transposition of i and j. Thus for example $S_1 = \{\mathrm{id}\}$, where id is the identity permutation, whereas $S_2 \setminus S_1 = \bigl\{(1\,2)\bigr\}$.

Let $w_0(\mathrm{id}) = 1$ and $w_l(\mathrm{id}) = 0$ for all nonzero $l \in \mathbb{Z}$. Extend the family $\{w_l\}_{l\in\mathbb{Z}}$ to all permutations by letting

$$w_l(\pi) = q^{n-\pi(n)-1} w_{l-1}\Bigl(\bigl(\pi(n)\, n\bigr) \circ \pi\Bigr) \qquad \text{for all } \pi \in S_n \setminus S_{n-1}$$

for all $n \geqslant 2$; this works because $\pi \in S_n \setminus S_{n-1}$ implies $\pi(n) < n$, which in turn implies that $\bigl(\pi(n)\,n\bigr) \circ \pi \in S_{n-1}$. By repeating this argument, one can show that every $\pi \in S_n$ has a unique decomposition

$$\begin{aligned} \pi = (i_l\, j_l) \circ (i_{l-1}\, j_{l-1}) \circ \cdots \circ (i_1\, j_1), \\ \text{where } 1 \leqslant i_k < j_k \text{ for } k = 1, \dots, l \\ \text{and } j_1 < j_2 < \cdots < j_l \leqslant n \end{aligned} \tag{C.33}$$

for some $l \in \mathbb{N}$. Thus an immediate application of the w_l functions is to recognise the permutations whose decomposition of this type has a given length l, as $w_l(\pi) \neq 0$ if and only if the (C.33) decomposition of π consists of exactly l transpositions.

The $\{w_l\}_{l\in\mathbb{Z}}$ also have the interesting property that

$$\begin{bmatrix} n \\ k \end{bmatrix}_q = \sum_{\pi \in S_n} w_{n-k}(\pi) \qquad \text{for all } n \in \mathbb{Z}^+ \text{ and } k \in \mathbb{N}, \tag{C.34}$$

which is trivially verified for $n = 1$. As $1 < j_1 < \cdots < j_l \leqslant n$ for every decomposition of some $\pi \in S_n$ of the type described above, it follows that $l < n$, and hence $w_n(\pi) = 0$ for all $\pi \in S_n$. This implies that $\sum_{\pi \in S_n} w_n(\pi) = 0$ which verifies (C.34) for $k = 0$. Thus for proving (C.34) in general it would suffice verifying that the right hand side obeys the recursion formula (C.28). To that end, just observe that

$$\begin{aligned} \sum_{\pi \in S_n} w_{n-k}(\pi) &= \sum_{\pi \in S_{n-1}} w_{n-k}(\pi) + \sum_{\pi \in S_n \setminus S_{n-1}} w_{n-k}(\pi) = \\ &= \sum_{\pi \in S_{n-1}} w_{(n-1)-(k-1)}(\pi) + \sum_{\pi \in S_{n-1}} \sum_{i=1}^{n-1} w_{n-k}\bigl((i\, n) \circ \pi\bigr) = \\ &= \sum_{\pi \in S_{n-1}} w_{(n-1)-(k-1)}(\pi) + \sum_{\pi \in S_{n-1}} \sum_{i=1}^{n-1} q^{n-1-i} w_{n-k-1}(\pi) = \\ &= \sum_{\pi \in S_{n-1}} w_{(n-1)-(k-1)}(\pi) + \{n-1\}_q \sum_{\pi \in S_{n-1}} w_{(n-1)-k}(\pi). \end{aligned}$$

Now what has all this to do with the number of cycles that a permutation has? Quite a lot, since the cycle structures of two permutations π and $(i\,j) \circ \pi$ are closely related—either i and j are in different cycles of π, and then $(i\,j) \circ \pi$ differ from π only in that these two cycles have been joined, or else i and j are in the same cycle of π, and then $(i\,j) \circ \pi$ differ from π

only in that this cycle has been split in two. As the decomposition (C.33) is set up so that each j_{k+1} is alone in its cycle in $(i_k\, j_k) \circ \cdots \circ (i_1\, j_1)$, it follows that each additional transposition decreases the number of cycles in the permutation by one. Therefore a permutation π has exactly l cycles fewer than the identity if and only if $w_l(\pi) \neq 0$. As the identity always has n cycles up to n, this implies that the $\pi \in S_n$ for which $w_{n-k}(\pi) \neq 0$ are exactly those which have k cycles up to n.

The rest is simple. When $q = 1$ each w_l only assumes the values 0 and 1, and thus $\sum_{\pi \in S_n} w_{n-k}(\pi)$ is simply the number of $\pi \in S_n$ which have $w_{n-k}(\pi) = 1$, i.e., the number of such permutations which have k cycles (up to n). Therefore $\left[{n \atop k}\right]$ is the number of permutations of an n-set which have k cycles.

If one uses the decomposition (C.33) to express the $\pi \in S_n$ for which $w_{n-k}(\pi) \neq 0$, one gets the rather hideous formula

$$\begin{bmatrix} n \\ k \end{bmatrix}_q = \sum_{\pi \in S_n} w_{n-k}(\pi) = \sum_{1 < j_1 < \cdots < j_l \leqslant n} \sum_{i_1=1}^{j_1-1} \cdots \sum_{i_l=1}^{j_l-1} \prod_{k=1}^{l} q^{j_k - i_k - 1}.$$

Some algebraic manipulations of this would however lead to the much nicer formula (C.37) in the following theorem, but it turns out there is a much quicker way of getting that result.

Theorem C.16 *Let* $[n] = \{1, \ldots, n\}$ *for all* $n \in \mathbb{Z}^+$ *and let* $[0] = \varnothing$. *For all* $n, k \in \mathbb{N}$,

$$\begin{bmatrix} n \\ k \end{bmatrix}_q = \sum_{\substack{S \subseteq [n] \\ |S| = n-k}} \prod_{s \in S} \{s-1\}_q, \tag{C.35}$$

$$\begin{Bmatrix} n \\ k \end{Bmatrix}_q = \sum_{\substack{\mathbf{b} \in \mathbb{N}^k \\ |\mathbf{b}| = n-k}} \prod_{i=1}^{k} \{i\}_q^{b_i}. \tag{C.36}$$

For $n \geqslant 1$ *in particular,*

$$\begin{bmatrix} n \\ k \end{bmatrix}_q = \sum_{\substack{S \subseteq [n-1] \\ |S| = n-k}} \prod_{s \in S} \{s\}_q. \tag{C.37}$$

Proof. If $k > n$ then $\left[{n \atop k}\right]_q = 0$ and the right hand side of (C.35) is an empty sum (no subsets of $[n]$ has a negative number of elements). For $k \leqslant n$

one has by definition that

$$\sum_{k=0}^{n} \begin{bmatrix} n \\ k \end{bmatrix}_q (-1)^{n-k} x^k = F_n(x;q) = \prod_{s=1}^{n} \left(x - \{s-1\}_q\right) =$$

$$= \sum_{S \subseteq [n]} \prod_{\substack{1 \leqslant s \leqslant n \\ s \in S}} \left(-\{s-1\}_q\right) \prod_{\substack{1 \leqslant s \leqslant n \\ s \notin S}} x =$$

$$= \sum_{k=0}^{n} (-1)^{n-k} \left(\sum_{\substack{S \subseteq [n] \\ |S| = n-k}} \prod_{s \in S} \{s-1\}_q \right) x^k.$$

Now (C.35) follows from identifying the coefficients of each x^k in the sums. (C.37) is simply (C.35) with all terms containing the factor $\{0\}_q = 0$ removed.

The formula (C.36) seems easiest to show using the recursion (C.29). First consider the case $k = 0$. The set $\mathbb{N}^0$ contains only one element, namely the empty vector $\mathbf{e} = ()$. As $|\mathbf{e}| = 0$, the sum in (C.36) is empty for all $n > 0$, and it has exactly one term for $n = 0$. That term is 1 since it consists of an empty product. Hence (C.36) agrees with (C.31) for $k = 0$.

In the case $k > 0$, (C.36) is proved by induction on n. In the base case $n = 0$, the sum in (C.36) is over all $\mathbf{b} \in \mathbb{N}^k$ such that $|\mathbf{b}| = n - k < 0$. As no vector in $\mathbb{N}^k$ satisfies that, the sum is empty and the right hand side is 0 in accordance with (C.31).

Now assume that (C.36) holds for some $n = m \in \mathbb{N}$ and all $k \in \mathbb{N}$. Consider it for $n = m + 1$. By (C.29),

$$\begin{Bmatrix} n \\ k \end{Bmatrix}_q = \begin{Bmatrix} n-1 \\ k-1 \end{Bmatrix}_q + \{k\}_q \begin{Bmatrix} n-1 \\ k \end{Bmatrix}_q =$$

$$= \sum_{\substack{\mathbf{b} \in \mathbb{N}^{k-1} \\ |\mathbf{b}| = n-k}} \prod_{i=1}^{k-1} \{i\}_q^{b_i} + \{k\}_q \sum_{\substack{\mathbf{b} \in \mathbb{N}^k \\ |\mathbf{b}| = n-1-k}} \prod_{i=1}^{k} \{i\}_q^{b_i} =$$

$$= \sum_{\substack{\mathbf{b} \in \mathbb{N}^k \\ |\mathbf{b}| = n-k \\ b_k = 0}} \prod_{i=1}^{k} \{i\}_q^{b_i} + \sum_{\substack{\mathbf{b} \in \mathbb{N}^k \\ |\mathbf{b}| = n-k \\ b_k > 0}} \prod_{i=1}^{k} \{i\}_q^{b_i} = \sum_{\substack{\mathbf{b} \in \mathbb{N}^k \\ |\mathbf{b}| = n-k}} \prod_{i=1}^{k} \{i\}_q^{b_i}.$$

Thus by the principle of induction, it follows that (C.36) holds for all $n \in \mathbb{N}$ and $k > 0$ as well. □

$n \backslash k$	0	1	2	3	4
0	1	0	0	0	0
1	0	1	0	0	0
2	0	1	1	0	0
3	0	$q+1$	$q+2$	1	0
4	0	q^3+2q^2+2q+1	q^3+3q^2+4q+3	q^2+2q+3	1

Table C.3 The q-Stirling numbers of the first kind $\left[{n \atop k}\right]_q$ for $k, n \leqslant 4$.

$n \backslash k$	0	1	2	3	4	5
0	1	0	0	0	0	0
1	0	1	0	0	0	0
2	0	1	1	0	0	0
3	0	1	$q+2$	1	0	0
4	0	1	q^2+3q+3	q^2+2q+3	1	0
5	0	1	q^3+4q^2+6q+4	$q^4+3q^3+7q^2+8q+6$	q^3+2q^2+3q+4	1

Table C.4 The q-Stirling numbers of the second kind $\left\{{n \atop k}\right\}_q$ for $k, n \leqslant 5$.

The formulae (C.36) and (C.37) indicate another interesting property of the Stirling numbers, namely that $\left\{{l+m \atop l}\right\}$ and $\left[{l+1 \atop l+1-m}\right]$ are the sums of all possible products of the first l positive integers, taken m at a time, with or without repetition respectively. During the nineteenth century this used to be the main reason for studying the Stirling numbers. This approach is quite different from that Stirling used, however. When he introduced the Stirling numbers, he used the same definition through coefficients in polynomials as we have done (except he didn't involve any q). A funny thing about Stirling's definition is that he defined the second kind numbers *first*, and only several pages later introduced the first kind numbers. The modern terminology here is thus historically incorrect (as well as mathematically rather void), but at least when it was introduced (in the early twentieth century) it had the merit that it made clear Stirling's connection with these numbers—something which had been almost entirely forgotten.

Despite their compactness the formulae (C.35)–(C.37) aren't very useful for other than small n and k, since the number of terms grows very rapidly, but at least they will serve in computing the small tables of q-Stirling numbers found in Tables C.3 and C.4. There exist formulae in which the number of terms grows much slower, and we will derive two of these, but for that we need to establish a few additional facts about the q-binomial coefficients.

Definition C.3 The **Gauss polynomials** $\{G_n(x;q)\}_{n=0}^{\infty}$ are the elements of $\mathcal{K}(q)[x]$ which are defined by

$$G_n(x;q) = \prod_{i=0}^{n-1}(x - q^i) \qquad \text{for all } n \in \mathbb{N}. \tag{C.38}$$

Note that the Gauss polynomials constitute a basis of the vector space $\mathcal{K}(q)[x]$.

Theorem C.17 *For all $n \in \mathbb{N}$,*

$$x^n = \sum_{k=0}^{n} \binom{n}{k}_q G_k(x;q), \tag{C.39}$$

$$G_n(x;q) = \sum_{k=0}^{n} (-1)^{n-k} q^{\binom{n-k}{2}} \binom{n}{k}_q x^k. \tag{C.40}$$

Proof. Both equations are proved by induction on n. For $n = 0$ they both reduce to $1 = 1$, and hence the base for the induction presents no problems. Assuming that (C.39) holds for n, it follows that

$$\begin{aligned}
x^{n+1} &= x\sum_{k=0}^{n} \binom{n}{k}_q G_k(x;q) = \sum_{k=0}^{n} \binom{n}{k}_q G_k(x;q)(x - q^k + q^k) = \\
&= \sum_{k=0}^{n} \binom{n}{k}_q \left(G_{k+1}(x;q) + q^k G_k(x;q)\right) = \\
&= \sum_{k=1}^{n+1} \binom{n}{k-1}_q G_k(x;q) + \sum_{k=0}^{n} q^k \binom{n}{k}_q G_k(x;q) = \\
&= \sum_{k=0}^{n+1} \left(\binom{n}{k-1}_q + q^k \binom{n}{k}_q\right) G_k(x;q) = \\
&= \sum_{k=0}^{n+1} \binom{n+1}{k}_q G_k(x;q)
\end{aligned}$$

as required for the induction step. Assuming that (C.40) holds for n, it follows that

$$\begin{aligned}
G_{n+1}(x;q) = G_n(x;q)(x-q^n) = \\
&= \sum_{k=0}^{n}(-1)^{n-k}q^{\binom{n-k}{2}}\binom{n}{k}_q x^k(x-q^n) = \\
&= \sum_{k=0}^{n}(-1)^{n-k}q^{\binom{n-k}{2}}\binom{n}{k}_q x^{k+1}+ \\
&\qquad - q^n\sum_{k=0}^{n}(-1)^{n-k}q^{\binom{n-k}{2}}\binom{n}{k}_q x^k = \\
&= \sum_{k=1}^{n+1}(-1)^{n-(k-1)}q^{\binom{n-(k-1)}{2}}\binom{n}{k-1}_q x^k+ \\
&\qquad + \sum_{k=0}^{n}(-1)^{n+1-k}q^{\binom{n-k}{2}+(n-k)}q^k\binom{n}{k}_q x^k = \\
&= \sum_{k=0}^{n+1}(-1)^{n+1-k}q^{\binom{n+1-k}{2}}\left(\binom{n}{k-1}_q + q^k\binom{n}{k}_q\right)x^k = \\
&= \sum_{k=0}^{n+1}(-1)^{n+1-k}q^{\binom{n+1-k}{2}}\binom{n+1}{k}_q x^k
\end{aligned}$$

as required for the induction step. Hence by the principle of induction, the theorem follows. □

We will also need a technique called inversion of sequences. Given two sequences $\{u_n\}_{n=0}^{\infty}$ and $\{v_n\}_{n=0}^{\infty}$ which are related by an equation of the form $v_n = \sum_{k=0}^{n} a_{n,k}u_k$ for some coefficients $\{a_{n,k}\}_{0\leqslant k\leqslant n}$, the inversion consists of finding coefficients $\{b_{n,k}\}_{0\leqslant k\leqslant n}$ such that $u_n = \sum_{k=0}^{n} a_{n,k}v_k$. The following lemma shows how one can use a relation between two sequences of polynomials to show that two families of coefficients $\{a_{n,k}\}$ and $\{b_{n,k}\}$ form such a pair.

Lemma C.18 *Let $\mathcal{L}$ be a field. Let $\{P_n\}_{n=0}^{\infty}, \{Q_n\}_{n=0}^{\infty} \subseteq \mathcal{L}[x]$ be two bases of $\mathcal{L}[x]$. If $\{a_{n,k}\}_{0\leqslant k\leqslant n} \subseteq \mathcal{L}$ and $\{b_{n,k}\}_{0\leqslant k\leqslant n} \subseteq \mathcal{L}$ are coefficients which satisfy*

$$P_n = \sum_{k=0}^{n} a_{n,k}Q_k \quad \textit{and} \quad Q_n = \sum_{k=0}^{n} b_{n,k}P_k \quad \textit{for all } n \in \mathbb{N}, \tag{C.41}$$

then

$$\delta_{n,k} = \sum_{j=k}^{n} a_{n,j} b_{j,k} \qquad \textit{and} \qquad \delta_{n,k} = \sum_{j=k}^{n} b_{n,j} a_{j,k} \tag{C.42}$$

—where $\delta_{n,k}$ denotes the Kronecker delta—for all $k, n \in \mathbb{N}$ such that $k \leqslant n$. Furthermore for all $\{u_n\}_{n=0}^{\infty}, \{v_n\}_{n=0}^{\infty} \subseteq \mathcal{L}$ the same conditions imply that

$$\forall\, n \in \mathbb{N}\colon\ u_n = \sum_{k=0}^{n} a_{n,k} v_k \quad \Longleftrightarrow \quad \forall\, n \in \mathbb{N}\colon\ v_n = \sum_{k=0}^{n} b_{n,k} u_k. \tag{C.43}$$

Proof. For each $n \in \mathbb{N}$, by (C.41),

$$P_n = \sum_{j=0}^{n} a_{n,j} Q_j = \sum_{j=0}^{n} a_{n,j} \sum_{k=0}^{j} b_{j,k} P_k =$$
$$= \sum_{0 \leqslant k \leqslant j \leqslant n} a_{n,j} b_{j,k} P_k = \sum_{k=0}^{n} \left(\sum_{j=k}^{n} a_{n,j} b_{j,k} \right) P_k.$$

From identifying coefficients of P_k in the left and right hand sides, it follows that $\sum_{j=k}^{n} a_{n,j} b_{j,k} = 0$ if $k < n$ and that $\sum_{j=n}^{n} a_{n,j} b_{j,n} = 1$. This has proved the first part of (C.42), and the second part is proved analogously.

To prove the equivalence (C.43), assume that $u_n = \sum_{k=0}^{n} a_{n,k} v_k$ for all $n \in \mathbb{N}$. Then by (C.42),

$$\sum_{k=0}^{n} b_{n,k} u_k = \sum_{j=0}^{n} b_{n,j} u_j = \sum_{j=0}^{n} b_{n,j} \sum_{k=0}^{j} a_{j,k} v_k =$$
$$= \sum_{k=0}^{n} \left(\sum_{j=k}^{n} b_{n,j} a_{j,k} \right) v_k = \sum_{k=0}^{n} \delta_{n,k} v_k = v_n$$

for every $n \in \mathbb{N}$. The proof of the reverse implication is identical. □

When (C.43) is applied, one usually starts with an equation that looks like

$$f(n) = \sum_{k=0}^{n} a_{n,k}\, g(k)$$

and which holds for all $n \in \mathbb{N}$. The equivalence then implies that

$$g(n) = \sum_{k=0}^{n} b_{n,k} f(k)$$

for all $n \in \mathbb{N}$. The functions f and g may well depend on other variables than the ones explicitly mentioned above, but it is important that the $g(k)$ factor in the sum does not depend on n.

Corollary C.19 *For all sequences $\{u_n\}_{n=0}^{\infty}, \{v_n\}_{n=0}^{\infty} \subseteq \mathcal{K}(q)$, the following equivalences hold:*

$$\forall n \in \mathbb{N}\colon\ u_n = \sum_{k=0}^{n} \left\{ {n \atop k} \right\}_q v_k \iff \forall n \in \mathbb{N}\colon\ v_n = \sum_{k=0}^{n} (-1)^{n-k} \begin{bmatrix} n \\ k \end{bmatrix}_q u_k, \tag{C.44}$$

$$\forall n \in \mathbb{N}\colon\ u_n = \sum_{k=0}^{n} \binom{n}{k}_q v_k \iff \forall n \in \mathbb{N}\colon\ v_n = \sum_{k=0}^{n} (-1)^{n-k} q^{\binom{n-k}{2}} \binom{n}{k}_q u_k. \tag{C.45}$$

Theorem C.20 *For all $k, n \in \mathbb{N}$ such that $k \leqslant n$,*

$$(1-q)^{n-k} \begin{bmatrix} n \\ k \end{bmatrix}_q = \sum_{j=0}^{n-k} (-1)^j q^{\binom{j}{2}} \binom{n}{j}_q \binom{n-j}{k}_1. \tag{C.46}$$

Proof. Notice that

$$\begin{aligned}(q-1)^n F_n(x;q) = \prod_{i=0}^{n-1} \Big((q-1)\big(x - \{i\}_q\big) \Big) &= \prod_{i=0}^{n-1} \big((q-1)x + 1 - q^i \big) = \\ &= G_n\big((q-1)x + 1; q \big)\end{aligned}$$

and

$$\sum_{k=0}^{n} \binom{n}{k}_1 x^k = \sum_{k=0}^{n} \binom{n}{k}_1 \big((x+1) - 1 \big)^k = \sum_{k=0}^{n} \binom{n}{k}_1 G_k(x+1;1) = (x+1)^n,$$

the latter of which is of course only a special case of the binomial theorem.

These observations yield

$$\sum_{k=0}^{n}\binom{n}{k}_1 (q-1)^k x^k = \big((q-1)x+1\big)^n =$$

$$= \sum_{i=0}^{n}\binom{n}{i}_q G_i\big((q-1)x+1;q\big) =$$

$$= \sum_{i=0}^{n}\binom{n}{i}_q (q-1)^i F_i(x;q) =$$

$$= \sum_{i=0}^{n}\binom{n}{i}_q (q-1)^i \sum_{k=0}^{i}\begin{bmatrix} i \\ k \end{bmatrix}_q (-1)^{i-k} x^k =$$

$$= \sum_{0\leqslant k\leqslant i\leqslant n}\binom{n}{i}_q (q-1)^i \begin{bmatrix} i \\ k \end{bmatrix}_q (-1)^{i-k} x^k =$$

$$= \sum_{k=0}^{n}\left(\sum_{i=k}^{n}\binom{n}{i}_q (-1)^{i-k}(q-1)^i \begin{bmatrix} i \\ k \end{bmatrix}_q\right) x^k .$$

From identifying the coefficients of x^k on both sides it follows that

$$\binom{n}{k}_1 (q-1)^k = \sum_{i=k}^{n}\binom{n}{i}_q (-1)^{i-k}(q-1)^i \begin{bmatrix} i \\ k \end{bmatrix}_q ,$$

hence since $\begin{bmatrix} i \\ k \end{bmatrix}_q = 0$ for $i < k$,

$$\binom{n}{k}_1 (1-q)^k = \sum_{i=0}^{n}\binom{n}{i}_q (1-q)^i \begin{bmatrix} i \\ k \end{bmatrix}_q .$$

Now by (C.45),

$$(1-q)^n \begin{bmatrix} n \\ k \end{bmatrix}_q = \sum_{i=0}^{n}(-1)^{n-i} q^{\binom{n-i}{2}}\binom{n}{i}_q (1-q)^k \binom{i}{k}_1 ,$$

from which (C.46) follows by observing that $\binom{i}{k}_1 = 0$ for $i < k$ and the substitution $j = n - i$. □

Theorem C.21 *For all $k, n \in \mathbb{N}$,*

$$q^{\binom{k}{2}}\{k\}_q! \left\{ \begin{matrix} n \\ k \end{matrix} \right\}_q = \sum_{i=0}^{k}(-1)^i q^{\binom{i}{2}}\binom{k}{i}_q \{k-i\}_q^n . \tag{C.47}$$

Proof. Notice that

$$F_k(\{n\}_q;q) = \prod_{i=0}^{k-1}(\{n\}_q - \{i\}_q) = \prod_{i=0}^{k-1} q^i\{n-i\}_q = q^{\binom{k}{2}}\{k\}_q! \binom{n}{k}_q.$$

Hence

$$\{k\}_q^n = \sum_{i=0}^{n} \left\{{n \atop i}\right\}_q F_i(\{k\}_q;q) = \sum_{i=0}^{n} \left\{{n \atop i}\right\}_q q^{\binom{i}{2}}\{i\}_q! \binom{k}{i}_q = \\ = \sum_{i=0}^{k} \binom{k}{i}_q \left\{{n \atop i}\right\}_q q^{\binom{i}{2}}\{i\}_q!. \qquad \text{(C.48)}$$

Now by (C.45),

$$\left\{{n \atop k}\right\}_q q^{\binom{k}{2}}\{k\}_q! = \sum_{i=0}^{k}(-1)^{k-i} q^{\binom{k-i}{2}} \binom{k}{i}_q \{i\}_q^n = \\ = \sum_{i=0}^{k}(-1)^{i} q^{\binom{i}{2}} \binom{k}{i}_q \{k-i\}_q^n.$$

□

There is a direct combinatorial interpretation of the quantity on the left hand side of (C.47), and we will indicate how this can be refined into an interpretation of $\left\{{n \atop k}\right\}_q$. In doing this, we will once again denote the set $\{1,\ldots,n\}$ by $[n]$ for all $n \in \mathbb{Z}^+$, and in particular write $[0]$ for $\varnothing$.

Following [77], we let $b(n,k,q)$ be the number of functions from $[n]$ to $\mathbb{F}_q^k \setminus \{0\}$ whose images span $\mathbb{F}_q^k$. Since all i-dimensional vector spaces over $\mathbb{F}_q$ are isomorphic, this means that there are $b(n,i,q)$ functions from $[n]$ to $\mathbb{F}_q^k \setminus \{0\}$ whose images span a given i-dimensional subspace of $\mathbb{F}_q^k$. We know from Section C.2 that there are $\binom{k}{i}_q$ such subspaces, and therefore there are in total $\binom{k}{i}_q b(n,i,q)$ functions from $[n]$ to $\mathbb{F}_q^k \setminus \{0\}$ whose images span some i-dimensional subspace. The total number of functions $f\colon [n] \longrightarrow \mathbb{F}_q^k \setminus \{0\}$ is $(q^k-1)^n$, but one can also get this number by summing over all i from 0 to k. Thus

$$(q^k-1)^n = \sum_{i=0}^{k} \binom{k}{i}_q b(n,i,q).$$

On the other hand, by (C.48),

$$(q^k - 1)^n = (q-1)^n \{k\}_q^n = \sum_{i=0}^{k} \binom{k}{i}_q \begin{Bmatrix} n \\ i \end{Bmatrix}_q q^{\binom{i}{2}} \{i\}_q! \, (q-1)^n.$$

Using (C.45) on these two equalities, one gets

$$b(n,k,q) = \sum_{i=0}^{k} (-1)^{k-i} q^{\binom{k-i}{2}} \binom{k}{i}_q (q^i - 1)^n = \begin{Bmatrix} n \\ k \end{Bmatrix}_q q^{\binom{k}{2}} \{k\}_q! \, (q-1)^n. \tag{C.49}$$

The factor $(q-1)^n$ here is due to that the span of the image of a function $f\colon [n] \longrightarrow \mathbb{F}_q^k \setminus \{0\}$ is the same as that of the function gf for all $g\colon [n] \longrightarrow \mathbb{F}_q \setminus \{0\}$. Therefore one can get rid of this factor by instead counting functions from $[n]$ to points in the projective geometry corresponding to $\mathbb{F}_q^k$, or what is the same thing, to the rank 1 elements in the subspace lattice of $\mathbb{F}_q^k$.

Milne [280] found a modification of this which gets rid of the other two factors as well and thus becomes a combinatorial interpretation of the q-Stirling numbers of the second kind $\left\{ {n \atop k} \right\}_q$, but working out the details of that is a bit beyond this appendix. In order to still present the general idea, we shall employ some of Milne's concepts in proving that $\left\{ {n \atop k} \right\}$ is the number of partitions of the set $[n]$ into k parts.

Define a *restricted growth function* to be a function $f\colon [n] \longrightarrow [n]$ which satisfies

$$f(1) = 1, \tag{C.50a}$$

$$f(i+1) \leqslant 1 + \max_{j \in [i]} f(j) \quad \text{for all } i \in [n-1]. \tag{C.50b}$$

Every restricted growth function f defines a partition of the domain $[n]$, namely

$$\{\, i \in [n] \mid f(i) = 1 \,\} \cup \{\, i \in [n] \mid f(i) = 2 \,\} \cup \dots \cup \{\, i \in [n] \mid f(i) = \max f \,\}.$$

The restricted growth condition immediately ensures that all these sets are nonempty. It is furthermore easy to see that every partition of $[n]$ can be realised as a partition of the above type for some restricted growth function f (enumerate the parts in order of their smallest element, and let f be the function which gives the number of the part an element is in), and that no two distinct restricted growth functions will produce the same partition of

$[n]$. Therefore there is a 1–1 correspondence between the partitions of $[n]$ and the restricted growth function on $[n]$.

Instead of directly counting partitions, one may therefore count restricted growth functions. Let $a(n,k)$ be the number of restricted growth functions on $[n]$ which have maximum k—this corresponds to partitions of $[n]$ with k parts. These restricted growth functions can be roughly classified in two categories: those functions f which have $f(n) = 1 + \max_{j\in[n-1]} f(j)$, and those functions f which have $f(n) \leqslant \max_{j\in[n-1]} f(j)$. In the first category $f(n) = k$, which implies $\max_{j\in[n-1]} f(j) = k-1$, and hence $f\big|_{[n-1]}$ must be one of the $a(n-1,k-1)$ restricted growth functions on $[n-1]$ which have maximum $k-1$. As conversely any such function can be extended to a restricted growth function on $[n]$ which is of the first category, one concludes that there are $a(n-1,k-1)$ restricted growth functions on $[n]$ which are of this first category. For functions of the second category $\max_{j\in[n-1]} f(j) = k$, and hence $f\big|_{[n-1]}$ for these functions must be one of the $a(n-1,k)$ restricted growth functions on $[n-1]$ which have maximum k. There are k different ways (k possible values for $f(n)$) of extending such a function to a restricted growth function on $[n]$ which is of the second category, and thus one concludes that there are $ka(n-1,k)$ restricted growth functions on $[n]$ which are of this second category. In summary this implies that $a(n,k) = a(n-1,k-1) + ka(n-1,k)$, which is precisely the recursion formula (C.29) for $q = 1$. The two base cases $n = 1$ and $k = 1$ are easily seen to match the values of $\left\{ {n \atop k} \right\}$, and thus $a(n,k) = \left\{ {n \atop k} \right\}$.

To get to $\left\{ {n \atop k} \right\}_q$, one must reformulate the restricted growth function concept in lattice-theoretical language. In that set-up, the codomain of the restricted growth functions should not be the positive integers, but the set of rank 1 elements of some lattice $\mathcal{L}$, and by choosing this lattice one sets the equivalent of $\max f$. In the $q = 1$ case the lattice is the subset lattice of the set $[k]$—where the rank 1 elements are $\{1\}$, $\{2\}$, …, and $\{k\}$—and in the $q > 1$ case the lattice is the subspace lattice of $\mathbb{F}_q^k$. The restricted growth condition becomes the rather curious condition that

$$\left\{ \bigvee_{j=1}^{i} f(j) \right\}_{i=0}^{n} = C,$$

where C is a fixed maximal chain (totally ordered subset) in $\mathcal{L}$ and of course $\bigvee_{j=1}^{0} f(j)$ should be interpreted as the minimum element in the lattice. The chain C that corresponds to the (C.50) restricted growth condition is

$$C = \{[0], [1], [2], \ldots, [k]\},$$

but the number of restricted growth functions will be the same no matter which maximal chain is used. In fact, every function f for which $\bigvee_{j=1}^{n} f(j)$ is the maximum element in $\mathcal{L}$ will be a restricted growth function for some maximal chain C.

As that is precisely the set of functions that are counted in the modified form of (C.49), there should be a factor that is the total number of chains in the lattice, and indeed there is; it was shown in Section C.2 that the number of such chains is $\{k\}_q!$. This leaves only the $q^{\binom{k}{2}}$ factor in (C.49) to explain. It turns out that in $\mathbb{F}_q^k$ the group of automorphisms which map each element in a given chain of subspaces onto itself is not trivial, but rather large. The elements of this group are linear transformations which shear the space, but they can also be seen as automorphisms of $\mathcal{L}$, and in that context the group for any given maximal chain C has exactly $q^{\binom{k}{2}}$ elements. Thus one can get rid of this final $q^{\binom{k}{2}}$ factor by saying that two restricted growth functions are equivalent if one can be turned into the other by shearing the underlying space $\mathbb{F}_q^k$ (without changing the chain C), and then counting equivalence classes of restricted growth functions.

For the exact formulation and for proofs, we refer to [280], which is where the result is from, or to [388], where the concept of modular binomial lattices (or q-lattices) is brought in as a background to the result.

C.4 Extending the q-combinatorial functions

All the q-combinatorial functions have been defined with certain restrictions on their integer parameters. These restrictions are by no means absolute; it just happens that we found them convenient for the purposes of this book. We therefore close this appendix with a few remarks on how to extend the q-combinatorial functions to all integers.

The q-natural numbers are the building blocks of all the other q-combinatorial functions, so we should start with them. Here it is logical to require that (C.4), i.e., $q^k\{n\}_q + \{k\}_q = \{n+k\}_q$, should hold for all $n, k \in \mathbb{Z}$. In particular this implies that

$$\{-n\}_q = \{-n\}_q - \{0\}_q = \{-n\}_q - \left(q^{-n}\{n\}_q + \{-n\}_q\right) = -q^{-n}\{n\}_q.$$

This is also in accordance with the convention to let $\sum_{i=m}^{n} a_i = -\sum_{i=n+1}^{m-1} a_i$ for $n < m-1$ (cf. the rule $\int_a^b f(x)\,dx = -\int_b^a f(x)\,dx$) that is sometimes applied, since one gets

$$\{-n\}_q = \sum_{i=0}^{-n-1} q^i = -\sum_{i=-n}^{-1} q^i = -q^{-n}\sum_{i=0}^{n-1} q^i = -q^{-n}\{n\}_q$$

for $n \in \mathbb{Z}^+$ if that convention is applied to the definition of the q-natural numbers. Therefore it makes sense to define $\{n\}_q$ for all $n \in \mathbb{Z}$.

The same can sadly not be said about the q-factorials $\{n\}_q!$, which do not extend to negative n in the same way because that would require dividing $\{0\}_q! = 1$ by $\{0\}_q = 0$ to get $\{-1\}_q!$. If the problem is instead considered in the slightly different context that $n, q \in \mathbb{C}$ then one could claim that the fact that you cannot extend the q-factorials to the negative integers is just due to bad luck. This is since the q-factorials can be seen as just (a shift of) the restriction to the integers of the q-Gamma function (yes, it's the q-analogue of Euler's Gamma function), and one property that shares with the ordinary Gamma function is that its poles happen to lie on the negative integers. Therefore it isn't very interesting to extend the q-factorials to precisely the integers, as that would only be an extension with the bad parts of the complex plane.

The reader who is accustomed to defining the binomials through factorials as in (C.11) might think that this gets the q-binomials in the same kind of trouble, but that is actually not the case. It is common to extend the ordinary binomials $\binom{n}{k}$ to $n \notin \mathbb{N}$ and $k \in \mathbb{N}$ through the identity

$$\binom{n}{k} = \frac{1}{k!}\prod_{k=0}^{k-1}(n-k),$$

especially since this makes $\binom{n}{k}$ a polynomial in n. The q-binomial coefficients $\binom{n}{k}_q$ are not in general polynomials in n, but one can still use (C.8) to extend them to the negative integers. For $k \geqslant 0$ and $n > 0$ this yields

$$\{k\}_q!\binom{-n}{k}_q = \prod_{j=-n-k+1}^{-n} \{j\}_q = \prod_{j=n}^{n+k-1} \{-j\}_q =$$
$$= \prod_{j=n}^{n+k-1} -q^{-j}\{j\}_q =$$

$$= (-1)^k q^{-kn} \prod_{j=n}^{n+k-1} q^{n-j}\{j\}_q =$$

$$= (-1)^k q^{-kn-\binom{k}{2}}\{k\}_q! \binom{n+k-1}{k}_q,$$

which implies

$$\binom{-n}{k}_q = (-1)^k q^{-kn-\binom{k}{2}} \binom{n+k-1}{k}_q. \tag{C.51}$$

Alternatively one can postulate that $\binom{n}{0}_q = 1$ for all $n \in \mathbb{Z}$ and $q \in \mathcal{K}$, and then use either of the recursion formulae (C.14). The definitions turn out to be equivalent.

All of that was about $\binom{n}{k}_q$ for $n < 0$ and $k \geqslant 0$, however. What about the case $n < 0$ and $k < 0$? It seems that there are two schools on how to make that extension for the ordinary binomials—one that uses the recursion formula $\binom{n-1}{k-1} + \binom{n-1}{k} = \binom{n}{k}$, and one that uses the identity $\binom{n}{k} = \binom{n}{n-k}$. These two definitions do *not* produce the same results, but either one can be used for extending the q-binomial coefficients. If the q-binomial coefficients are defined through a recursion, it turns out that $\binom{n}{k}_q = 0$ for all $k < 0$. Clearly this doesn't satisfy the symmetry $\binom{n}{k}_q = \binom{n}{n-k}_q$ for $n < 0$. If on the other hand that symmetry and (C.51) is taken as the definition, one gets the result

$$\binom{-n}{-k}_q = \binom{-n}{k-n}_q = (-1)^{k-n} q^{\binom{n}{2}-\binom{k}{2}} \binom{k-1}{k-n}_q =$$

$$= (-1)^{k-n} q^{\binom{n}{2}-\binom{k}{2}} \binom{k-1}{n-1}_q$$

for $0 < n \leqslant k$ and

$$\binom{-n}{-k}_q = (-1)^{-k} q^{kn-\binom{k+1}{2}} \binom{n-k-1}{-k}_q = 0$$

for $0 < k < n$. This fails to satisfy the recursion formulae (C.14), but only in the case $n = k = 0$. Which extension one prefers is usually a matter of taste.

The q-Stirling numbers, finally, have a really beautiful extension to all integers. If the recursion formulae of Theorem C.13 are postulated to hold for all $n, k \in \mathbb{Z}$ and also completed with $\left[\begin{smallmatrix}-n\\0\end{smallmatrix}\right]_q = \left[\begin{smallmatrix}0\\-n\end{smallmatrix}\right]_q = \left\{\begin{smallmatrix}-n\\0\end{smallmatrix}\right\}_q = \left\{\begin{smallmatrix}0\\-n\end{smallmatrix}\right\}_q = 0$ for all $n \in \mathbb{Z}^+$, then it turns out that

$$\begin{bmatrix}-n\\-k\end{bmatrix}_q = q^{n-k}\begin{Bmatrix}k\\n\end{Bmatrix}_{q^{-1}} \qquad \text{and} \qquad \begin{Bmatrix}-n\\-k\end{Bmatrix}_q = q^{n-k}\begin{bmatrix}k\\n\end{bmatrix}_{q^{-1}} \tag{C.52}$$

for all $n, k \in \mathbb{Z}$. Furthermore $\left[\begin{smallmatrix}n\\k\end{smallmatrix}\right]_q = \left\{\begin{smallmatrix}n\\k\end{smallmatrix}\right\}_q = 0$ for all $k > n$ and all $(-n, -k) \in \mathbb{N}^2 \setminus \{(0,0)\}$. For $n \geqslant 1$, this is exactly what (C.34) and (C.37) says for $k < 0$, and likewise for $k \geqslant 0$ this is exactly what (C.36) says for $n < 0$.

The third major approach to the q-Stirling numbers that was treated in Section C.3 is that through sequences of polynomials which was used as definition. It too generalises to arbitrary $n, k \in \mathbb{Z}$ if $F_k(x;q)$ is defined for $k < 0$ through

$$F_k(x;q) = \prod_{i=k}^{-1}\left(x - \{i\}_q\right)^{-1} = \prod_{i=1}^{-k}\left(x + q^{-i}\{i\}_q\right)^{-1},$$

but for negative n the sums in general become series, and thus convergence problems complicates the treatment. The reformulations of (C.23) and (C.24) are that

$$x^n = \sum_{k\in\mathbb{Z}}\begin{Bmatrix}n\\k\end{Bmatrix}_q F_k(x;q), \tag{C.53}$$

$$F_n(x;q) = \sum_{k\in\mathbb{Z}}\begin{bmatrix}n\\k\end{bmatrix}_q (-1)^{n-k}x^k \tag{C.54}$$

for all $n \in \mathbb{Z}$. As an example of how all this works, and of the problems related to convergence which might appear, it is instructive to consider (C.53) for $n = -1$:

$$\frac{1}{x} = \sum_{k=1}^{\infty}\begin{Bmatrix}-1\\-k\end{Bmatrix}_q F_{-k}(x;q) = \sum_{k=1}^{\infty} q^{k-1}\begin{bmatrix}k\\1\end{bmatrix}_{\frac{1}{q}} F_{-k}(x;q) =$$

$$= \sum_{k=1}^{\infty}\frac{q^{k-1}\{k-1\}_{q^{-1}}!}{\prod_{i=1}^{k}\left(x + q^{-i}\{i\}_q\right)} = \sum_{k=1}^{\infty}\frac{\prod_{i=1}^{k-1} q^{-1}\{i\}_{q^{-1}}}{\prod_{i=1}^{k}\left(x + q^{-1}\{i\}_{q^{-1}}\right)}.$$

(It might be hard at first sight to get any impression of how the right hand side behaves, but it helps to consider the identity

$$\frac{1}{x} = \sum_{k=1}^{n} \frac{\prod_{i=1}^{k-1} a_i}{\prod_{i=1}^{k} (x + a_i)} + \frac{\prod_{i=1}^{n} a_i}{x \prod_{i=1}^{n} (x + a_i)}$$

which is easy to show for any commutative variable x and ditto constants $a_1, \ldots, a_n$.)

In summary all the q-combinatorial functions except the q-factorial can be generalised to all integers, and the majority of their properties continue to hold. One that does not is that of the functions being polynomials in q, instead they become polynomials in q and q^{-1}. Finally, there are formulae given above for calculating their values for arbitrary integers using only their values for the natural numbers.

Appendix D

Notes on notations

Arthur slightly smiled. "It seems a paradox, does it not," he went on, "that the image formed on the Retina should be inverted?"

"It *is* puzzling," she candidly admitted. "Why is it we do not *see* things upside-down?"

"You have never heard of the Theory, then, that the *Brain* also is inverted?"

"No *indeed*! What a *beautiful* fact! But how is it *proved*?"

"*Thus*," replied Arthur, with all the gravity of ten Professors rolled into one. "What we call the *vertex* of the Brain is really its *base*; and what we call its *base* is really its *vertex*; it is simply a question of *nomenclature*."

This last polysyllable settled the matter.

— from *Sylvie and Bruno* by L. CARROLL

What follows is a brief list of some notations we fear might cause trouble for the reader.

$\mathbb{N}$	The set $\{ n \in \mathbb{Z} \mid n \geqslant 0 \}$.
$\mathbb{Z}^+$	The set $\{ n \in \mathbb{Z} \mid n > 0 \}$.
$\mathcal{K}^*$	The set $\mathcal{K} \setminus \{0\}$, seen as a group.
S_n	The group of permutations of $\{1, 2, \ldots, n\}$.
$\mathrm{End}(V)$	The algebra of endomorphisms of V (homomorphisms $V \longrightarrow V$), where V is any linear space.
$\lfloor x \rfloor$	The integer $\max\{ n \in \mathbb{Z} \mid n \leqslant x \}$, where x may be any real number.
$[\alpha, \beta]$	The *commutator* of α and β, i.e., the same thing as $\alpha\beta - \beta\alpha$. In particular one has that $[\alpha, \beta] = 0$ if and only if α and β commute.

$O(f(n))$ If $g(n) = O(f(n))$ then there exists an integer n_0 and a positive constant c such that

$$\left|\frac{g(n)}{f(n)}\right| < c \tag{D.1}$$

for all $n > n_0$.

$\Theta(f(n))$ If $g(n) = \Theta(f(n))$ then there exists an integer n_0 and two positive constants c_1 and c_2 such that

$$c_1 < \left|\frac{g(n)}{f(n)}\right| < c_2$$

for all $n > n_0$.

Originally, the notation $O(f(n))$ merely meant "some unspecified term that stays bounded when divided by $f(n)$", but it—and its relatives $o(f(n))$, $\Omega(f(n))$, $\omega(f(n))$, and $\Theta(f(n))$—has grown to become much more than that. To the modern algebraicist, it probably makes most sense to view $O(f(n))$ as a set of functions—the set of functions which satisfy (D.1) for all $n > n_0$, for some n_0 and c. It is an elementary exercise to show that $\{O(n^k)\}_{k\in\mathbb{N}}$ is an ascending $\mathbb{N}$-filtration in the algebra of real-valued functions in n. Adopting this view requires that it is understood that $=$ in conjunction with expressions containing O-expressions in most cases really means $\in$, $\subseteq$, or something similar, but it also makes it much easier to understand why for example $O(n^2) + f(n) = O(n^2)$ does not imply $f(n) = 0$, but merely $f(n) = O(n^2)$.

Bibliography

References by author

W. H. Abdi [1, 2]
C. R. Adams [3–6]
M. Adler [7]
A. K. Agarwal [9]
R. P. Agarwal [8]
W. A. Al-Salam [10–13]
S. A. Amitsur [14]
G. E. Andrews [15, 16]
K. Aomoto [17–19]
M. Arik [20–22]
R. Askey [23]
M. Aspenberg [24]
S. S. Avancini [25]
J. A. de Azcárraga [26, 104, 105]
Yu. A. Bahturin [27]
Yu. M. Berezanskiĭ [28–30]
R. Berger [31]
G. M. Bergman [32]
I. N. Bernstein [33]
J. Bertrand [34]
L. C. Biedenharn [35, 36]
G. D. Birkhoff [37]
J.-E. Björk [38]
D. Bonatsos [39, 40]
F. Bonechi [41]
M. J. Boussinesq [42]
M. Bożejko [43–45]
G. Brodimas [201, 202]
J. L. Burchnall [46–48]
A. C. Cadavid [49, 50]
L. Carlitz [51]
R. C. Carlson [52]
R. D. Carmichael [53]
U. Carow-Watamura [54, 55]
B. L. Cerchiai [56]
M. Chaichian [57–63]
F. L. Chan [64, 65]
Z. Chang [66]
S. Chaturvedi [67, 68]
T. W. Chaundy [46–48, 69]
A. R. Chowdhury [70, 291]
W.-S. Chung [71, 72]
N. Ciccoli [41]
J. Cigler [73–81]
S. Codriansky [82]
A. Connes [83]
D. D. Coon [22]
T. L. Curtright [84]
J.-H. Dai [85]
E. V. Damaskinsky [86–89]
C. Daskaloyannis [40]
C. Delbecq [90, 91]
A. P. Demichev [57, 58]
L. A. Dickey [92, 154, 155]
J. F. van Diejen [93]
A. Dimakis [94]

J. Dixmier [95]
V. K. Dobrev [96–99]
H. D. Doebner [98]
V. G. Drinfeld [100]
M. Dubois-Violette [101]
B. A. Dubrovin [102, 103]
R. S. Dunne [104, 105]
S. Durand [106]
K. Dykema [107]
D. W. Ebner [324]
F. H. L. Essler [108]
H. Exton [109]
L. D. Faddeev [402]
A. Faessler [40]
D. B. Fairlie [110, 111, 156]
H.-Y. Fan [112, 113, 205, 206]
Ph. Feinsilver [114–126]
R. J. Finkelstein [49, 50, 64, 65, 127]
G. Fiore [128, 129]
D. I. Fivel [130]
H. Flanders [131]
N. Fleury [132]
G. Floquet [133]
R. Floreanini [134–146]
E. Frenkel [147]
I. B. Frenkel [148]
B. Fuglede [149]
P. Furlan [150]
L. Gårding [151, 152]
G. Gasper [153]
I. M. Gelfand [154–158, 384]
F. Gesztesy [159]
R. Giachetti [41]
V. Ya. Golodets [160]
F. R. Gonzalez [59]
K. R. Goodearl [52]
V. I. Gorbachuk [161]
H. W. Gould [162]
M. I. Graev [157, 384]
R. L. Graham [163]
O. W. Greenberg [164, 165]
H. Grosse [60]
H.-Y. Guo [66, 85]
N. D. Gupta [70]
L. K. Hadjiivanov [150]
W. Hahn [166–174]
L. Haine [175]
J. Harnad [176]
T. Hayashi [177]
A. Hebecker [178, 179]
E. Heine [180, 181]
L. Hellström [182–184]
P. A. Hendriks [185]
R. Hinterding [56]
E. Horikawa [186]
E. Horozov [7]
B. Y. Hou [187]
J. Hruby [188]
E. C. Ihrig [189]
P. Iliev [175, 190–193]
K. N. Ilinski [194]
M. Irac-Astaud [34]
E. H. Ismail [11]
M. E. H. Ismail [189]
F. H. Jackson [195–199]
R. Jagannathan [67]
A. Jannussis [200–203]
O. R. Jensen [204]
S.-C. Jing [112, 113, 205–207]
A. Joseph [208]
P. E. T. Jørgensen [209–219]
D. Kahaner [335]
G. V. Kalinin [194]
E. G. Kalnins [9]
G. Kaniadakis [220]
M. V. Karasev [221, 222]
C. Kassel [223, 224]
Y. Kato [19]
A. Kempf [225–227]
R. Kerner [101]
B. Khesin [228]
M. Klimek [229, 230]
A. U. Klimyk [72, 231, 386]
D. E. Knuth [163, 232]
A. N. Kochubei [233]
W. Koepf [234]
Yu. G. Kondrat'ev [29]
B. G. Konopelchenko [235]

T. H. Koornwinder [236, 237]
D. Korteweg [238]
B. S. Kostadinov [99]
N. Kravitsky [256]
G. Krein [25]
I. M. Krichever [239–247]
P. P. Kulish [58, 61, 62, 86–89, 248]
R. Kumar [340]
V. V. Kuryshkin [249]
E. C. Lance [251]
L. Lapointe [252]
A. Lavagno [220]
P. D. Lax [253, 254]
J. Le Caine [250]
P. Lesky [255]
J. LeTourneux [136, 137]
M. S. Livšic [256, 257]
M. A. Lohe [36]
J. Lukierski [62]
V. Lyubashenko [228]
L. Ma [258, 259]
Z. Q. Ma [187]
I. G. Macdonald [264]
A. J. Macfarlane [26, 104, 105, 265]
G. W. Mackey [267]
J. Madore [56]
S. Majid [268–270]
S. V. Manakov [403]
Yu. I. Manin [271, 272]
V. A. Marchenko [273]
E. Marcus [127]
A. S. Markus [256]
F. Marotte [274]
V. P. Maslov [221, 275]
T. E. Mason [276]
V. B. Matveev [103]
V. Mazorchuk [277]
R. J. McDermott [260–263]
H. P. McKean [266]
H. G. Meijer [367]
U. Meyer [278]
A. A. Mikhalev [27]
W. Miller [9, 279]
S. C. Milne [280]
K. Mimachi [281–286]
M. N. Mnatsakanova [63]
P. van Moerbeke [7, 287]
O. I. Mokhov [288]
R. T. More [214]
H. Morikawa [289]
J. Moser [290]
I. Mukhopadhaya [291]
M. Mulase [292]
F. Müller-Hoissen [94]
D. Mumford [287, 293]
H. C. Myung [294]
J. S. Møller [295]
V. E. Nazaikinskii [296]
M. H. A. Newman [297]
A. Nica [107]
M. Nishizawa [298, 299]
N. E. Nörlund [300]
M. Noumi [286]
S. P. Novikov [103, 244–247, 301, 403]
E. Novikova [222]
A. Odlyzko [335]
V. Oganesyan [65]
V. L. Ostrovskyĭ [30, 302–305]
L. Papaloucas [201–203]
E. Papp [306–308]
O. Patashnik [163]
G. K. Pedersen [309]
S. Pedersen [215, 310]
J. C. Pérez Bueno [104, 105]
V. M. Petrogradsky [27]
M. Pillin [311]
L. P. Pitayevsky [403]
G. Post [312, 374]
P. Prešnajder [59, 60]
A. Pressley [313]
E. Previato [314, 315]
D. P. Proskurin [216, 316]
W. Pusz [317, 318]
C. R. Putnam [319]
P. Quarati [220]
C. Quesne [90, 91, 320]
M. Rahman [153]
M. Rasetti [321]

M. Rausch de Traubenberg [322, 323]
N. Yu. Reshetikhin [148]
V. S. Retakh [157]
V. Rittenberg [108]
S. Rodriguez-Romo [324]
C. Roger [228]
L. J. Rogers [325]
M. Roitman [326]
S. Roman [327–332]
A. L. Rosenberg [333, 334]
G.-C. Rota [332, 335]
F. Ryde [336]
Yu. S. Samoĭlenko [30, 161, 216, 302–304, 337, 382]
R. M. Santilli [294, 338, 339]
R. K. Saxena [340]
A. Schirrmacher [341]
M. Schlieker [54]
L. M. Schmitt [217, 218]
K. Schmüdgen [231, 342–344]
S. Schreckenberg [179]
I. Schur [345]
J. Schwenk [179, 346–349]
J. Scott-Russell [350]
G. Segal [313, 351]
S. V. Shabanov [352, 353]
A. B. Shabat [404, 405]
V. E. Shatalov [296]
P. Siafarikas [201–203]
S. D. Silvestrov [24, 183, 184, 305, 354–358]
S. Skorik [359]
V. V. Sokolov [360, 361]
A. I. Solomon [260–263, 362]
E. Sorace [41]
D. Sourlas [201, 202]
R. Speicher [43–45, 363, 364]
V. P. Spiridonov [134, 135, 359, 365]
R. Sridhar [67]
V. Srinivasan [67, 68]
G. W. Starcher [366]
A. S. Stepanenko [194]
B. Yu. Sternin [296]
A. Streclas [201]
T. Striker [94]
R. F. Swarttouw [367]
Z. Tang [258, 259]
M. Tarlini [41]
J. Thomae [368–370]
J. M. Thuswaldner [371]
I. T. Todorov [150]
W. J. Trjitzinsky [372]
N. Tsangas [201]
A. Turbiner [132, 373, 374]
L. B. Turowska [277, 357]
R. Twarock [98, 375, 376]
M. R. Ubriaco [377]
K. Ueno [299]
M. Upadhyay [378–380]
G. F. Us [161]
N. Vansteenkiste [320]
E. Ye. Vaysleb [381, 382]
J.-L. Verdier [383]
A. Verma [12, 13]
Yu. S. Vernov [63]
A. M. Vershik [384]
A. P. Veselov [385]
N. Ya. Vilenkin [158, 386]
L. Vinet [134–146, 252]
V. Vinnikov [256]
O. V. Viskov [387]
K. Vlachos [202]
G. de Vries [238]
C. G. Wagner [388]
L. L. Waksman [257]
G. Wallenberg [389]
H. Wallin [358]
R. Wallisser [390]
S. Watamura [54, 55]
W. Weich [178, 179]
R. Weikard [159]
R. F. Werner [217–219]
J. Wess [56, 179, 349, 391–394]
A. Wightman [151, 152]
G. Wilson [351, 395–397]
J. Wilson [23]
S. L. Woronowicz [318, 398]
J. J. Xu [207]

H. Yan [66, 85]
K.-W. Yang [399]
C. K. Zachos [111, 400]
D. Zagier [401]
M. V. Zaicev [27]
V. E. Zakharov [402–405]
C.-Z. Zha [406]
C. Zhang [274, 407]
Y.-D. Zhang [258, 259]
W.-Z. Zhao [406]
A. S. Zhedanov [408, 409]
B. Zumino [393, 394, 410]

References by topic

The following list contains abbreviations for some subjects considered in the references which have been mentioned in the introduction and other parts of the book. All references are listed in the Bibliography and abbreviations for those primary subjects which are considered in each reference are placed at the end of the corresponding entry in the Bibliography. These subjects are interrelated in a complicated way, and many of the references treat several of these subjects. Thus in order to reach some useful classification we had to make some own judgements on the primary subjects of each article. We hope that this classification together with the introduction to the book would assist the reader in his exploration of the related literature.

BGLA braided geometry and braided Lie algebras [208, 224, 231, 269, 270]

CDO commuting differential operators as separate subject [14, 46–48, 52, 69, 95, 131, 133, 183, 184, 345, 389]

COAG commuting operators and algebraic geometry [100, 102, 103, 159, 239–247, 256, 257, 266, 271, 287, 292, 293, 301, 313–315, 351, 360, 361, 383, 396, 397, 403–405]

COCT commuting operators in colligation theory, investigations on non-self-adjoint operators, and representations of semigroups [256, 257]

CONDE commuting operators in investigation of solutions to non-linear differential equations [42, 92, 100, 102, 103, 154, 155, 159, 176, 235, 238–247, 253, 254, 266, 271, 273, 287, 288, 290, 292, 293, 301, 313–315, 350, 351, 360, 361, 383, 385, 395–397, 402–405]

CORT commuting operators in representation theory [28–30, 83, 93, 149, 151, 152, 158, 160, 161, 209–211, 213–215, 222, 251, 252, 264, 267, 286, 304, 305, 309, 310, 319, 337, 342, 354, 355, 357, 358, 381, 382, 384]

DHO deformed harmonic oscillators [20, 26, 34–36, 39–41, 49, 50, 54, 55, 57–66, 71, 72, 84, 86–89, 104, 105, 110–113, 117–127, 134–137, 141, 200, 206, 207, 220, 248, 258–263, 265, 291, 324, 352, 353, 359, 362, 365, 377, 400, 408, 409]

DL Diamond Lemma for rings and algebras [27, 31, 32, 184, 297, 312]

DQ deformation quantization [156, 212]

H∗-A C^* and $*$-algebras associated with q-deformed Heisenberg relations and their representations by bounded and unbounded operators on a Hilbert space [107, 216–219, 302–304, 333, 334, 343, 344]

LAA Lie-admissible algebras [200–203, 294, 338, 339]

NCM non-canonical mechanics [200–203]

NGC noncommutative geometry and noncommutative differential and integral calculus [26, 58, 67, 68, 84–89, 104, 105, 111, 187, 208, 221, 223, 224, 229–231, 248, 269, 270, 272, 275, 296, 346, 377, 393, 394, 398, 410]

QA quadratic algebras [108, 272, 302, 303]

q-A q-analogues of important algebras [39–41, 49, 50, 57, 58, 61, 62, 64–66, 71, 84–89, 108, 110, 111, 117–127, 134, 135, 138, 139, 142–146, 148, 150,

156, 177, 187, 194, 216, 218, 223, 224, 248, 265, 270, 277, 278, 346, 375–377, 398, 400, 406, 408, 410]

q-DE q-difference equations and q-analysis [1–6, 8–13, 15–20, 23, 37, 53–55, 73–76, 79–81, 96–99, 109, 114–116, 138–140, 142–145, 148, 153, 157, 166–174, 180, 181, 185, 186, 195–199, 205, 229, 230, 250, 255, 274, 276, 278–286, 291, 298–300, 306–308, 311, 325, 336, 340, 341, 366–372, 375, 376, 378–380, 386, 388, 390, 399, 407]

q-FA q-Fourier analysis [72, 236, 237, 347, 348]

QGQS quantum groups and quantum spaces [20, 34–36, 39–41, 49, 50, 54, 55, 57, 58, 61, 62, 64, 65, 67, 68, 71, 84–89, 96–99, 117–127, 138–140, 150, 177, 186, 187, 208, 224, 231, 236, 248, 265, 269, 270, 278, 321, 324, 341, 346, 377, 398, 400, 410]

q-H q-analogues of various objects from homological algebra [18, 19, 187, 223, 224]

q-HARA representations and applications of q-deformed Heisenberg algebras [14, 20–22, 24–26, 38–41, 43–45, 49, 50, 56, 58–66, 72, 82, 84, 86–91, 104–106, 108, 110–131, 133, 141, 151, 152, 156, 160, 164, 165, 177–179, 184, 188, 194, 204, 206, 212, 220, 225–227, 233, 248, 249, 258–263, 265, 277, 289, 295, 316–318, 320, 322, 323, 345, 349, 352, 353, 359, 362–365, 391, 392, 398, 401, 408–410]

q-NDE nonlinear q-difference equations [7, 147, 175, 190–193, 228]

q-OP q-orthogonal polynomials [23, 61, 62, 66, 72–76, 79–81, 84, 86–89, 136–146, 280–286, 347, 348]

q-SF q-special functions [51, 61, 62, 72–81, 84, 86–89, 109, 114–116, 136–140, 142–146, 148, 153, 157, 162, 163, 166–174, 180, 181, 186, 195–199, 232, 234, 236, 237, 280–286, 325, 368–370, 378–380, 386, 388, 408]

RF reordering formulas in q-deformed Heisenberg algebras [11, 31, 32, 73–76, 79–81, 114–116, 132, 182–184, 236, 237, 295, 312, 356, 373, 374, 387]

UC umbral calculus [75, 94, 189, 327–332, 335]

References

[1] W. H. Abdi: *On certain q-difference equations and q-Laplace transform*, Proceedings of the national academy of sciences, India A, (1962), 1–15. (q-DE)

[2] ———: *Certain inversion and representation formulae for q-Laplace transforms*, Math. Z. **83** (1964), 238–249. (q-DE)

[3] C. R. Adams: *The general theory of a class of linear partial q-difference equations*, Trans. Amer. Math. Soc. **26** (1924), 283–312. (q-DE)

[4] ———: *Note on the existence of analytic solutions of the non-homogeneous linear q-difference equation, ordinary and partial*, Ann. of Math. **27** (1925), 73–83. (q-DE)

[5] ———: *On the linear ordinary q-difference equation*, Ann. of Math. **30** (1929), 195–205. (q-DE)

[6] ———: *Linear q-difference equations*, Bull. Amer. Math. Soc. **37** (1931), 361–400. (q-DE)

[7] M. Adler, E. Horozov, P. van Moerbeke: *The solution to the q-KdV equation*, Phys. Lett. A. **242**, 3 (1998), 139–151. (q-NDE)

[8] R. P. Agarwal: *Certain fractional q-integrals and q-derivatives*, Proc. Camb. Phil. Soc. **66** (1969), 365–370. (q-DE)

[9] A. K. Agarwal, E. G. Kalnins, W. Miller: *Canonical equations and symmetry techniques for q-series*, SIAM J. Math. Anal. **18**, 6 (1987), 1519–1538. (q-DE)

[10] W. A. Al-Salam: *Some fractional q-integrals and q-derivatives*, Proc. Edin. Phil. Soc. **15** (1966), 135–140. (q-DE)

[11] W. A. Al-Salam, E. H. Ismail: *Some operational formulas*, J. Math. Anal. Appl. **51** (1975), 208–218. (RF, q-DE)

[12] W. A. Al-Salam, A. Verma: *A fractional Leibniz q-formula*, Pacific J. Math. **60** (1975), 1–9. (q-DE)

[13] ———: *Remarks on fractional q-integrals*, Bul. Soc. Roy. Sci. Liège, **44** (1975), 600–607. (q-DE)

[14] S. A. Amitsur: *Commutative linear differential operators*, Pacific J. Math. **8** (1958), 1–10. (CDO, q-HARA)

[15] G. E. Andrews: *On q-difference equations for certain well-poised basic hypergeometric series*, Quart. J. Math. (Oxford) **19**, 2 (1968), 433–447. (q-DE)

[16] ———: *On the foundations of combinatorial theory V, Eulerian differential operators*, Studies in Appl. Math. **50** (1971), 345–375. (q-DE)

[17] K. Aomoto: *A note on holonomic q-difference system*, in *Algebraic analysis. I*, (M. Kashiwara, T. Kawai, eds.), Academic Press, New York, 1988, 25–28. (q-DE)

[18] ———: *Finiteness of a cohomology associated with certain Jackson integrals*, Tôhoku Math. J. **43** (1991), 75–101. (q-DE, q-H)

[19] K. Aomoto, Y. Kato: *A q-analogue of de Rham cohomology associated with Jackson integrals*, in *Special functions*, (Okayama, 1990), ICM-90 Satell. Conf. Proc., Springer, Tokyo, 1991, 30–60. (q-DE, q-H)

[20] M. Arik: *The q-difference operator, the quantum hyperplane, Hilbert spaces of analytic functions and q-oscillators*, Z. Phys. C, **51**, 4 (1991), 627–632. (DHO, q-HARA, QGQS, q-DE)

[21] ______: *From Q-oscillators to quantum groups*, in *Symmetries in Science VI* (edited by B. Gruber), Plenum Press, New York, 1993. (q-HARA)

[22] M. Arik, D. D. Coon: *Hilbert spaces of analytic functions and generalized coherent states*, J. Math. Phys. **17**, 4 (1976), 524–527. (q-HARA)

[23] R. Askey, J. Wilson: *Some basic orthogonal polynomials that generalize Jacobi polynomials*, Mem. AMS **54**, 319 (1985). (q-DE, q-OP)

[24] M. Aspenberg, S. D. Silvestrov: *Complex two-dimensional representations of noncommutative quadratic curves*, Preprint, TRITA-MAT-1997-MA **35**, Dep. of Math., Royal Institute of Technology, (1997), 126 pp. (q-HARA)

[25] S. S. Avancini, G. Krein: *Many-body problems with composite particles and q-Heisenberg algebras*, J. Phys. A: Math. Gen. **28**, 3 (1995), 685–691. (q-HARA)

[26] J. A. de Azcárraga, A. J. Macfarlane: *Group theoretical foundations of fractional supersymmetry*, J. Math. Phys. **37**, 3 (1996), 1115–1127. (q-HARA, NGC, DHO, NGC)

[27] Yu. A. Bahturin, A. A. Mikhalev, V. M. Petrogradsky, M. V. Zaicev: *Infinite Dimensional Lie Superalgebras*, Walter de Gruyter, Berlin, 1992. (DL)

[28] Yu. M. Berezanskiĭ: *Selfadjoint operators in spaces of functions of infinitely many variables*, Trans. Math. Monographs **63**, AMS, Providence, 1986. (Transl. to English from Russian edn.: Naukova Dumka, Kiev, 1978.) (CORT)

[29] Yu. M. Berezanskiĭ, Yu. G. Kondrat'ev: *Spectral methods in infinite-dimensional analysis*, Kluwer Acad. Publ., Dordrecht, 1992. (Transl. to English from Russian edn.: Naukova Dumka, Kiev, 1988.) (CORT)

[30] Yu. M. Berezanskiĭ, V. L. Ostrovskyĭ, Yu. S. Samoĭlenko: *Expansion in eigenfunctions of families of commuting operators, and representations of commutation relations*, Ukrain. Mat. Zh. **40**, 1 (1988), 106–109. (CORT)

[31] R. Berger: *Quantification de l'identité de Jacobi*, C. R. Acad. Sci. Paris, **312**, Sér. I (1991), 171–174. (RF, DL)

[32] G. M. Bergman: *The diamond lemma for ring theory*, Adv. Math. **29** (1978), 178–218. (RF, DL)

[33] I. N. Bernstein: *The analytic continuation of generalised functions with respect to a parameter*, Funktz. Anal. Priloz. **6**, 4 (1972), 26–40.

[34] J. Bertrand, M. Irac-Astaud: *Invariant differential calclus on a deformation of the Weyl-Heisenberg algebra*, in *Modern Group Theoretical Methods in Physics*, Kluwer Acad. Publ. (1995), 37–49. L873–L878. (DHO, QGQS)

[35] L. C. Biedenharn: *The quantum group* $\mathrm{SU}_q(2)$ *and a q-analogue of the Boson operators*, J. Phys. A: Math. Gen. **22** (1989), L873–L878. (DHO, QGQS)

[36] L. C. Biedenharn, M. A. Lohe: *A concise introduction to quantum groups*, in *Proceedings of the International School of Physics "Enrico Fermi"*, (L. Castellani, J. Wess, eds.), Italian Physical Society, IOS Press, 1996, 1–52. (DHO, QGQS)

[37] G. D. Birkhoff: *The generalized Riemann problem for linear differential equations and the allied problems for linear difference and q-difference equations*, Proc. Amer. Acad. Arts Sci. **49** (1913), 521–568 . (q-DE)

[38] J.-E. Björk: *Rings of Differential Operators*, North-Holland Publishing Company, 1979. (q-HARA)

[39] D. Bonatsos: *Are q-bosons suitable for the description of correlated fermion pairs?*, J. Phys. A: Math. Gen. **25** (1992), L101–L108. (DHO, q-HARA, q-A, QGQS)

[40] D. Bonatsos, C. Daskaloyannis, A. Faessler: *Quantum algebraic description of the pairing correlations in a single-j nuclear shell*, J. Phys. A: Math. Gen. **27** (1994), 1299–1306. (DHO, q-HARA, q-A, QGQS)

[41] F. Bonechi, N. Ciccoli, R. Giachetti, E. Sorace, M. Tarlini: *Free q-Schrödinger equation from homogeneous spaces of the 2-dim Euclidean quantum group*, Comm. Math. Phys. **175** (1996), 161–176. (DHO, q-HARA, q-A, QGQS)

[42] M. J. Boussinesq: *Théorie des ondes et des remous qui se propagent le long d'un canal rectangulaire horizontal, en communiquant au liquide contenu dans ce canal des vitesses sensiblement pareilles de la surface au fond*, J. Math. Pures et Appl. **27** (1872), 55–108. (CONDE)

[43] M. Bożejko, R. Speicher: *An example of a generalised Brownian motion*, Commun. Math. Phys. **137** (1991), 519–531. (q-HARA)

[44] ______: *An example of a generalised Brownian motion II*, in *Quantum Probability and Related Topics VII*, (L. Accardi, ed.), World Scientific, Singapore, 1992, 67–77. (q-HARA)

[45] ______: *Completely positive maps on Coxeter groups, deformed commutation relations, and operator spaces*, Math. Ann. **300** (1994), 97–120. (q-HARA)

[46] J. L. Burchnall, T. W. Chaundy: *Commutative ordinary differential operators*, Proc. London Math. Soc. (Ser. 2) **21** (1922), 420–440. (CDO)

[47] ______: *Commutative ordinary differential operators*, Proc. Roy. Soc. London **A 118** (1928), 557–583. (CDO)

[48] ______: *Commutative ordinary differential operators. II. — The Identity* $P^n = Q^m$, Proc. Roy. Soc. London **A 134** (1932), 471–485. (CDO)

[49] A. C. Cadavid, R. J. Finkelstein: Integration over the group space of $\mathrm{SU}_q(2)$, Clebsch-Gordan coefficients, and related identities. Lett. Math. Phys. 33 (1995), no. 3, 221–229. (DHO, q-A, QGQS, q-HARA)

[50] ______: *The q-Coulomb problem in configuration space*, J. Math. Phys. **37**, 8 (1996), 3675–3683. (DHO, q-A, QGQS, q-HARA)

[51] L. Carlitz: *q-Bernoulli numbers and polynomials*, Duke. Math. J. **15** (1948), 987–1000. (q-SF)

[52] R. C. Carlson, K. R. Goodearl: *Commutants of ordinary differential operators*, J. Differential Equations, **35** 3 (1980), 339–365. (CDO)

[53] R. D. Carmichael: *The general theory of linear q-difference equations*, Amer. J. Math. **34** (1912), 147–168. (q-DE)

[54] U. Carow-Watamura, M. Schlieker, S. Watamura: *$SO_q(N)$ covariant differential calculus on quantum space and quantum deformation of Schrödinger equation*, Z. für Phys. C, Particles and fields, **49** (1991), 439–446. (DHO, QGQS, q-DE)

[55] U. Carow-Watamura, S. Watamura: *The quantum group as a symmetry. The Schrödinger equation of the N-dimensional q-deformed harmonic oscillator*, Progress of theoretical physics supplement, **118** (1995), 375–389. (DHO, QGQS, q-DE)

[56] B. L. Cerchiai, R. Hinterding, J. Madore, J. Wess: *A calculus based on a q-deformed Heisenberg algebra*, Eur. Phys. J. C Part. Fields, **8**, 3 (1999), 547–558. (q-HARA)

[57] M. Chaichian, A. P. Demichev: *Introduction to quantum groups*, World Sci. Publishing, 1996. (QGQS, DHO, q-A)

[58] M. Chaichian, A. P. Demichev, P. P. Kulish: *Quasiclassical limit in q-deformed systems, noncommutativity and the q-path integral*, Phys. Lett. A, **233**, 4–6 (1997), 251–260. (DHO, q-HARA, NGC, q-A, QGQS)

[59] M. Chaichian, F. R. Gonzalez, P. Prešnajder: *Unitary irreducible representations of covariant q-oscillators*, J. Phys. A: Math. Gen. **28**, 8, (1995), 2247–2253. (q-HARA, DHO)

[60] M. Chaichian, H. Grosse, P. Prešnajder: *Unitary representations of the q-oscillator algebra*, J. Phys. A: Math. Gen. **27**, (1994), 2045–2051. (q-HARA, DHO)

[61] M. Chaichian, P. P. Kulish: *Quantum superalgebras, q-oscillators and applications*, in *Nonperturbative methods in low dimensional quantum field theories*: Proceedings of 14th Johns Hopkins Workshop on Current Problems in Particle Theory, Debrecen, Hungary, 1990, (edited by G. Domokos, Z. Horvath, S. Kovesi-Domokos), World Scientific, Singapore, 1991, 214–236. (DHO, q-A, QGQS, q-HARA, q-OP, q-SF)

[62] M. Chaichian, P. P. Kulish, J. Lukierski: *Supercovariant systems of q-oscillators and q-supersymmetric hamiltonians*, Phys. Lett. B. **262**, 1 (1991), 43–48. (DHO, q-A, QGQS, q-HARA, q-OP, q-SF)

[63] M. Chaichian, M. N. Mnatsakanova, Yu. S. Vernov: *Invariance properties of the q-oscillator algebra: q-analogue of von Neumann's theorem*, J. Phys. A: Math. Gen. **27**, (1994), 2053–2059. (DHO, q-HARA)

[64] F. L. Chan, R. J. Finkelstein: *q-Deformation of the Coulomb problem*, J. Math. Phys. **35**, 7 (1994), 3273–3284. (DHO, q-A, QGQS, q-HARA)

[65] F. L. Chan, R. J. Finkelstein, V. Oganesyan: *The q-isotropic oscillator,* J. Math. Phys. **38**, 5 (1997), 2132–2147. (DHO, q-A, QGQS, q-HARA)

[66] Z. Chang, H.-Y. Guo, H. Yan: *The q-Hermite polynomial and the representations of Heisenberg and quantum Heisenberg algebras*, J. Phys. A: Math. Gen. **25**, 6 (1992), 1517–1525. (DHO, q-HARA, q-A, q-OP)

[67] S. Chaturvedi, R. Jagannathan, R. Sridhar, V. Srinivasan: *Nonrelatvistic quantum mechanics in a noncommutative space*, J. Phys. A: Math. Gen. **26**, 3 (1993), L105–L112. (NGC, QGQS)

[68] S. Chaturvedi, V. Srinivasan: *Aspects of q-oscillator quantum mechanics*, Phys. Rev. A (3) **26**, 12 (1991), 8020–8023. (NGC, QGQS)

[69] T. W. Chaundy: *The differential calculus*, Oxford, 1935. (CDO)

[70] A. R. Chowdhury, N. D. Gupta: *Commutative differential operator and a class of solutions for the BKP equation on the singular rational curve*, J. Math. Phys. **32**, 12 (1991), 3473–3475.

[71] W.-S. Chung: *Some realization of $gl_q(n)$-covariant oscillator algebra and $gl_q(n)$-covariant q-Virasoro algebra with q a root of unity*, Preprint, q-alg/9610001, (1996), 9 pp. (DHO, q-A, QGQS)

[72] W.-S. Chung, A. U. Klimyk: *On position and momentum operators in the q-oscillator algebras*, J. Math. Phys. **37**, 2 (1996), 917–932. (DHO, q-HARA, q-FA, q-OP, q-SF)

[73] J. Cigler: *Operatormethoden für q-Identitäten*, Monatsh. Math. **88** (1979), 87–105. (RF, q-DE, q-OP, q-SF)

[74] ______: *Operatormethoden für q-Identitäten II: q-Laguerre-Polynome*, Monatsh. Math. **91** (1981), 105–117. (RF, q-DE, q-OP, q-SF)

[75] ______: *Operatormethoden für q-Identitäten III: Umbrale Inversion and die Lagrangesche Formel*, Archiv der Math. **35** (1980), 533–543. (RF, q-DE, q-OP, q-SF, UC)

[76] ______: *Elementare q-Identitäten*, Publication de l'institute de recherche Mathématique avancée (1982) 23–57. (RF, q-DE, q-OP, q-SF)

[77] ______: *Über die Anzahl erzeugender Mengen in endlichen Vektorräumen*, Anz. Österreich. Akad. Wiss. Math.-Natur. Kl. 129 (1992), 27–33. (q-SF)

[78] ______: *A new q-analog of Stirling Numbers*, Österreich. Akad. Wiss. Math.-Natur. Kl. Sitzungsber. II **201**, no. 1-10, (1992), 97–109. (q-SF)

[79] ______: *Operatormethoden für q-Identitäten IV: Eine Klasse von q-Gould-Polynomen*, Sitzungsber. Abt. II, 205 (1996), 169–174. (RF, q-DE, q-OP, q-SF)

[80] ______: *Operatormethoden für q-Identitäten V: q-Catalan-Bäume*, Sitzungsber. Abt. II, 205 (1996), 175–182. (RF, q-DE, q-OP, q-SF)

[81] ______: *Operatormethoden für q-Identitäten VI: Geordnete Wurzelbäume und q-Catalan-Zahlen*, Sitzungsber. Abt. II, 206 (1997), 253–266. (RF, q-DE, q-OP, q-SF)

[82] S. Codriansky: *Localized states in deformed quantum mechanics*, Phys. Lett. A, 184 (1994), 381–384. (q-HARA)

[83] A. Connes: *Noncommutative Geometry*, Academic Press, 1994. (Extended transl. to English from French edn.: InterEditions, 1990.) (CORT)

[84] T. L. Curtright: *Deformations, Coproducts, and U*, in *Quantum Groups*, Proceedings of the Argonne Workshop, Argonne National Laboratory, 1990, (T. Curtright, D. Fairlie, C. Zachos, eds.), World Scientific, Singapore, 1991, 72–96. (DHO, q-HARA, q-OP, q-SF, q-A, NGC, QGQS)

[85] J.-H. Dai, H.-Y. Guo, H. Yan: *The q-deformed differential operator algebra, a new solution to the Yang-Baxter equation and quantum plane*, J. Phys. A: Math. Gen. **24**, 8 (1991), L409–L414. (NGC, QGQS, q-A)

[86] E. V. Damaskinsky, P. P. Kulish: *On the q-oscillator and the quantum algebra $su_q(1,1)$*, J. Phys. A: Math. Gen. **23** (1990), L415–L419. (DHO, q-HARA, q-OP, q-SF, q-A, NGC, QGQS)

[87] ______: *Deformed oscillators and their applications*, Zap. Nauch. Semin. LOMI **189** (1991), 37–74, (in Russian). (DHO, q-HARA, q-OP, q-SF, q-A, NGC, QGQS)

[88] ______: *Quantum groups, deformed oscillators and their interrelations*, Preprint q-alg/9501006, (1995), 14 pp. (DHO, q-HARA, q-OP, q-SF, q-A, NGC, QGQS)

[89] ______: *Irreducibe representations of deformed oscillator algebra and q-special functions*, Intern. J. Math. Phys. **12**, 1 (1997), 153–158. (DHO, q-HARA, q-OP, q-SF, q-A, NGC, QGQS)

[90] C. Delbecq, C. Quesne: *A cubic deformation of $su(2)$*, Modern Phys. Lett. **8**, 10 (1993), 961–966. (q-HARA)

[91] ______: *Representation theory and q-boson realizations of Witten's $su(2)$ and $su(1,1)$ deformations*, Phys. Lett. B. **300**, 3 (1993), 227–233. (q-HARA)

[92] L. A. Dickey: *Soliton Equations and Hamiltonian Systems*, World Scientific, Singapore, 1991. (CONDE)

[93] J. F. van Diejen: *Commuting difference operators with polynomial eigenfunctions*, Compositio Math. **95** (1995), 183–233. (CORT)

[94] A. Dimakis, F. Müller-Hoissen, T. Striker: *Umbral calculus, discretization, and quantum mechanics on a lattice*, J. Phys. A: Math. Gen. **29** (1996), 6861–6876. (UC)

[95] J. Dixmier: *Sur les algèbres de Weyl*, Bull. Soc. math. France, **96** (1968), 209–242. (CDO)

[96] V. K. Dobrev: *New q-Minkowski space-time and q-Maxwell equations hierarchy from q-conformal invariance*, Phys. Lett. B. **341** (1994), 133–138. (q-DE, QGQS)

[97] ______: *q-Conformally covariant q-Minkowski space-time and invariant equations*, Czechoslovak J. Phys. **47**, 11 (1997), 1107–1114. (q-DE, QGQS)

[98] V. K. Dobrev, H. D. Doebner, R. Twarock: *Quantisation of kinematics and dynamics on S^1 with difference operators and a related q-deformation of the Witt algebra*, J. Phys. A: Math. Gen. **30** (1997), 6841–6859. (q-DE, QGQS)

[99] V. K. Dobrev, B. S. Kostadinov: *Non-commutative quantum conformal plane wave deformation*, Phys. Lett. B. **439** (1998), 337–344. (q-DE, QGQS)

[100] V. G. Drinfeld: *On commutative subrings of some non-commutative rings*, Funktz. Anal. Priloz. **11**, 1 (1977), 11–14. (CONDE, COAG)

[101] M. Dubois-Violette, R. Kerner: *Universal q-differential calculus and q-analogue of homological algebra*, Acta Math. Univ. Comenianae, LXV, 2 (1996), 175–188.

[102] B. A. Dubrovin: *Completely integrable Hamiltonian systems associated with matrix operators and Abelian varieties*, Funktz. Anal. Priloz. **11**, 4 (1977), 28–41. (CONDE, COAG)

[103] B. A. Dubrovin, V. B. Matveev, S. P. Novikov: *Non-linear equations of Korteweg–de Vries type, finite-zone linear operators, and Abelian varieties*, Uspekhi Mat. Nauk **31**, 1 (1976), 55–136. (CONDE, COAG)

[104] R. S. Dunne, A. J. Macfarlane, J. A. de Azcárraga, J. C. Pérez Bueno: *The q-calculus for generic q and q a root of unity*, Czechoslovak J. Phys. **46**, 12 (1996), 1235–1242. (q-HARA, DHO, NGC)

[105] ______: *Geometrical foundations of fractional supersymmetry*, Internat. J. Modern Phys. A, **12**, 19 (1997), 3275–3305. (q-HARA, DHO, NGC)

[106] S. Durand: *Fractional supersymmetry and quantum mechanics*, Phys. Lett. B, 312 (1993), 115–120. (q-HARA)

[107] K. Dykema, A. Nica: *On the Fock representation of the q-commutation relations*, J. reine angew. Math. **440** (1993), 201–212. (H$*$-A)

[108] F. H. L. Essler, V. Rittenberg: *Representations of the quadratic algebra and partially assymmetric diffusion with open boundaries*, J. Phys. A: Math. Gen. **29**, 13 (1996), 3375–3407. (QA, q-HARA, q-A)

[109] H. Exton: *q-Hypergeometric functions and applications*, Ellis Horwood Limited, Chichester, 1983. (q-DE, q-SF)

[110] D. B. Fairlie: *Polynomial algebras with q-Heisenberg operators*, in *Quantum Groups*, Proceedings of the Argonne Workshop, Argonne National Laboratory, 1990, (T. Curtright, D. Fairlie, C. Zachos, eds.), World Scientific, Singapore, 1991, 133–142. (DHO, q-A, q-HARA)

[111] D. B. Fairlie, C. K. Zachos: *Multiparameter associative generalizations of canonical commutation relations and quantized planes*, Phys. Lett. B. **256**, 1 (1991), 43–49. (q-HARA, DHO, NGC, q-A)

[112] H.-Y. Fan, S.-C. Jing: *Integration within an ordered product for q-deformed bosons*, Phys. Lett. A. **179**, 6 (1993), 379–384. (DHO, q-HARA)

[113] ______: *q-Deformed IWOP technique and its applications*, Commun. Theoret. Phys. **20**, 3 (1993), 347–352. (DHO, q-HARA)

[114] Ph. Feinsilver: *Commutators, anti-commutators and Eulerian calculus*, Rocky Mountain J. Math. **12**, 1 (1982), 171–183. (RF, q-SF, q-DE, q-HARA)

[115] ______: *Discrete analogues of the Heisenberg-Weyl algebra*, Monatsh. Math. **104** (1987), 89–108. (RF, q-SF, q-DE, q-HARA)

[116] ______: *Elements of q-harmonic analysis*, J. Math. Anal. Appl. **141** (1989), 509–526. (RF, q-SF, q-DE, q-HARA)

[117] ______: *q-Deformations of the oscillator*, J. Math. Phys. **33**, 12 (1992), 4259–4266. (DHO, q-A, QGQS, q-HARA)

[118] ______: *The orthogonality relations for* $\mathrm{SU}_q(2)$, J. Math. Phys. **35**, 11 (1994), 6123–6143. (DHO, q-A, QGQS, q-HARA)

[119] ______: *q-Field theory*, Lett. Math. Phys. **34**, 2 (1995), 169–176. (DHO, q-A, QGQS, q-HARA)

[120] ______: *q-Gravity*, Lett. Math. Phys. **38**, 1 (1996), 53–62. (DHO, q-A, QGQS, q-HARA)

[121] ______: *q-Deformation of the Lorentz group*, J. Math. Phys. **37**, 2 (1996), 953–964. (DHO, q-A, QGQS, q-HARA)

[122] ______: *q-Gauge theory*, Internat. J. Modern Phys. A **11**, 4 (1996), 733–746. (DHO, q-A, QGQS, q-HARA)

[123] ______: *The q-Coulomb problem*, J. Math. Phys. **37**, 6 (1996), 2628–2636. (DHO, q-A, QGQS, q-HARA)

[124] ______: *A note on q-transformation theory* Lett. Math. Phys. **41**, 2 (1997), 195–206. (DHO, q-A, QGQS, q-HARA)

[125] ______: *q-Uncertainty relations*, Internat. J. Modern Phys. A **13**, 11 (1998), 1795–1803. (DHO, q-A, QGQS, q-HARA)

[126] ______: *Observable properties of q-deformed physical systems*, Lett. Math. Phys. **49**, 2 (1999), 105–114. (DHO, q-A, QGQS, q-HARA)

[127] R. J. Finkelstein, E. Marcus: *Transformation theory of the q-oscillator*, J. Math. Phys. **36**, 6 (1995), 2652–2672. (DHO, q-A, QGQS, q-HARA)

[128] G. Fiore: *Braided chains of q-deformed Heisenberg algebrae*, J. Phys. A: Math. Gen. **31**, 23 (1998), 5289–5298. (q-HARA)

[129] ______: *Embedding q-deformed Heisenberg algebras into undeformed ones*, Rep. Math. Phys. **43**, 1–2 (1999), 101–108. (q-HARA)

[130] D. I. Fivel: *Interpolation between Fermi and Bose statistics using generalized commutators*, Phys. Rev. Lett. **65**, 27 (1990), 3361–3364; Errata: Phys. Rev. Lett. **69**, 13 (1992), 2020. (q-HARA)

[131] H. Flanders: *Commutative linear differential operators*, Technical Report No. 1, Dep. Math., Univ. of California, Berkeley, (1955), 39 pp. (CDO, q-HARA)

[132] N. Fleury, A. Turbiner: *Polynomial relations in the Heisenberg algebra*, J. Math. Phys. **35**, 11 (1994), 6144–6149. (RF)

[133] G. Floquet: *Sur la théorie des équations différentielles linéaires*, Ann. Sci École. Norm. Sup. **8** (1879), Suppl. 1–132. (CDO, q-HARA)

[134] R. Floreanini, V. P. Spiridonov, L. Vinet: *Bosonic realizations of the quantum superalgebra* $\mathrm{osp}_q(1,2n)$, Phys. Lett. B. **242**, 3–4 (1990), 383–386. (DHO, q-A)

[135] ______: *q-Oscilator realizations of the quantum superalgebras* $\mathrm{sl}_q(m,n)$ *and* $\mathrm{osp}_q(m,2n)$, Comm. Math. Phys. **137**, 1 (1991), 149–160. (DHO, q-A)

[136] R. Floreanini, J. LeTourneux, L. Vinet: *More on the q-oscillator algebra and q-orthogonal polynomials*, J. Phys. A. **28** (1995), L287-L293. (DHO, q-OP, q-SF)

[137] ______: *A q-deformed e(4) and continuous q-Jacobi polynomials*, J. Math. Phys. **37**, 8 (1996), 4135–4149. (DHO, q-OP, q-SF)

[138] R. Floreanini, L. Vinet: *Representations of quantum algebras and q-special functions*, in *Quantum symmetries*, (Clausthal, 1991), World Sci. Publishing, 1993, 264–284. (QGQS, q-A, q-SF, q-OP, q-DE)

[139] ______: *Quantum algebras and q-special functions*, Ann. Phys. **221**, 1 (1993), 53–70. (QGQS, q-A, q-SF, q-OP, q-DE)

[140] ______: *On the quantum group and quantum algebra approach to q-special functions*, Lett. Math. Phys. **27** (1993), 179–190. (q-SF, q-OP, QGQS, q-DE)

[141] ______: *Automorphisms of the q-oscillator algebra and basic orthogonal polynomials*, Phys. Lett. A. **180**, 6 (1993), 393–401. (DHO, q-HARA, q-OP)

[142] ______: *Symmetries of the q-difference heat equation*, Lett. Math. Phys. **32**, 1 (1994), 37–44. (q-DE, q-SF, q-OP, q-A)

[143] ______: *Quantum symmetries of q-difference equations*, J. Math. Phys. **36**, 6 (1995), 3134–3156. (q-DE, q-SF, q-OP, q-A)

[144] ______: *Symmetries of the q-deformed heat equations*, in *Nonlinear, deformed and irreversible quantum systems*, (Clausthal, 1994), World Sci. Publishing, 1995, 385–400. (q-DE, q-SF, q-OP, q-A)

[145] ______: *Quantum algebra approach to q-Gegenbauer polynomials*, J. Comput. Appl. Math. **57**, 1–2 (1995), 123–133. (q-DE, q-SF, q-OP, q-A)

[146] ______: *q-Gamma and q-beta functions in quantum algebra representation theory*, J. Comput. Appl. Math. **68**, 1–2 (1996), 57–68. (q-SF, q-OP, q-A)

[147] E. Frenkel: *Deformations of the KdV hierarchy and related soliton equations*, IMRN (Internat. Math. Res. Notices) **2** (1996), 55–76. (q-NDE)

[148] I. B. Frenkel, N. Yu. Reshetikhin: *Quantum Affine Algebras and Holonomic difference equations*, Commun. Math. Phys. **146**, 1 (1992), 1–60. (q-DE, q-A, q-SF)

[149] B. Fuglede: *Commuting self-adjoint partial differential operators and a group theoretic problem*, J. Funct. Anal. **16** (1974), 101–121. (CORT)

[150] P. Furlan, L. K. Hadjiivanov, I. T. Todorov: *Quantum deformation of the ladder representations of* $U(1,1)$ *for* $|q| = 1$, J. Math. Phys. **33**, 12 (1992), 4255–4258. (q-A, QGQS)

[151] L. Gårding, A. Wightman: *Representations of the anticommutation relations*, Proc. Nat. Acad. Sci. USA **40** (1954), 617–622. (CORT, q-HARA)

[152] ______: *Representations of the commutation relations*, Proc. Nat. Acad. Sci. USA **40**, 9 (1954), 623–626. (CORT, q-HARA)

[153] G. Gasper, M. Rahman: *Basic Hypergeometric Series*, Cambridge University Press, 1990. (q-SF, q-DE)

[154] I. M. Gelfand, L. A. Dickey: *Asymptotic behaviour of the resolvent of Sturm–Liouville equations and the algebra of the Korteweg–de Vries equations*, Uspekhi Mat. Nauk **30**, 5 (1975), 67–100. (CONDE)

[155] ______: *Fractional powers of operators and Hamiltonian systems*, Funktz. Anal. Priloz. **10**, 4 (1976), 13–29. (CONDE)

[156] I. M. Gelfand, D. B. Fairlie: *The Algebra of Weyl Symmetrised Polynomials and Its Quantum Extension*, Commun. Math. Phys. **136** (1991), 487–499. (q-HARA, DQ, q-A)

[157] I. M. Gelfand, M. I. Graev, V. S. Retakh: *Difference analogues and q-analogues of general hypergeometric systems of differential equations*, Dokl. Akad. Nauk **325**, 2 (1992), 215–220, (Russian). (q-DE, q-SF)

[158] I. M. Gelfand, N. Ya. Vilenkin: *Generalized functions*, **4**, Academic Press, New York, 1964, Transl. to English from Russian edn.: Fizmatgiz, Moscow, 1961. (CORT)

[159] F. Gesztesy, R. Weikard: *Elliptic algebro-geometric solutions of the KdV and AKNS hierarchies — an analytic approach*, Bull. American Math. Soc. **35**, 4 (1998), 271–317. (CONDE, COAG)

[160] V. Ya. Golodets: *Classification of representations of the anti-commutation relations*, Russ. Math. Surveys, **24**, (1969), 1–63. (CORT, q-HARA)

[161] V. I. Gorbachuk, Yu. S. Samoĭlenko, G. F. Us: *Spectral theory of selfadjoint operators, and infinite-dimensional analysis*, (in Russian), Uspehi Mat. Nauk. **31**, 1 (187) (1976), 203–216. (CORT)

[162] H. W. Gould: *The q-Stirling numbers of first and second kinds*, Duke. Math. J. **28** (1961), 281–289. (q-SF)

[163] R. L. Graham, D. E. Knuth, O. Patashnik: *Concrete mathematics*, second edition, Addison-Wesley Publishing Company, Reading, MA, 1994. (q-SF)

[164] O. W. Greenberg: *Particles with small violations of Fermi or Bose statistics*, Phys. Rev. D. **43** (1991), 4111–4120. (q-HARA)

[165] ______: *Q-mutators and violations statistics*, in *Quantum Groups*, Proceedings of the Argonne Workshop, Argonne National Laboratory, 1990, (T. Curtright, D. Fairlie, C. Zachos, eds.), World Scientific, Singapore, 1991, 166–180. (q-HARA)

[166] W. Hahn: *Über Orthogonalpolynome, die q-Differenzengleichungen genügen*, Math. Nachr. **2** (1949), 4–34. (q-DE, q-SF)

[167] ______: *Über Polynome, die gleichzeitig zwei verschiedenen Orthogonal systemen angehören*, Math. Nachr. **2** (1949), 263–278. (q-DE, q-SF)

[168] ______: *Beiträge zur Theorie der Heineschen Reihen, die 24 integrale der hypergeometrischen q-Differenzengleichung, das q-Analogon der Laplace-Transformation*, Math. Nachr. **2** (1949), 340–379. (q-DE, q-SF)

[169] ______: *Über die höheren Heineschen Reihen und eine einheitliche Theorie der sogenannten speziellen Funktionen*, Math. Nachr. **3** (1950), 257–294. (q-DE, q-SF)

[170] ______: *Über uneigentliche Lösungen linearer geometrischer Differenzengleichungen*, Math. Annalen, **125** (1952), 67–81. (q-DE, q-SF)

[171] ______: *Die mechanische Deutung einer geometrischen Differenzgleichung*, Z. angew. Math. Mech. **33** (1953), 270–272. (q-DE, q-SF)

[172] ______: *Über geometrische Differenzengleichungen mit einer Einheitswurzel als Parameter*, Bollettino U.M.I. **4**, 11 (1975), 95–99. (q-DE, q-SF)

[173] ______: *On Nonlinear Geometric difference equations*, Jour. Math. Phy. Sci. (India), **11**, 1 (1977), 89–94. (q-DE, q-SF)

[174] ______: *Lineare geometrische Differenzengleichungen*, Forschungszentrum Graz Math. Stat. Section, Bericht **169** (1981). (q-DE, q-SF)

[175] L. Haine, P. Iliev: *The bispectral property of q-deformation of Schur polynomials and q-KdV hierarchy*, J. Phys. A: Math. Gen. **30**, 20 (1997), 7217–7227. (q-NDE)

[176] J. Harnad: *Isospectral flow and Liouville–Arnol'd integration in loop algebras*, 1–42, in *Geometric and Quantum Aspects of Integrable Systems*, (G. F. Helminck, ed.), Lecture Notes in Physics 424, Springer-Verlag, Berlin, 1993. (CONDE)

[177] T. Hayashi: *q-Analogues of Clifford and Weyl algebras—spinor and oscillator representations of quantum enveloping algebras*, Comm. Math. Phys. **127**, 1 (1990), 129–144. (q-A, QGQS, q-HARA)

[178] A. Hebecker, W. Weich: *Free particle in q-deformed configuration space*, Letters Math. Phys. **26** (1992), 245–258. (q-HARA)

[179] A. Hebecker, S. Schreckenberg, J. Schwenk, W. Weich, J. Wess: *Representations of a q-deformed Heisenberg algebra*, Z. Phys. C **64**, 2 (1994), 355–359. (q-HARA)

[180] E. Heine: *Über die Reihe ...*, J. Reine Angew. Math. **32**, (1846), 210–212. (q-DE, q-SF)

[181] ______: *Untersuchungen über die Reihe ...*, J. Reine Angew. Math. **34**, (1847), 285–328. (q-DE, q-SF)

[182] L. Hellström: *On centralisers in the q-deformed Heisenberg algebra*, master thesis, Department of Mathematics, Umeå University, Umeå, 1996. (RF)

[183] L. Hellström, S. D. Silvestrov: *On centralisers in q-deformed Heisenberg algebras*, Czechoslovak J. Phys. **47**, 11 (1997), 1163–1169. (CDO, RF)

[184] ______: *Centralisers in q-deformed Heisenberg algebras*, Research Reports **2**, Dep. Math., Umeå University, (1999), 215 pp. (CDO, DL, q-HARA, RF)

[185] P. A. Hendriks: *An algorithm for computing a standard form for second-order linear q-difference equations*, J. Pure Appl. Algebra, **117, 118** (1997), 331–352. (q-DE)

[186] E. Horikawa: *Contiquity relations for q-hypergeometric function and related quantum group*, Proc. Japan Acad. **68**, Ser. A (1992), 157–160. (q-DE, q-SF, QGQS)

[187] B. Y. Hou, B. Y. Hou, Z. Q. Ma: *q-Deformed Chern class, Chern–Simons and cocycle hierarchy*, J. Phys. A: Math. Gen. **28**, 3 (1995), 543–558. (NGC, QGQS, q-H, q-A)

[188] J. Hruby: *On the q-deformed Heisenberg uncertainty relations and discrete time*, Rendicinti del Circolo Matematico di Palermo, Serie II, Suppl. **96**, 133–152. (q-HARA)

[189] E. C. Ihrig, M. E. H. Ismail: *A q-umbral calculus*, J. Math. Anal. Appl. **84**, 1 (1981), 178–207. (UC)

[190] P. Iliev: *Solutions to Frenkel's deformation of the KP hierarchy*, J. Phys. A: Math. Gen. **31**, 12 (1998), L241–L244. (q-NDE)

[191] ______: *Tau function solutions to a q-deformation of the KP hierarchy*, Lett. Math. Phys. **44**, 3 (1998), 187–200. (q-NDE)

[192] ______: *q-KP hierarchy, bispectrality and Calogero–Moser systems*, preprint (1999). (q-NDE)

[193] ______: *Algèbres commutatives d'operateurs aux q-différences et systèmes de Calogero-Moser*, C. R. Acad. Sci. Paris, Sér. I, **329**, 10 (1999), 877–882. (q-NDE)

[194] K. N. Ilinski, G. V. Kalinin, A. S. Stepanenko: *q-functional field theory for particles with exotic statistics*, Phys. Lett. A, **232**, 6 (1997), 399–408. (q-HARA, q-A)

[195] F. H. Jackson: *On q-functions and a certain difference operator*, Trans. Roy. Soc. Edin. **46** (1908), 253–281. (q-DE, q-SF)

[196] ______: *A q-form of Taylor's theorem*, Mess. Math. **38** (1909), 62–64. (q-DE, q-SF)

[197] ______: *A q-series corresponding to Taylor's series*, Mess. Math. **39** (1909), 26–28. (q-DE, q-SF)

[198] ______: *On q-definite integrals*, Quart. J. Pure Appl. Math. **41** (1910), 193–203. (q-DE, q-SF)

[199] ______: *q-Difference equations*, American J. Math. **32** (1910), 305–314. (q-DE, q-SF)

[200] A. Jannussis: *Some properties of q-analysis and applications to non-canonical mechanics*, J. Phys. A: Math. Gen. **26** (1993), L233–L237. (NCM, DHO, LAA)

[201] A. Jannussis, G. Brodimas, D. Sourlas, A. Streclas, P. Siafarikas, L. Papaloucas, N. Tsangas: *Foundations of the Lie-admissible Fock space of the hadronic mechanics*, Hadronic Journal **5** (1982), 1923–1947. (NCM, LAA)

[202] A. Jannussis, G. Brodimas, D. Sourlas, K. Vlachos, P. Siafarikas, L. Papaloucas: *Some properties of q-analysis and applications to non-canonical mechanics*, Hadronic Journal **6** (1983), 1653–1686. (NCM, LAA)

[203] A. Jannussis, L. Papaloucas, P. Siafarikas: *Eigenfunctions and eigenvalues of the q-differential operators*, Hadronic Journal **3** (1980), 1622–1632. (NCM, LAA)

[204] O. R. Jensen: *Construction of vacuum for the positive energy-representation and its Bose counterpart*, Math. Scand. **80** (1997), 101–106. (q-HARA)

[205] S.-C. Jing, H.-Y. Fan: *q-Taylor's formula with its q-remainder*, Comm. Theoret. Phys. **23**, 1 (1995), 117–120. (q-DE)

[206] ______: *q-deformed binomial state*, Phys. Rev. A (3) **49**, 4 (1994), 2277–2279. (DHO, q-HARA)

[207] S.-C. Jing, J. J. Xu: *Comment on the q-deformed fermionic oscillator*, J. Phys. A: Math. Gen. **24** (1991), L891–L894. (DHO)

[208] A. Joseph: *Quantum groups and their primitive ideals*, Springer-Verlag, 1995. (BGLA, NGC, QGQS)

[209] P. E. T. Jørgensen: *Selfadjoint extension operators commuting with an algebra*, Math. Z. **169** (1979), 41–62. (CORT)

[210] ———: *Unbounded operators: perturbations and commutativity problems*, J. Funct. Anal. **39**, 3 (1980), 281–307. (CORT)

[211] ———: *Spectral theory of finite volume domains in $\mathbb{R}^n$*, Adv. Math. **44** (1982), 105–120. (CORT)

[212] ———: *Quantization and deformation of Lie algebras*, Contemp. Math. **160** (1994), 141–149. (q-HARA, DQ)

[213] ———: *Operators and Representation Theory,* North-Holland, Amsterdam, 1988. (CORT)

[214] P. E. T. Jørgensen, R. T. More: *Operator Commutation Relations,* D. Reidel Publishing Company, Dordrecht, 1984. (CORT)

[215] P. E. T. Jørgensen, S. Pedersen: *Spectral theory for Borel sets in $\mathbb{R}^n$ of finite measure*, J. Funct. Anal. **107**, 1 (1992), 72–104. (CORT)

[216] P. E. T. Jørgensen, D. P. Proskurin, Yu. S. Samoĭlenko: *The kernel of Fock representations of Wick algebras with braided operator of coefficients*, to appear in Pacific J. Math. (H$*$-A, q-A)

[217] P. E. T. Jørgensen, L. M. Schmitt, R. F. Werner: *q-Canonical Commutation Relations and stability of the Cuntz algebra*, Pacific J. Math. **165** (1994), 131–151. (H$*$-A)

[218] ———: *Positive representations of general commutation relations allowing Wick ordering*, J. Funct. Anal. **134** (1995), 33–99. (H$*$-A, q-A)

[219] P. E. T. Jørgensen, R. F. Werner: *Coherent States of the q-Canonical Commutation Relations*, Commun. Math. Phys. **164** (1994), 455–471. (H$*$-A)

[220] G. Kaniadakis, A. Lavagno, P. Quarati: *Kinetic model for q-deformed bosons and fermions*, Phys. Lett. A. **227**, 3–4 (1997), 227–231. (q-HARA, DHO)

[221] M. V. Karasev, V. P. Maslov: *Non-Linear Poisson Brackets. Geometry and Quantization*, Nauka, Moscow, 1991. (Transl. to English: Transl. Math. Monographs 119, AMS, Providence, Rhode Island, 1993). (NGC)

[222] M. V. Karasev, E. Novikova: *Non-Lie Permutation Relations, Coherent States, and Quantum Embedding*, in *Coherent Transform, Quantization, and Poisson Geometry*, (M. V. Karasev, ed.), Transl. Math. Monographs (2), 187, AMS, Providence, Rhode Island, 1998. (CORT)

[223] C. Kassel: *Cyclic homology of differential operators, the Virasoro algebra and a q-analogue*, Commun. Math. Phys. **146** (1992), 343–356. (NGC, q-A, q-H)

[224] ———: *Quantum groups*, Springer-Verlag, 1995. (QGQS, NGC, BGLA, q-A, q-H)

[225] A. Kempf: *Uncertainty relation in quantum mechanics with quantum group symmetry*, J. Math. Phys. **35**, 9 (1994), 4483–4496. (q-HARA)

[226] ______: *Algebraic q-integral and Fourier theory on quantum and braided spaces*, J. Math. Phys. **35**, 12 (1994), 6802–6837. (q-HARA)

[227] ______: *On path integration on noncommutative geometries*, in *Quantum Groups and Quantum Spaces*, Banach Center Publ. **40**, Inst. Math. Polish Acad. Sci., Warszawa, (1997), 379–386. (q-HARA)

[228] B. Khesin, V. Lyubashenko, C. Roger: *Extensions and contractions of the Lie algebra of q-pseudodifferential symbols on the circle*, J. Funct. Anal. **143**, 1 (1997), 55–97. (q-NDE)

[229] M. Klimek: *The conservation laws and integrals of motion for a certain class of equations in discrete models*, J. Phys. A: Math. Gen. **29** (1996), 1747–1758. (q-DE, NGC)

[230] ______: *Integrals of motion for some equations on q-Minkowski space*, Czechoslovak J. Phys. **47**, 12 (1997), 1199–1206. (q-DE, NGC)

[231] A. U. Klimyk, K. Schmüdgen: *Quantum groups and their representations*, Springer-Verlag, Berlin, 1997. (BGLA, NGC, QGQS)

[232] D. E. Knuth: *Two Notes on Notation*, Amer. Math. Month. **99** (1992), 403–422. (q-SF)

[233] A. N. Kochubei: *p-adic commutation relations*, J. Phys. A: Math. Gen. **29** (1996), 6375–6378. (q-HARA)

[234] W. Koepf: *Hypergeometric summation. An algorithmic approach to summation and special function identities*, Friedr. Vieweg & Sohn, Braunschweig, 1998. (q-SF)

[235] B. G. Konopelchenko: *Introduction to Multidimentional Integrable Equations. The Inverse Spectral Transform in* $2+1$ *Dimensions*, Plenum Press, 1992. (CONDE)

[236] T. H. Koornwinder: *Compact quantum groups and q-special functions*, European school of group theory, Pitman Research Notes 311, 1994. (QGQS, RF, q-SF, q-FA)

[237] ______: *Special functions and q-commuting variables*, Report No. 1, Mittag-Leffler Institute, Stockholm, (1995/1996). (q-SF, q-FA, RF)

[238] D. Korteweg, G. de Vries: *On the change of form of long waves advancing in a rectangular canal, and on a new type of long stationary wave*, Phylos. Mag. **35** (1895), 422–443. (CONDE)

[239] I. M. Krichever: *Algebraic curves and commuting matrix differential operators*, Funktz. Anal. Priloz. **10**, 2 (1976), 75–76. (CONDE, COAG)

[240] ______: *Integration of non-linear equations by the methods of algebraic geometry*, Funktz. Anal. Priloz. **11**, 1 (1977), 15–31. (CONDE, COAG)

[241] ______: *Methods of algebraic geometry in the theory of nonlinear equations*, Uspekhi. Mat. Nauk **32**, 6 (1977), 183–208. (CONDE, COAG)

[242] ______: *Commutative rings of ordinary linear differential operators*, Funktz. Anal. Priloz. **12**, 3 (1978), 20–31. (CONDE, COAG)

[243] ______: *Algebraic curves and non-linear difference equations*, Comm. Moscow Math. Soc. **33** (1978). (CONDE, COAG)

[244] I. M. Krichever, S. P. Novikov: *Holomorphic bundles over Riemann surfaces and the Kadomcev–Petviašvili equation. I*, Funktz. Anal. Priloz. **12**, 4 (1978), 41–52. (CONDE, COAG)

[245] ______: *Holomorphic bundles and nonlinear equations. Finite-zone solutions of rank 2*, Dokl. Akad. Nauk. SSSR **247** (1979), 33–36. (CONDE, COAG)

[246] ______: *Holomorphic bundles over algebraic curves, and nonlinear equations*, Uspekhi. Mat. Nauk. **35**, 6 (216) (1980), 47–68. (CONDE, COAG)

[247] ______: *The inverse problem method and holomorphic bundles on Riemann surfaces*, in V. G. Drinfeld, I. M. Krichever, Yu. I. Manin, S. P. Novikov: *Methods of algebraic geometry in contemporary mathematical physics*, Soviet Sci. Rev. Sect. C: Math. Phys. Rev. **1**, (1980), 1–54. (CONDE, COAG)

[248] P. P. Kulish: *Quantum Group Covariant Algebras*, in *Proceedings of the International School of Physics "Enrico Fermi"*, (L. Castellani, J. Wess, eds.), Italian Physical Society, IOS Press, 1996, 203–214. (QGQS, DHO, NGC, q-A, q-HARA)

[249] V. V. Kuryshkin: *On some generalisation of creation and annihilation operators in quantum theory (μ-quantisation)*, VINITI Dep. 3936–76 (in Russian), (1976); Ann. Found. L. de Broglie, **5**, (1980). (q-HARA)

[250] J. Le Caine: *The linear q-difference equation of the second order*, Amer. J. Math. **65** (1943), 585–600. (q-DE)

[251] E. C. Lance: *Finitely-presented C^*-algebras*, in *Operator algebras and applications*, (A. Katavolos, ed.), NATO ASI Ser. C: Math. Phys. Sci. **495**, Kluwer Acad. Publ., (1997), 255–266. (CORT)

[252] L. Lapointe, L. Vinet: *Creation operators for the Macdonald and Jack polynomials*, Lett. Math. Phys. **40** (1997), 269–286. (CORT)

[253] P. D. Lax: *Integrals of Nonlinear Equations of Evolution and Solitary Waves*, Comm. Pure Appl. Math. **21** (1968), 467–490. (CONDE)

[254] ______: *Periodic solution of the KdV equation*, Comm. Pure Appl. Math. **28** (1975), 141–188. (CONDE)

[255] P. Lesky: *Zweigliederige Rekursionen für die Koeffizienten von Polynomlösungen Sturm-Liouvillescher q-Differenzengleichungen*, Zeitung angewandte Mathematik und Mechanik, **74**, 10 (1994), 497–500. (q-DE)

[256] M. S. Livšic, N. Kravitsky, A. S. Markus, V. Vinnikov: *Theory of Commuting Nonselfadjoint Operators*, Kluwer Acad. Publ., Dordrecht, 1995. (COAG, COCT)

[257] M. S. Livšic, L. L. Waksman: *Commuting Nonselfadjoint Operators in Hilbert Space*, Lecture Notes in Mathematics 1272, Springer-Verlag, 1987. (COAG, COCT)

[258] L. Ma, Z. Tang, Y.-D. Zhang: *The q-deformed normal-ordered product*, Nuovo Cimento B (11) **108**, 10 (1993), 1075–1080. (DHO, q-HARA)

[259] ______: *Coherent states in the form of a quantum group*, Internat. J. Theoret. Phys. **32**, 4 (1993), 545–553. (DHO, q-HARA)

[260] R. J. McDermott, A. I. Solomon: *An analogue of the unitary displacement operator for q-oscillator*, J. Phys. A: Math. Gen. **27**, 6 (1994), 2037–2043. (DHO, q-HARA)

[261] ______: *Double squeezing in generalized q-coherent states*, J. Phys. A: Math. Gen. **27**, 2 (1994), L15–L19. (DHO, q-HARA)

[262] ______: *Squeezed states parametrized by elements of noncommutative algebras*, Czechoslovak J. Phys. **46**, 2–3 (1996), 235–241. (DHO, q-HARA)

[263] ______: *q-Correlated states and noncommutative algebras*, in Symmetry methods in physics, (Dubna, 1995), Vol. **2**, Inst. Nuclear Res., Dubna, (1996), 416–422. (DHO, q-HARA)

[264] I. G. Macdonald: *Symmetric Functions and Hall Polynomials*, (Second Edition), Oxford Mathematical Monographs, Clarendon Press, Oxford, 1995. (CORT)

[265] A. J. Macfarlane: *On q-analogues of the quantum harmonic oscilator and the quantum group* $\mathrm{SU}(2)_q$, J. Phys. A: Math. Gen. **22**, (1989), 4581–4588. (DHO, q-A, QGQS, q-HARA)

[266] H. P. McKean: *Integrable systems and algebraic curves*, in *Global Analysis*, Proceedings of the Biennial seminar of the Canadian Mathematical Congress, Calgary, Alberta, 1978, (edited by M. Grmela, J. E. Marsden), Lecture Notes in Mathematics 755, Springer-Verlag, (1979), 83–200. (CONDE, COAG)

[267] G. W. Mackey: *Imprimitivity for representations of locally compact groups*, Proc. Nat. Acad. Sci. USA, **35**, 9 (1949), 537–545. (CORT)

[268] S. Majid: *Free braided differential calculus, braided binomial theorem, and the braided exponential map*, J. Math. Phys. **34**, 10 (1993), 4843–4856.

[269] ______: *Foundations of Quantum Group Theory*, Cambridge University Press, 1995. (BGLA, NGC, QGQS)

[270] ______: *Introduction to braided geometry and q-Minkowski space*, in *Proceedings of the International School of Physics "Enrico Fermi"*, (L. Castellani, J. Wess, eds.), Italian Physical Society, IOS Press, 1996, 267–345. (BGLA, NGC, q-A, QGQS)

[271] Yu. I. Manin: *Algebraic aspects of nonlinear differential equations*, in Itogi nauki i tekhniki, ser. Sovremennye problemy Mat., **11**, (1978) , 5–152. (CONDE, COAG)

[272] ______: *Quantum Groups and Non-Commutative Geometry*, Les publications du Centre de Recherches Mathématiques, Université de Montréal, 1988. (NGC, QA)

[273] V. A. Marchenko: *Nonlinear Equations and Operator Algebras*, D. Reidel Publishing Company, Dordrecht, 1988, Transl. to English from Russian edn.: Naukova Dumka, Kiev, 1986. (CONDE)

[274] F. Marotte, C. Zhang: *Sur la sommabilité des séries entières solutions formelles d'une équation aux q-différences. II*, C. R. Acad. Sci. Paris, Sér. I, **327**, 8 (1998), 715–718. (q-DE)

[275] V. P. Maslov: *Operational Methods*, Mir, Moscow, 1976. (NGC)

[276] T. E. Mason: *On properties of the solutions of linear q-difference equations with entire function coefficients*, Amer. J. Math. **37** (1915), 439–444. (q-DE)

[277] V. Mazorchuk, L. B. Turowska: *$*$-Representations of twisted generalized Weyl constructions*, Preprint, TRITA-MAT-1998-MA **4**, Dep. of Math., Royal Institute of Technology, (1998), 22 pp. (q-A, q-HARA)

[278] U. Meyer: *Wave equations on q-Minkowski space*, Commun. Math. Phys. **174** (1996), 457–475. (q-DE, q-A, QGQS)

[279] W. Miller: *Lie theory and q-difference equations*, SIAM J. Math. Anal. **1** (1970), 171–188. (q-DE)

[280] S. C. Milne: *A q-analog of restricted growth functions, Dobinski's equality, and Charlier polynomials*, Trans. Amer. Math. Soc. **245** (1978), 89–118. (q-DE, q-OP, q-SF)

[281] K. Mimachi: *Connection problem in holonomic q-difference system associated with a Jackson integral of Jordan-Pochhammer type*, Nagoya Math. J. **116** (1989), 149–161. (q-DE, q-SF, q-OP)

[282] ______: *Holonomic q-difference system associated with the basic hypergeometric series $_{n+1}\phi_n$*, Tôhoku Math. J. (2) **45**, 4 (1993), 485–490. (q-DE, q-SF, q-OP)

[283] ______: *A solution to quantum Knizhnik Zamolodchikov equations and its application to eigenvalue problems of the Macdonald type*, Duke Math. J. **85**, 3 (1996), 635–658. (q-DE, q-SF, q-OP)

[284] ______: *A solution of the quantum Knizhnik Zamolodchikov equation of type C_n*, Comm. Math. Phys. **197**, 1 (1998), 229–246. (q-DE, q-SF, q-OP)

[285] ______: *Eigenfunctions of Macdonald's q-difference operator for the root system of type C_n*, J. Funct. Anal. **163**, 2 (1999), 272–278. (q-DE, q-SF, q-OP)

[286] K. Mimachi, M. Noumi: *An integral representation of eigenfunctions for Macdonald's q-difference operators*, Tôhoku Math. J. **49**, (1997), 517–525. (CORT, q-DE, q-SF, q-OP)

[287] P. van Moerbeke, D. Mumford: *The spectrum of difference operators and algebraic curves*, Acta Mathematica, **143**, (1979), 93–154. (CONDE, COAG)

[288] O. I. Mokhov: *Commuting differential operators of rank 3, and nonlinear differential equations*, Izv. Akad. Nauk. SSSR, Ser. Mat. **53**, 6 (1989). (Transl. to English: Math USSR Izvestiya, **35**, 3 (1990), 629–655.) (CONDE)

[289] H. Morikawa: *ζ_n-Weyl algebra $W_r(\zeta_n, \mathbb{Z})$*, Nagoya Math. J. **113** (1989), 153–159. (q-HARA)

[290] J. Moser: *Integrable Hamiltonian Systems and Spectral Theory*, Lezioni Fermiane, Accademia Nazionale Dei Lincei Scuola Normale Superiore, Pisa, 1981. (CONDE)

[291] I. Mukhopadhaya, A. R. Chowdhury: *q-Deformed dressing operators and modified integrable hierarchies* J. Nonlinear Math. Phys. **1**, 4 (1994), 414–419. (DHO, q-DE)

[292] M. Mulase: *Algebraic geometry of soliton equations* Proc. Japan Acad. Ser. A Math. Sci. **59**, 6 (1983), 285–288. (CONDE, COAG)

[293] D. Mumford: *An algebro-geometric construction of commuting operators and of solutions to the Toda lattice equation, Korteweg–de Vries equation and related non-linear equations*, Proc.Int. Symp. on Algebraic Geometry, Kyoto, 1978, 115–153. (CONDE, COAG)

[294] H. C. Myung, R. M. Santilli: *Foundations of the Hadronic generalization of the atomic Mechanics, II: modular-isotopic Hilbert space formulation of the exterior strong problem*, Hadronic Journal **5** (1982), 1277–1366. (LAA)

[295] J. S. Møller: *Second quantization in a quon-algebra*, J. Phys. A: Math. Gen. **26**, 18 (1993), 4643–4652. (RF, q-HARA)

[296] V. E. Nazaikinskii, V. E. Shatalov, B. Yu. Sternin: *Methods of Noncommutative Analysis. Theory and Applications*, De Gruyter studies in mathematics 22, De Gruyter, Berlin, New York, 1976. (NGC)

[297] M. H. A. Newman: *On theories with a combinatorial definition of "equivalence"*, Ann. of Math., **43** (1942), 223–243. (DL)

[298] M. Nishizawa: *On a solution of a q-difference analogue of Lauricella's D-type hypergeometric equation with* $|q| = 1$, Publ. Res. Inst. Math. Sci. **34**, 3 (1998), 277–290. (q-DE)

[299] M. Nishizawa, K. Ueno: *Integral solutions of q-difference equations of the hypergeometric type with* $|q| = 1$, 1997. (q-DE)

[300] N. E. Nörlund: *Differenzenrechnung*, Berlin, 1924. (q-DE)

[301] S. P. Novikov: *The periodic problem for the Korteweg–de Vries equation*, Funktz. Anal. Priloz. **8**, 3 (1974), 54–66. (CONDE, COAG)

[302] V. L. Ostrovskyĭ, Yu. S. Samoĭlenko: *Structure theorems for a pair of unbounded selfadjoint operators satisfying a quadratic relation*, Adv. Sov. Math. **9** (1992), 131–149. (QA, H∗-A)

[303] ______: *On Pairs of self-adjoint operators*, Seminar Sophus Lie **3**, 2 (1993), 185–218. (QA, H∗-A)

[304] ______: *Introduction to the Theory of Representations of Finitely Presented ∗-Algebras. I. Representations by bounded operators*, The Gordon and Breach Publ. Group, London, 1999. (CORT, H∗-A)

[305] V. L. Ostrovskyĭ, S. D. Silvestrov: *Representations of the real forms of the graded analogue of a Lie algebra*, Ukrain. Mat. Zh. **44**, 11 (1992), 1518–1524. (Transl. to English in : Ukrainian Math. J. **44**, 11 (1993), 1395–1401). (CORT)

[306] E. Papp: *Explicit polynomial solutions to the q-difference form of the Harper-equation*, Modern Phys. Lett. B **11**, 18 (1997), 773–778. (q-DE)

[307] ______: *Derivation of q-analogs for the radial Schrödinger equation in N space-dimensions*, Nova J. Theor. Phys. **5**, 4 (1997), 227–242. (q-DE)

[308] ______: *Non-polynomial solutions to the q-difference form of the Harper equation*, J. Phys. A: Math. Gen. **31**, 12 (1998), 2881–2887. (q-DE)

[309] G. K. Pedersen: *C^*-algebras and their Automorphism Groups*, London Math. Soc. Monographs, **4**, Academic Press, 1979. (CORT)

[310] S. Pedersen: *Spectral theory of commuting self-adjoint partial differential operators*, J. Funct. Anal. **73** (1987), 122–134. (CORT)

[311] M. Pillin: *q-deformed relativistic wave equations*, J. Math. Phys. **35**, 6 (1994), 2804–2817. (q-DE)

[312] G. Post: *Operator identities, representations of algebras and the problem of normal ordering*, Lett. Math. Phys. **36** (1996), 109–116. (RF, DL)

[313] A. Pressley, G. Segal: *Loop Groups*, Cambridge University Press, 1984. (CONDE, COAG)

[314] E. Previato: *Geometry of the modified KdV equation*, in *Geometric and Quantum Aspects of Integrable Systems*, (G. F. Helminck, ed.), Lecture Notes in Physics 424, Springer-Verlag, (1993), 43–65. (CONDE, COAG)

[315] ______: *Seventy Years of Spectral Curves: 1923–1993*, in R. Donagi, B. Dubrovin, E. Frenkel, E. Previato, M. Francaviglia (ed.), S. Greco (ed.), *Integrable Systems and Quantum Groups*, Lecture Notes in Mathematics 1620, Springer-Verlag, (1996), 419–481. (CONDE, COAG)

[316] D. P. Proskurin: *Homogeneouse ideals in Wick $*$-algebras*, Proc. Amer. Math. Soc. **126**, 11 (1998), 3371–3376. (q-HARA)

[317] W. Pusz: *Twisted canonical anticommutation relations*, Rep. Math. Phys. **27**, 3 (1989), 349–360. (q-HARA)

[318] W. Pusz, S. L. Woronowicz: *Twisted second quantization*, Rep. Math. Phys. **27**, 3 (1989), 231–257. (q-HARA)

[319] C. R. Putnam: *Commutation properties of Hilbert space operators and related topics*, Ergebnisse der Mathematik und ihrer Grenzgebiete, **36**, Springer-Verlag, 1967. (CORT)

[320] C. Quesne, N. Vansteenkiste: *Representation theory of deformed oscillator algebras*, Helv. Phys. Acta, **69**, 2 (1996), 141–157. (q-HARA)

[321] M. Rasetti: *Quantum Groups in Condensed-Matter Physics*, in *Proceedings of the International School of Physics "Enrico Fermi"* (L. Castellani, J. Wess, eds.), Italian Physical Society, IOS Press, 1996, 359–395. (QGQS)

[322] M. Rausch de Traubenberg: *Clifford algebras of polynomials, generalized Grassmann algebras and q-deformed Heisenberg algebras*, Adv. Appl. Clifford Algebras, **4**, 2, (1994), 131–144. (q-HARA)

[323] ______: *Algebrèbres de Clifford, Supersymètrie et Symètries $\mathbb{Z}_n$, Applications en Thèorie des Champs*, Habilitation a diriger des recherches, Laboratoire de Physique Thèorique, Univ. Louis Pasteur, 1997. (q-HARA)

[324] S. Rodriguez-Romo, D. W. Ebner: *Some Aspects of the q-Deformed Oscillator Algebras as Quantum Groups*, *Geometric and Quantum Aspects of Integrable Systems*, (G. F. Helminck ed.), Lecture Notes in Physics 424, Springer-Verlag, (1993), 213–222. (DHO, QGQS)

[325] L. J. Rogers: *On the expansion of some infinite products*, Proc. London Math. Soc. (2), **16** (1893), 337–352. (q-DE, q-SF)

[326] M. Roitman: *On Free Conformal and Vertex Algebras*, J. Algebra **217** (1999), 496–527.

[327] S. Roman: *The theory of the umbral calculus I*, J. Math. Anal. Appl. **87**, 1 (1982), 58–115. (UC)

[328] ______: *The theory of the umbral calculus II*, J. Math. Anal. Appl. **89**, 1 (1982), 290–314. (UC)

[329] ______: *The theory of the umbral calculus III*, J. Math. Anal. Appl. **95**, 2 (1983), 528–563. (UC)

[330] ______: *The Umbral Calculus*, Academic Press, 1984. (UC)

[331] ______: *More on the Umbral calculus, with emphasis on the q-Umbral calculus*, J. Math. Anal. Appl. **107** (1985), 222–254. (UC)

[332] S. Roman, G.-C. Rota: *The umbral calculus*, Adv. Math. **27** (1978), 95–188. (UC)

[333] A. L. Rosenberg: *The spectrum of the algebra of skew differential operators and the irreducible representations of the quantum Heisenberg algebra*, Commun. Math. Phys. **142** (1991), 567–588. (H*-A)

[334] ______: *The unitary irreducible representations of the quantum Heisenberg algebra*, Commun. Math. Phys. **144** (1992), 41–51. (H*-A)

[335] G.-C. Rota, D. Kahaner, A. Odlyzko: *On the foundations of combinatorial theory. VIII. Finite operator calculus* J. Math. Anal. Appl. **42** (1973), 684–760. (UC)

[336] F. Ryde: *A contribution to the theory of linear homogeneous geometric difference equations (q-difference equations)*, Dissertation, Lund, 1921. (q-DE)

[337] Yu. S. Samoĭlenko: *Spectral theory of families of self-adjoint Operators*, Kluwer Acad. Publ., Dordrecht, 1991. (Extended transl. to English from Russian edn.: Naukova Dumka, Kiev, 1984.) (CORT)

[338] R. M. Santilli: *Foundations of the hadronic generalization of the atomic Mechanics, I: generalization of Heisenberg's and Schrödinger's representations*, Hadronic Journal **5** (1982), 1194–1276. (LAA)

[339] ______: *Foundations of Theoretical Mechanics, II. Birkhoffian Generalization of Hamiltonian Mechanics*, Texts and Monographs in Physics, Springer-Verlag, 1983. (LAA)

[340] R. K. Saxena, R. Kumar: *Fractional q-differentiation and basic hypergeometric functions of two variables*, Bull. Math. Soc. Sci. Math. Roumanie (N.S.) **35** (83), 3–4 (1991), 287–294. (q-DE)

[341] A. Schirrmacher: *Quantum groups. Quantum space-time, and Dirac equation* in *Low-dimensional topology and quantum field theory*, NATO Adv. Sci. Inst. Ser. B Phys., 315, (1993), 221–230. (QGQS, q-DE)

[342] K. Schmüdgen: *Unbounded operator algebras and representation theory*, Birkhäuser, Basel, 1990. (CORT)

[343] ______: *Integrable Operator Representations of* $\mathbf{R}_q^2$, $X_{q,\gamma}$ *and* $\mathrm{SL}_q(2,\mathbf{R})$, Commun. Math. Phys. **159** (1994), 217–237. (H*-A)

[344] ______: *Operator representations of a q-deformed Heisenberg algebras*, J. Math. Phys. **40**, 9 (1999), 4596–4605. (H*-A)

[345] I. Schur: *Über vertauschbare lineare Differentialausdrücke*, der Berliner Math. Gesell., Sitzungsber. **4** (1905), 2–8. (CDO, q-HARA)

[346] J. Schwenk: *Differential calculus for the n-dimensional quantum plane*, in *Quantum Groups*, Proceedings of the Argonne Workshop, Argonne National Laboratory, 1990, (T. Curtright, D. Fairlie, C. Zachos, eds.), World Scientific, Singapore, 1991, 53–61. (NGC, q-A, QGQS)

[347] ______: *q-Deformed Fourier theory*, Preprint, MPI-PhT/94-36, hep-th-9406168, (1994). (q-FA, q-OP)

[348] ______: *Fourier transformation for the q-deformed Heisenberg algebra*, in *Proceedings of the International School of Physics "Enrico Fermi"*, (L. Castellani, J. Wess, eds.), Italian Physical Society, IOS Press, 1996, 525–540. (q-FA, q-OP)

[349] J. Schwenk, J. Wess: *A q-deformed quantum mechanical toy model*, Phys. Letters B **291**, 3 (1992), 273–277. (q-HARA)

[350] J. Scott-Russell: *Report on waves*, Proc. Roy. Soc. Edinburgh, (1844), 319–320. (CONDE)

[351] G. Segal, G. Wilson: *Loop groups and equations of KdV type*, Publ. Math. IHES, **61** (1985), 5–65. (CONDE, COAG)

[352] S. V. Shabanov: *q-Oscillators, (non-)Kähler manifolds, and constrained dynamics*, Modern Phys. Lett. A. **10**, 12 (1995), 941–948. (q-HARA, DHO)

[353] ______: *Quantum and classical mechanics of q-deformed systems*, J. Phys. A: Math. Gen. **26** (1993), 2583–2606. (q-HARA, DHO)

[354] S. D. Silvestrov: *Representations of Commutation Relations. A Dynamical Systems Approach*, Hadronic Journal Supplement, **11**, 1 (1996), 1–116. (CORT)

[355] ______: *Hilbert space representations of the graded analogue of the Lie algebra of the group of plane motions*, Studia Mathematica, **117**, 2 (1996), 195–203. (CORT)

[356] ______: *On a class of rings generalizing commutativity*, Czechoslovak J. Phys. **48**, 11 (1997), 1495–1500. (RF)

[357] S. D. Silvestrov, L. B. Turowska: *Representations of the q-deformed Lie algebra of the group of motions of the Euclidean plane*, J. Funct. Anal. **160**, (1998), 79–114. (CORT)

[358] S. D. Silvestrov, H. Wallin: *Representations of algebras associated with a Möbius transformation*, J. Nonlin. Math. Physics, **3**, 1–2 (1996), 202–213. (CORT)

[359] S. Skorik, V. P. Spiridonov: *Self-similar potentials and the q-oscillator algebra at roots of unity*, Lett. Math. Phys. **28**, 1 (1993), 59–74. (q-HARA, DHO)

[360] V. V. Sokolov: *Examples of commutative rings of differential operators*, (in Russian), Funktz. Anal. Priloz. **12**, 1 (1978), 82–83. (CONDE, COAG)

[361] ______: *Birationally isomorphic commutative rings of differential operators*, (in Russian), Funktz. Anal. Priloz. **12**, 3 (1978), 88–89. (CONDE, COAG)

[362] A. I. Solomon: *Quantum group analogues of squeezed states*, in *Symmetries in Science VI*, (B. Gruber, ed.), Plenum Press, New York, 1993. (DHO, q-HARA)

[363] R. Speicher: *A new example of 'independence' and 'white noise'*, Prob. Th. Rel. Fields, **84** (1990), 141–159. (q-HARA)

[364] ———: *Generalized Statistics of Macroscopic Fields*, Lett. Math. Phys. **27** (1993), 97–104. (q-HARA)

[365] V. P. Spiridonov: *Coherent states of the q-Weil algebra*, Lett. Math. Phys. **35**, 2 (1995), 179–185. (q-HARA, DHO)

[366] G. W. Starcher: *On identities arising from solutions of q-difference equations and some interpretations in number theory*, Amer. J. Math. **53** (1931), 801–816. (q-DE)

[367] R. F. Swarttouw, H. G. Meijer: *A q-analogue of the Wronskian and a second solution of the Hahn-Exton q-Bessel difference equation*, Proc. Amer. Math. Soc. **120**, 3 (1994), 855–864. (q-DE)

[368] J. Thomae: *Beiträge zur Theorie der durch die Heinische Reihe ...* , J. Rreine Angew. Math. **70** (1869), 258–281. (q-DE, q-SF)

[369] ———: *Über die höheren hypergeometrischen Reihen, insbesondere die Reihe ...*, Math. Ann., Bd.2, (1870), 427–444. (q-DE, q-SF)

[370] ———: *Les séries Heinéennes supérieures, ou les séries de la forme ...*, Ann. di Mat. Pura ed Appl. **4** (1871), 105–138. (q-DE, q-SF)

[371] J. M. Thuswaldner: *q-Difference equations and their applications to combinatorial problems*, Grazer Math. Ber. **328** (1996), 103–112. (q-DE)

[372] W. J. Trjitzinsky: *Analytic theory of linear q-difference equations*, Acta Math. **61** (1933), 1–89. (q-DE)

[373] A. Turbiner: *Invariant identities in the Heisenberg algebra*, (in Russian), Funktz. Anal. Priloz. **29**, 4 (1995), 88–91. (RF)

[374] A. Turbiner, G. Post: *Operator identities, representations of algebras and the problem of normal ordering*, J. Phys. A: Math. Gen. **27** (1994), L9–L13. (RF)

[375] R. Twarock: *A descrete quantum kinematics based on q-derivatives*, Arnold Sommerfeld Institut für Mathematische Physik, Clausthal, (1995). (q-DE, q-A)

[376] ———: *Derivation of a q-Schrödinger equation on S^1 via q-deformation of the kinematical algebra ...* , Arnold Sommerfeld Institut für Mathematische Physik, Clausthal, (1996). (q-DE, q-A)

[377] M. R. Ubriaco: *Non-commutative differential calculus and q-analysis*, J. Phys. A: Math. Gen. **25**, (1992), 169–173. (DHO, NGC, q-A, QGQS)

[378] M. Upadhyay: *q-Fractional differentiation and basic hypergeometric transformations*, Ann. Polon. Math. **25** (1971/72), 109–124. (q-DE, q-SF)

[379] ———: *Fractional q-integration and integral representations of basic double hypergeometric series of higher order*, Gaṇita, **23**, 1 (1972), 19–36. (q-DE, q-SF)

[380] ______: *Certain fractional q-integral operators*, Pure Appl. Math. Sci. **4**, 1–2 (1976), 141–149. (q-DE, q-SF)

[381] E. Ye. Vaysleb: *Representations of relations which connect a family of commuting operators with non-selfadjoint one*, Ukrain. Mat. Zh. **42** (1990), 1258–1262. (CORT)

[382] E. Ye. Vaysleb, Yu. S. Samoĭlenko: *Representations of operator relations by unbounded operators and multi-dimensional dynamical systems*, Ukrain. Mat. Zh. **42**, 9 (1990), 1011–1019. (CORT)

[383] J.-L. Verdier: *Equations différentielles algébriques*, in *Seminaire Bourbaki, vol. 1977/78, Exposés 507–524*, Lecture Notes in Mathematics 710, Springer-Verlag, (1979), 101–122. (CONDE, COAG)

[384] A. M. Vershik, I. M. Gelfand, M. I. Graev: *A commutative model of the representations of current group* $\mathrm{SL}(2,\mathbb{R})^X$ *related to the unipotent subgroup*, (in Russian), Funktz. Anal. Priloz. **17**, 2 (1983), 70–72. (CORT)

[385] A. P. Veselov: *Hamiltonian formalism for the Novikov–Krichever equations for the commutativity of two operators*, (in Russian), Funktz. Anal. Priloz. **13**, 1 (1979), 1–7. (CONDE)

[386] N. J. Vilenkin, A. U. Klimyk: *Representations of Lie groups and special functions*, Vol. 3, Kluwer Academic Publishers, 1992. (q-DE, q-SF)

[387] O. V. Viskov: *A commutative-like noncommutative identity*, Acta Sci. Math. (Szeged) **59** (1994), 585–590. (RF)

[388] C. G. Wagner: *Generalized Stirling and Lah numbers*, Discrete Mathematics **160** (1996), 199–218. (q-DE, q-SF)

[389] G. Wallenberg: *Über die Vertauschbarkeit homogener linearer Differentialausdrücke*, Archiv Math. Phys. **4** (1903), 252–268. (CDO)

[390] R. Wallisser: *Über ganze Funktionen, die in einer geometrischen Folge ganze Werte annehmen*, Monatsh. für Math. **100** (1985), 329–335. (q-DE)

[391] J. Wess: *q-Deformed Heisenberg Algebra*, in *Supersymmetry and Quantum Field Theory*, Proc. of the D. V. Volkov Memorial Seminar, Kharkov, Ukraine, 1997, (J. Wess, V. P. Akulov, eds.), Lecture Notes in Physics 509, Springer-Verlag, 1998. (q-HARA)

[392] ______: *q-Deformed Heisenberg Algebras*, Preprint, math-ph/9910013, (1999). (q-HARA)

[393] J. Wess, B. Zumino: *Covariant differential calculus of the quantum plane*, Nucl. Phys. B **4** (1990), 252–268. (NGC)

[394] ______: *Covariant differential calculus on the quantum hyperplane*, Nucl. Phys. B (Proc. Suppl.) **18** B (1990), 302. (NGC)

[395] G. Wilson: *Commuting flows and conservation lows for Lax equations*, Math. Proc. Camb. Phil. Soc. **86** (1979), 131–143. (CONDE)

[396] ______: *Hamiltonian and algebro-geometric integrals of stationary equations of KdV type*, Math. Proc. Camb. Phil. Soc. **87** (1980), 295–305. (CONDE, COAG)

[397] ______: *Algebraic curves and soliton equations*, in *Geometry Today*, (E. Arbarello, C. Procesi, E. Strickland, eds.), Birkhäuser, Boston, 1985, 303–329. (CONDE, COAG)

[398] S. L. Woronowicz: *Twisted* SU(2) *group. An example of a non-commutative differential calculus*, Publ. RIMS, **23** (1987), 117–181. (NGC, q-A, QGQS, q-HARA)

[399] K.-W. Yang: *Solution of q-difference equations*, Bull. London Math. Soc. **20**, 1 (1988), 1–4. (q-DE)

[400] C. Zachos: *Elementary paradigms of quantum algebras*, Cont. Math. **134** (1992), 351–377. (DHO, q-A, QGQS)

[401] D. Zagier: *Realizability of a Model in Infinite Statistics*, Comm. Math. Phys. **147** (1992), 199–210. (q-HARA)

[402] V. E. Zakharov, L. D. Faddeev: *The Korteweg–de Vries equation is a completely integrable Hamiltonian system*, (in Russian), Funktz. Anal. Priloz. **5**, 4 (1971), 18–27. (CONDE)

[403] V. E. Zakharov, S. V. Manakov, S. P. Novikov, L. P. Pitayevsky: *Theory of solitons*, (in Russian), Nauka, Moscow, 1980. (CONDE, COAG)

[404] V. E. Zakharov, A. B. Shabat: *A scheme for integrating the nonlinear equations of mathematical physics by the method of the inverse scattering problem. I*, (in Russian), Funktz. Anal. Priloz. **8**, 3 (1974), 43–53. (CONDE, COAG)

[405] ______: *Integration of the nonlinear equations of mathematical physics by the method of the inverse scattering transform. II*, (in Russian), Funktz. Anal. Priloz. **13**, 3 (1979), 13–22. (CONDE, COAG)

[406] C.-Z. Zha, W.-Z. Zhao: *The q-deformation of super high-order Virasoro algebra*, J. Math. Phys. **36**, 2 (1995), 967–979. (q-A)

[407] C. Zhang: *Sur la sommabilité des séries entières solutions formelles d'une équation aux q-différences. I*, C. R. Acad. Sci. Paris, Sér. I, **327**, 4 (1998), 349–352. (q-DE)

[408] A. S. Zhedanov: *'Non-classical' q-oscillator realization of the quantum* SU(2), J. Phys. A: Math. Gen. **25**, (1992), L713–L717. (DHO, q-HARA, q-A, q-SF)

[409] ______: *The Weyl shift for q-oscillator and a q-polynomial*, (in Russian), Teoret. Mat. Fiz. **94**, 2 (1993), 307–315. (DHO, q-HARA)

[410] B. Zumino: *Deformation of the quantum mechanical phase space with bosonic or fermionic coordinates*, Modern Phys. Lett. A, **6**, 13 (1991), 1225–1235. (q-HARA, NGC, q-A, QGQS)

[297] ———, [illegible] and [illegible], [illegible] Boston, [illegible]

[298] S. L. [illegible] equations [illegible], Dokl. [illegible] SSSR [illegible]

[299] [illegible] Spectra with differential [illegible], Comm. [illegible] Math. [illegible]

[300] [illegible] stochastic [illegible] equations [illegible] Math. [illegible]

[301] [illegible] stability of a [illegible], Comm. Math. Phys. [illegible]

[302] [illegible] [illegible] Hamiltonian systems [illegible]

[303] [illegible] Zaslavsky, [illegible] Sagdeev, [illegible] Chernikov, [illegible] Cambridge [illegible]

[304] V. E. Zakharov and A. B. Shabat, A scheme for integrating the nonlinear equations of mathematical physics by the method of the inverse scattering problem, [illegible] Funkts. Anal. [illegible]

[305] ———, Integration of the nonlinear equations of mathematical physics by the method of the inverse scattering problem [illegible] Funct. Anal. [illegible]

[306] [illegible]

[307] [illegible] [illegible] Trans. [illegible]

[308] [illegible]

[309] ———, [illegible]

[310] [illegible]

Index

$\binom{n}{k}_q$, 184
$\langle q\rangle$, 77
$\left[{n \atop k}\right]_q$, 197
$[\alpha,\beta]$, 221
$\lfloor x\rfloor$, 221
$\sqcap$, 64
$\sqsubseteq$, 64
$\left\{{n \atop k}\right\}_q$, 197
$\{n\}_q$, 184
$\{n\}_q!$, 184

1, 36, 163

A, 38
  A_j, 37
Alg, 36
α_s, 142
Artinian, 145

B, 38
  B_j, 37
Bernstein filtration, 160

Cen, 74
  Cen(α), 74
  Cen(d,f,α), 74
  Cen(F,α), 74
  Cen(f,α), 74
  Cen$\big([m,n]{:}j,\alpha\big)$, 74
  Cen$(n{:}j,\alpha)$, 74
centraliser, 74
chain, 62
chain function, 65
chain function in j, 65
chain space, 62
χ, 65
  χ_j, 65
  $\overline{\chi}_j$, 65
  $\underline{\chi}_j$, 65
compatible, 146

d_V, 159
d_v, 159
DCC, 145
deg, 54, 159
  $\deg_{A_j}$, 54
  $\deg_{B_j}$, 54
  $\deg_j$, 54
degree function, 159
δ_{ij}, 17
descending chain condition, 145
direct free type, 78
direct torsion type, 78
direct type, 77

End, 221

$F_n(x;q)$, 197
$\mathbb{F}_q$, 191

$\mathcal{F}(J, \mathcal{K})$, 36
faithful, 135
filtration, 156
free type, 77
fully homogeneous, 64

$G_n(x; q)$, 207
Gauss polynomials, 207
generalised polynomials, 129
generated
  subalgebra, 36
generating set, 156, 163
gradation, 179
graded, 179

$\mathcal{H}(q)$, 38
$\mathcal{H}(\mathbf{q}, J)$, 38
$\mathcal{H}(\mathbf{q}, J, \mathcal{K})$, 37
$\mathcal{H}(\mathbf{q}, n)$, 38
Heisenberg algebra, 20
homogeneous in j, 64

I, 37
$\mathcal{I}(S)$, 142
id
  id, 143
inclusion ambiguity, 144
intersection, 64
Irr, 143
irreducible, 143

K_f, 62
K_n^j, 62
$\mathcal{K}$, 37
$\mathcal{K}^*$, 221

lower chain functions, 65

M_i, 158
  $M_i(V)$, 162
  $M_i(v)$, 162
$M_{-\infty}$, 158
Maj, 169
Mon, 170
monomial, 156
monomial decomposition, 170
μ_s, 142

$\mathbb{N}$, 221
normal form, 41, 143

$O(f(n))$, 222
order, 77
overlap ambiguity, 144

P_i, 157
  $P_i(V)$, 162
  $P_i(v)$, 162
  P_i^n, 156
  $P_i^n(V)$, 162
  $P_i^n(v)$, 162
$\mathcal{P}(\alpha_0, \ldots, \alpha_k)$, 118
persistently reducible, 143
pr_2, 170

$\mathbf{q}$, 37
$\mathcal{Q}(J, \mathcal{K})$, 86, 98
q-analogue, 183
q-binomial coefficients, 184
q-deformed falling factorial
  polynomials, 197
q-factorial, 184
q-natural number, 184
q-special functions, 183
q-Stirling numbers
  of the second kind, 197
q-Stirling numbers
  of the first kind, 197

$r(n)$, 107
Red, 143
reduction system, 142
reductions, 143
resolvable, 144
rule, 142

S_n, 221
$S(\mathbf{q}, J)$, 37

semigroup partial order, 145
strictly direct type, 77
strongly nonzero, 90

$T(S)$, 143
  $T_0(S)$, 143
  $T_1(S)$, 143
  $T_n(S)$, 143
$t_{\lambda s \nu}$, 142
t^S, 143
$\Theta(f(n))$, 222
torsion type, 77

uniquely reducible, 143
upper chain functions, 65

weakly nonzero, 90

$\mathbb{Z}(J)$, 62
$\mathbb{Z}^+$, 221
$\hat{\mathbb{Z}}$, 159